Animal Dispersal

Animal Dispersal

Small mammals as a model

Edited by
NILS CHR. STENSETH
Department of Biology, University of Oslo, Norway

and

WILLIAM Z. LIDICKER, JR
Museum of Vertebrate Zoology, University of California, Berkeley, USA

CHAPMAN & HALL
London · New York · Tokyo · Melbourne · Madras

Published by Chapman & Hall, 2–6 Boundary Row, London SE1 8HN

Chapman & Hall, 2–6 Boundary Row, London SE1 8HN, UK

Chapman & Hall, 29 West 35th Street, New York NY10001, USA

Chapman & Hall Japan, Thomson Publishing Japan, Hirakawacho Nemoto Building, 7F, 1-7-11 Hirakawa-cho, Chiyoda-ku, Tokyo 102, Japan

Chapman & Hall Australia, Thomas Nelson Australia, 102 Dodds Street, South Melbourne, Victoria 3205, Australia

Chapman & Hall India, R. Seshadri, 32 Second Main Road, CIT East, Madras 600 035, India

First edition 1992

Typeset in 10/12pt Sabon by Graphicraft Typesetters Ltd, Hong Kong
Printed in Great Britain by Hartnolls Ltd, Cornwall

ISBN 0 412 29330 7

A catalogue record for this book is available from the British Library

Library of Congress Cataloging-in-Publication data

Animal dispersal: small mammals as a model/edited by Nils Chr. Stenseth and William Z. Lidicker, Jr.

p. cm.

Includes bibliographical references and index.

ISBN 0–412–29330–7

1. Animals—Dispersal. 2. Mammals—Dispersal. I. Stenseth, Nils Chr. II. Lidicker, William Zander, 1932–

QH543.3.A55 1992

574.5—dc20

91–31135

CIP

Contents

Contributors

N.H. BARTON Institute of Cell, Animal and Population Biology, University of Edinburgh, UK

C.A. BRANDT Department of Environmental Sciences, Battelle Pacific Northwest Laboratories, Richland, Washington, USA

R.L. CALDWELL Department of Integrative Biology, University of California, Berkeley, USA

A. COCKBURN Department of Zoology, Australian National University, Canberra, Australia

J. GLIWICZ Department of Wildlife Management, Agricultural University of Warsaw, Poland

L. HANSSON Department of Wildlife Ecology, Swedish University of Agricultural Science, Uppsala, Sweden

C.J. KREBS Department of Zoology, University of British Columbia, Vancouver, Canada

W.Z. LIDICKER, Jr Museum of Vertebrate Zoology, University of California, Berkeley, USA

D.M. MADISON Department of Biological Sciences, State University of New York Binghamton, USA

W.J. MCSHEA Conservation Center, Smithsonian Institution, Front Royal, Virginia, USA

A.E. PUSEY Department of Ecology and Behavioral Biology, University of Minnesota, Minneapolis, USA

G.K. RODERICK Department of Entomology, University of Maryland, College Park, USA

A.R.E. SINCLAIR Department of Zoology, University of British Columbia, Vancouver, Canada

N.C. STENSETH Department of Biology, University of Oslo, Norway

A. WATSON Institute of Terrestrial Ecology, Kincardineshire, Scotland, UK

Acknowledgements

We thank William A. Fuller, Chairman of the Congress Organizing Committee, for his copious contributions toward making the Fourth International Theriological Congress (ITC IV) a big success. He did far more than find a place and a time for the various symposia. He gave detailed thought and attention to the various scientific and logistic aspects of the program, including our symposium. International meetings like ITC are ideal places for merging ideas originating in different parts of the world and under the influence of different scientific traditions. Bill Fuller paid special attention to this fact; hence he encouraged both of us to organize the ITC symposium on dispersal that forms the basis for this book. Also, the support and co-operation of the Senior Editors at Chapman and Hall, A. Crowden, T. Hardwick and R.C.J. Carling are much appreciated and were absolutely essential to our task. M.L. Rosenzweig attended our symposium and encouraged us to put it into book form. He also gave us some helpful editorial feedback at an early stage. Finally, Louise N. Lidicker cheerfully typed and re-typed most of the chapters and was helpful with numerous other editorial chores, all of which greatly eased our workloads.

In spite of its many merits, transatlantic (and transcontinental) editorial collaboration has its disadvantages as well. Travel and telephone expenses quickly exceeded our expectations and budgets. However, we gratefully acknowledge support from the Department of Biology, University of Oslo, the Fulbright Foundation, ITC IV, the Museum of Vertebrate Zoology, University of California, Berkeley, the Norwegian National Science Foundation (NAVF), and The Scandinavian-American Association. Their help made our efforts possible.

Nils Chr. Stenseth
William Z. Lidicker, Jr

Preface

The literature on animal dispersal is extensive; in fact it is overwhelming. Early in the 1980s, for example, the BIOSIS database already had 18 539 entries on dispersal and spacing behaviour covering the previous 15 years. No single person can master such an extensive literature. Even specialists in this field will, therefore, need reviews of large portions of the literature, and certainly the non-specialist will need to make use of such resources. A moderate number of book-length reviews on dispersal do already exist, but the field is progressing rapidly and moreover there are still major areas which remain unsynthesized. We intend this book to be a contribution toward alleviating this need, as well as offering speculation into future research directions.

The term *presaturation dispersal* was proposed in 1975. The idea that individuals might disperse even if their home situation was not intolerable was not new, but it certainly was not a pervasive view at that time. The notion turned out to attract considerable attention, and led, we think, to an energetic and profitable period of dispersal research, especially among students of small mammal species. Ten years later, we felt it was appropriate to review the literature on dispersal in small mammal species and attempt a synthesis. Accordingly, we organized a symposium on this subject for the Fourth International Theriological Congress (ITC IV) held in Edmonton, Canada in August 1985. We asked a group of biologists who worked on small mammal dispersal to review various aspects of the subject. Speakers were requested to review both existing data and ideas, and, most importantly, to try to integrate the two. In our view the ITC symposium was a success. Certainly it was very well attended by Congress participants, and generated lively discussions.

Because the symposium speakers did such masterful jobs of reviewing their assigned subjects, and because their presentations were relatively well integrated, we were encouraged to use the ITC symposium as the starting point for a book on dispersal. Therefore, we asked the speakers in Edmonton to write review chapters based on their talks. All but two responded to this challenge. We wanted, however, to do more than to review the literature on small mammal dispersal. Although this alone would represent a major accomplishment, we wished to explore the potential for viewing small mammals as a model system for studying dispersal in animals generally. Hence, we wanted to broaden the scope of the book considerably relative to the ITC symposium. We had two options: (1) to ask the authors of the small mammal chapters to broaden their reviews and include references to other groups of organisms, or (2) to ask a group of additional authors who worked on dispersal in other kinds of organisms to comment on the core chapters of the book from their perspectives. We chose the latter option. By so doing, we would, we believed, achieve two important objectives. We could both go into depth with small mammals, exploiting the various authors' specialities and backgrounds, and still enjoy and profit from the enlarged perspectives provided by authors with experiences far beyond small mammals. Moreover, by adopting the 'commentary strategy' we hoped to produce a more cohesive, comprehensive, and broadly based product than we might have otherwise expected.

The result, the present book, has therefore developed into a treatise on animal dispersal which goes into depth in the area of small mammal biology, but claims a much wider scope. In writing this book we have had both the professional researcher and beginning graduate students in mind. For both we have provided a synthetic and broad review of the literature. We hope that readers whose experiences lie outside the small mammal area will find items of interest and profit in this enterprise as well. For the students particularly, we have provided in addition a guide to the conceptual framework of animal dispersal and a methodological guide (both with respect to experimental design and data analysis) for studying dispersal in small mammals. We hope that this book will be especially useful to anyone about to begin a project involving dispersal behaviour. For everyone we have provided some speculations on directions for the future. In science, it is not enough to know where we are and where we have been, we want also to consider where we go from here.

We have purposely not intended this book to be viewed merely as a symposium proceedings. Even though all but two of the papers presented at the ITC symposium are included in the book, several additional chapters have been written specifically for the book. In addition, the symposium chapters have been extensively revised in order to give them greater depth of coverage and to better connect the parts.

We have enjoyed working on this book, and have found the collaboration with the authors to be professionally very rewarding; we thank them for their co-operation. We hope that readers will find the book as useful as we have found it fulfilling to produce.

Nils Chr. Stenseth
William Z. Lidicker, Jr

PART ONE

Introductory chapters

We begin with three chapters that attempt to provide a foundation and a context for our subject. All three are general in the sense that they are not specific to small mammals. In fact they treat dispersal as a phenomenon common to all organisms, just as reproduction and death are also general features of all life on this planet. Starting with the fundamentals is appropriate to one of our objectives in this book, namely, to help bridge the gaps among the various taxon-oriented approaches to the study of dispersal. We believe these chapters achieve this goal, and we are hopeful that students of dispersal in any organism will find much that is helpful here.

Chapter 1 presents a conceptual guide to dispersal that sets the ecological and evolutionary stage for the subject. It is brief but presents both verbal and quantitative guidelines for what is to follow. This chapter was not part of the symposium at Edmonton, but was prepared in response to the perceived need for more general introductory material. In the second chapter, we tackle some difficult definitions and discuss the characterization of dispersers, their 'motivations', and the environmental circumstances that favour dispersal. We propose a classification system for dispersal that may prove too complex to be practical, but is conceptually as realistic as we can make it given the current state of our knowledge.

In Chapter 3, Nick Barton reviews the population genetics of dispersal. Except for the contributors of the invited commentaries (Part VI), Barton is the only author in this book who has not done empirical research on dispersal in small mammals. Accordingly, his non-taxon oriented approach serves us well in providing a general perspective to the genetic aspects of dispersal. This chapter is a guide to how to apply genetic information about populations and individuals to infer dispersal patterns and social organization. We are convinced that students of dispersal, who often are ecologically or behaviourally oriented, will profit from reading Chapter 3.

– 1 –

The study of dispersal: a conceptual guide

N.C. STENSETH AND W.Z. LIDICKER, JR

1.1 INTRODUCTION

Probably, the best known example of dispersal is the Biblical story of the Israelites traversing the desert in search of a new land. Since then many examples have been recorded in the literature. Today we know dispersal – or movement from one home site to another – as a phenomenon of potentially great importance to the demographic and evolutionary dynamics of populations. The most obvious effect of dispersal is that it changes the spatial distribution of individuals, i.e. dispersal can lead to changes in dispersion (*sensu* Odum, 1953; Southwood, 1978). These changes may cause increased clumping, greater randomness, or more even spacing. Unfortunately, considerable confusion persists in the literature over the terms 'dispersal' and 'dispersion' (in fact in some languages, e.g. Spanish and Norwegian, only a single word is available for both concepts). It is therefore necessary to be alert to this problem in reading the literature.

1.2 TERMINOLOGY

Circumstances in which individuals leave their existing home ranges and do not return, at least not in the short-term as they would after brief excursions, constitute our notion of dispersal movements (see also Chapter 2; Lidicker and Caldwell, 1982). Currently, this phenomenon has become a subject of profound interest. At least in part, this is because it comprises a 'glue', joining together ecology, population genetics, ethology, and evolution. Moreover, it draws these disciplines together with wildlife management, conservation biology, and epidemiology. Epidemiology is largely dispersal biology as applied to pathogens, because without dispersal there would be no such discipline, and conservation biology is increasingly preoccupied with movements among habitat patches (e.g. Chapter 8). Movements among habitat discontinuities, furthermore, link dispersal to the rapidly growing field of landscape ecology (Forman and Godron, 1986; Lidicker, 1988b).

Table 1.1 Contrasting terminologies for various movement phenomena (travels) as advocated by a leading entomologist (Taylor, 1986) and by us in this book. The first phenomenon includes the other three, and the last may or may not be a subset of the second

Phenomenon	*Taylor*	*This book*
Individual and/or population movements	Migration	Movements (travels)
One-way extra home range movements	Migration	Dispersal
Round-trip travelling (usually seasonal)	Migration	Migration
Socially mediated infra-dispersion	Dispersal	Spacing behaviour

In all of this interdisciplinary ferment, species of small mammals have played critically important roles, and should, we believe, continue to do so in the future. Primarily this is because population densities of these species often change dramatically on time scales which are amenable for study, and because it is sometimes possible to follow the activities of individuals over significant fractions of their lifespans. This book addresses the state of our knowledge and understanding of such movements in populations of small mammals.

While we define dispersal in a broad-based way and consequently establish the boundaries of our subject for this book, the reader should be aware that not all investigators in this field agree on a standard terminology (Table 1.1). The fluidity of terminology that exists reflects the diversity of organisms and of disciplines that are involved. While we lament these disagreements, we feel they serve beneficially to inhibit premature solidification of concepts. Especially among entomologists, there is a tendency to use 'migration' as the broad-based, generic term for movements (e.g. Kennedy, 1961; Dingle, 1972, 1980; Taylor and Taylor, 1977; Taylor, 1986; Chapter 13). This seems to us an unfortunate preemption of a general term with many meanings, and one that in popular usage (as well as used consistently by vertebrate biologists) refers to two-way movements generally over relatively long distances and organized seasonally (e.g. Chapter 10). In this latter sense, 'migration' is a subset of our concept of dispersal (Table 1.1). The case for the generic use of 'migration' has recently been espoused by Taylor (1986). In Table 1.1 we compare Taylor's terminology for a selected group of movement phenomena with ours. Perhaps because in insects the home range (or site) concept is more weakly developed, Taylor extends his concept of migration to the absurdity of claiming (p. 13) that a swallow laying eggs in a nest a few meters from the one in which it was hatched has thereby 'migrated' those few meters. In our opinion, he further confuses the process of dispersal with the population property of dispersion which, as explained above, is the spatial arrangement of individuals. Taylor (1986) defines dispersal as socially motivated movements leading toward greater infra-dispersion (or even spacing), such as the scattering of a school of fish (p. 7). Consequently, Taylor's concept of 'dispersal' has little to do with our subject here.

1.3 DISPERSAL OR RESIDENCY

Our view of a patchy environment (Figure 1.1), which serves as the framework for our discussions of dispersal, owes much to the pioneering contributions of Anderson (1970) and Hansson (1977). In non-sessile species an individual is continuously faced with the choice (option) of staying put in the home site it currently occupies or leaving in order to search for a different place to live. If it leaves (disperses), it faces a variety of potential risks as well as possible gains (summarized in Table 1.2). If the choice is made to stay at home, a similar set of advantages and disadvantages must be weighed. Finally, the net risks of dispersal can be compared with the net risks of philopatry, and a decision made. To make the correct decision is critically important as the issues are pervasive; consequences range from premature death without reproducing to the discovery of an unexploited patch of excellent habitat complete with the presence of an ideal mate(s). The solution to the balance of potential gains and losses for any given individual depends on a variety of factors which are discussed throughout the chapters of this book. In Chapter 2 we discuss more fully the complex of proximate motivations which lead individuals to initiate dispersal movements.

Bengtsson (1978) provides a model which is useful when considering the relative risks of dispersal versus residency. It is useful because it emphasizes the importance of incorporating both the conditions of the home site and those that would be encountered by a prospective disperser. Others have taken this approach from a more intuitive perspective (e.g. Southwood, 1977; Solbreck, 1978; Alerstam and Enckell, 1979). Bengtsson's model was devised to find an evolutionarily stable strategy (ESS; Maynard Smith, 1982) among strategies with different dispersal behavior, but it can be thought of as applying to the decision-making process of individual organisms in which alternative behaviours are possible.

The most widely perceived risk of not dispersing is inbreeding. This potential risk must be balanced against the risks of dispersing. And it is exceedingly difficult to assess either side of the equation for most situations. One needs to determine the expected genetic contribution of an individual whether it disperses and outbreeds, or remains philopatric and inbreeds. This analysis is further complicated by the necessity of considering 'inclusive fitness' components (Hamilton, 1964). That is, an in-

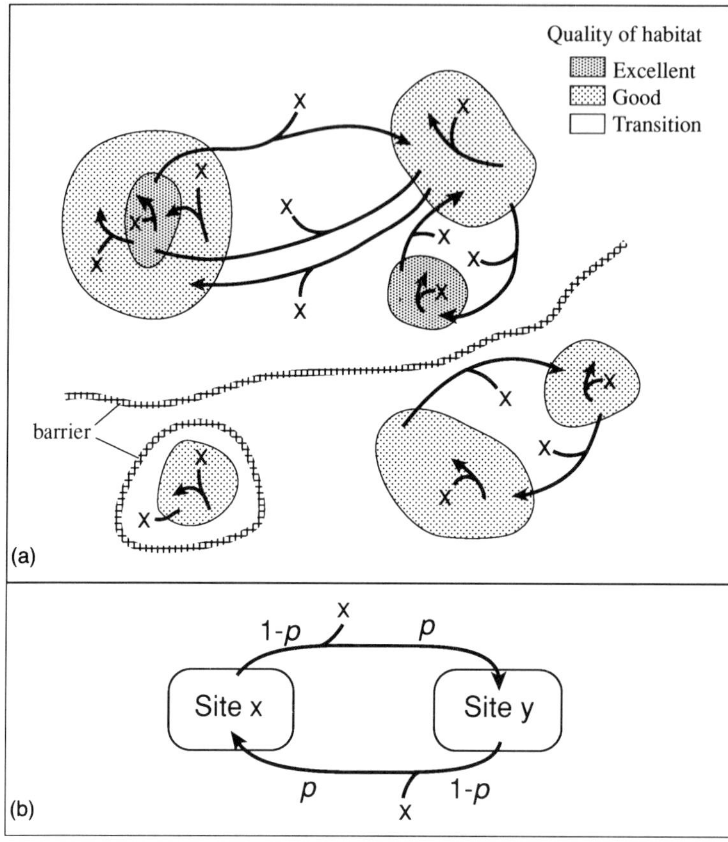

Figure 1.1 (a) Map of hypothetical meta-population (*sensu* Levins, 1970) with six habitat patches of varying quality and two dispersal barriers. Transition habitat (*sensu* Stenseth, 1983) is not suitable for sustained survival and reproduction. Barriers can be of variable resistance, and can be assigned transit probabilities. (b) Conceptual model applicable to any of the inter-patch or intra-patch dispersal events shown in (a). p is the probability of successfully transferring from one site to another. The p value for within-patch movements can be expected to be higher than that for between-patch movements. Very hostile transition habitats would generate lower p values than less hostile situations. Species with tight social structure into which newcomers may have difficulties in entering would be characterized by relatively low p values. Solid lines indicate dispersal trajectories; × indicates death during dispersal phase.

Table 1.2 A dispersal balance sheet

Types of factors	*Potential advantages*	*Potential disadvantages*
Environmental	Escape from unfavourable conditions (economic, physical, social) Reduced exposure to predators or competitors Reduced exposure to population crashes	Uncertainties of finding food, shelter, appropriate social milieu Greater exposure to predators or competitors
Genetic		
Quantitative	May find uninhabited or lower density area of suitable quality	May not find any empty habitat
	Promiscuity	Uncertainty of finding a mate
	Frequency-dependent selection may favour rare phenotypes	Strange phenotypes may be avoided
Qualitative	Heterosis and avoidance of inbreeding	Less viable offspring may be produced (breakdown of co-adapted systems; disadvantageous recombinations)
	Greater chance for new and advantageous recombinations occurring	

dividual passes on its genes both directly by successfully reproducing itself, and indirectly through the reproductive efforts of close relatives.

In order to show how this argument can be developed within the framework of Bengtsson's (1978) model, we simplify our conditions to full sibling (brother–sister) matings versus outbreeding (mating with non-relatives). Consider, for example, a species where adolescent males disperse and outbreed. There is a probability, m, that such a male will either die in the process or fail to find a mate. Thus, m is the cost of dispersal, or the cost of outbreeding (a quantity we refer to in Chapter 9 as $1 - p$). Females in our simplified model are assumed not to disperse, and hence do not incur the costs of dispersal (m). Hence, the adult sex ratio will be $1 - m$ males: 1 female, and every surviving male will on average mate with $(1 - m)^{-1}$ females.

Following Bengtsson's scheme (1978), we now assume that a behavioural mutant occurs which causes its male carriers not to disperse. They thus avoid the costs of dispersal and instead remain at home to mate with sisters. The cost associated with this new behaviour is that the male's fitness is decreased from 1 to $1 - i$ through inbreeding depression. We can often simply interpret i as that fraction of the male's offspring that die because of inbreeding complications, while the rest of the offspring are normal. Bengtsson then asks under what conditions can such a mutation be expected to spread through the population, assuming, of course, that it is genetic. The way we answer this question is analogous to the process of making the correct decision regarding dispersal of residency with which we began this chapter.

We need to define and compare the reproductive contribution of an outbreeding male with one that inbreeds. The philopatric male will, we assume, produce offspring only through matings with his sisters with whom he shares (on average) half his genes. Therefore, his offspring will have 0.75 proportion of his genes and there will be $(1 - i)$ of these for a total contribution of $0.75(1 - i)$. The outbreeding male, on the other hand, will contribute to the next generation in two ways. First, through his own breeding his genetic contribution will be 0.50, but his chances of success are only $1 - m$. Second, such a male will contribute through his philopatric and outbred sisters. He will share 0.25 genes with these offspring, so that the total contribution expected is $0.25 + 0.50(1 - m)$.

Now we can see that the mating strategy which makes the greatest contribution to future generations will depend on whether or not $0.25 + 0.50(1 - m)$ is greater or less than $0.75(1 - i)$. In particular, our new (and rare) philopatric mutant will spread if and only if

$$0.75(1 - i) > 0.25 + 0.50(1 - m),$$

which is equivalent to

$$m > (3/2)i. \tag{1.1}$$

We can see in this simplified situation, close inbreeding will never be favoured by natural selection if $i > 2/3$ no matter how risky dispersal might be. This is equivalent to saying that if inbreeding depression is as severe as 67%, a male would be more successful dispersing and not breeding, while allowing his genes to be passed on by his outbred sisters, than remaining philopatric.

This reasoning can be generalized to a situation where there are two dispersal strategies (named α and β), each with its own specific costs of inbreeding (i) and costs of dispersal (m). Assuming f to be the average coefficient of consanguinity among potential males (see, for example,

Crow and Kimura, 1970; Maynard Smith, 1989)*, it can be shown that the α-strategy will be favoured by natural selection if, and only if

$$(1 - m_\alpha)[1 - i_\alpha(1 + 2f_\alpha)] > (1 - m_\beta)[1 - i_\beta(1 + 2f_\beta)]. \quad (1.2)$$

This relation reduces to (1.1) if we put $m_\alpha = m$, $i_\alpha = f_\alpha = 0$, and $m_\beta = 0$, $i_\beta = i$ and $f_\beta = 0.25$. If there is no strategy β which beats the existing strategy α, then α is an ESS (Maynard Smith, 1982).

Although this approach oversimplifies a complex phenomenon, it indicates how the variables in Table 1.2 can be incorporated into models for analysing and clarifying the issues involved. Moreover, both intra-patch and inter-patch dispersal (Figure 1.1a) can be handled conceptually in the same way (Figure 1.1b).

The idea of a 'dispersal sink' (Lidicker, 1975) has been widely utilized in the recent literature on dispersal (see Chapter 4 for a discussion), and is an important heuristic component of the spatial paradigm we are describing (Figure 1.1). A dispersal sink is defined as habitat that is accessible to a potential disperser, and that is at least good enough to support temporary survival. It thus could be merely habitat through which a disperser moves (transition habitat), habitat that supports survival and at times reproduction as well (marginal habitat), or even good habitat into which it is possible to immigrate (survival habitat). Without the availability of such a sink (see the habitat patch in Figure 1.1a surrounded by a dispersal barrier), individuals motivated to disperse become thwarted in their efforts; that is dispersal becomes frustrated and is referred to as 'frustrated dispersal' (Lidicker, 1975).

1.4 A MATTER OF SCALE

In this book we concentrate on 'small mammals' (*sensu* Bourlière, 1975; i.e. body weight <5 kg). Even though there are many well-known difficulties in working with such species (e.g. their secretiveness and cleverness in avoiding being caught), there are many advantages as well (e.g. the relative ease with which adequate samples can be obtained, their ubiquitousness, and the diversity and complexity of their life history traits). This has led to an extensive literature on dispersal in small mammals. Our task in this

* The coefficient of consanguinity, *f*, of two individuals is defined as the probability that two homologous genes drawn at random, one from each of the two individuals, will be identical by descent. This is the same as the inbreeding coefficient of progeny produced by these two individuals. Hence, the inbreeding coefficient of an individual is the same as the coefficient of consanguinity of its parents (Crow and Kimura, 1970).

book is to review these data and to attempt its synthesis. In addition, we have striven to broaden its scope by including a series of commentaries on dispersal in taxa other than small mammals.

Whenever comparisons are made between small mammals and other organisms, a problem of differences in spatial scale must often be confronted. Insects are usually smaller than small mammals and other mammals are often considerably larger. The concept of 'habitat grain' (*sensu* Levins, 1968) is useful in this context. As expressed by Pianka (1988), organisms that encounter and use resources in the same proportions in which they actually occur utilize the environment in a fine-grained manner. In contrast, organisms that spend disproportionate amounts of time in one or a few patches are said to use the environment in a coarse-grained way. This means that among two organisms of about the same size, one could be a habitat (or patch) specialist and hence be more coarse-grained, and the other might be more of a generalist and therefore relatively fine-grained (MacArthur, 1972).

From the perspective of a human observer, small mammals experience a coarse-grained environment and large mammals a more fine-grained environment. This is because small species usually live their lives within one or a few habitat types (as defined by human observers), whereas larger species readily move across numerous habitat types. An example of the latter would be a raptor searching for prey over many habitats, whereas the former could be thought of as a thrush limiting most of its daily activities to one habitat patch. However, compared to a collembolan, the thrush would experience its environment in a more fine-grained fashion. Thus, we must keep in mind that organisms of greatly varying size may experience their environments as equally heterogeneous, but they do so on quite different spatial scales. It is a challenge for us to find comparable units of measure when working across spatial scales in comparing species of vastly different size. Some of the difficulties of communication between those working on different organisms can be, we believe, attributed to these inevitable problems of scale.

1.5 DISPERSAL: A CONCEPTUAL 'GLUE'

Assuming that we can accept dispersal to have the broad-based meaning of one-way movements from a home site, how can we claim for it the integrative and synthetic role that we have espoused? The reasons are as direct as they are profound. In considering demographic processes, movements of individuals into and out of populations are potentially as critical as births and deaths. And, the impact is not merely on numbers (although this in itself can be substantial), but also on the sex ratios, age structure, social dynamics, and genetics of populations. We feel that in the past investigators have focused too much on the quantity of dispersers and that the current trend (e.g. Krebs, 1979) toward exploring quality of individuals is a favourable development. Attention needs also to be directed toward possible changes in quality of both the disperser's source and recipient populations. The dynamics of the genetic composition of populations (i.e. their genetic quality) as well as their resulting chances for extinction and reproduction are clearly the stuff of evolutionary change. Moreover, these changes can be strongly influenced by gene flow among populations (demes). Although imprecise thinking about group versus individual selection has troubled this field as well as others (e.g. Maynard Smith, 1976), it remains an interesting and important area for discussion (e.g. Van Valen, 1971; Wilson, 1977, 1980; Wade and McCauley, 1980; Wade and Breden, 1987; Lidicker, 1988a; Chapter 12).

Furthermore, ecological, evolutionary, and behavioural issues merge inextricably in a consideration of a number of important issues relating dispersal to population processes. It is after all the ecological context of populations that guides evolutionary changes, and dispersal, while being a part of this nexus, is fundamentally a behavioural phenomenon. For example, can dispersal influence the demographic stability of populations by disrupting co-adapted gene complexes? This has been suggested as an explanation for multi-annual density cycles in microtine rodents (Abramsky and Tracy, 1979; Rosenzweig and Abramsky, 1980). Alternatively, does dispersal provide a stabilizing influence on numbers by equalizing densities throughout a meta-population (Hastings, 1983; Stenseth, 1986, 1988)? Or, is it possible that dispersal actually contributes to density instability by disrupting stabilizing effects of social interactions (Petrusewicz, 1957, 1963; DeLong, 1967; Bailey, 1969; Charnov and Finerty, 1980; Hestbeck, 1982; Warkowska-Dratnal and Stenseth, 1985)? A related issue is whether increased heterozygosity, perpetrated by dispersal, can increase the vigour of a population sufficiently to enhance population growth (Turner, 1960; Smith *et al.*, 1975). Another question concerns the kind of spatial and genetic structure of populations that leads to the evolution of dispersal behaviour (e.g. Bull *et al.*, 1987). Brown (1985) provides a good example from her studies of mosquitofish (*Gambusia affinis*) relating the demographic and genetic bases of dispersal. A further question is how do social factors encourage or discourage dispersal behaviour? To what extent is dispersal genetically programmed? Is there any relationship between social structure and genetic structure (Chepko-Sade and Halpin, 1987; Lidicker and Patton, 1987)? Finally, what is the basis for sex differences in dispersal? These and other questions are discussed at length in the various chapters in this book.

An important conceptual correlate of the recent emphasis on the role of dispersal is the increasing appreciation for the patchy nature of most environments (Hansson, 1977; Anderson, 1980; Lidicker, 1985, 1988a,b; Chapter 4). The world is spatially uneven in its properties and this is true on the scale of movements for most organisms. Even sessile creatures 'feel' this unevenness on a temporal basis. Where there is unevenness across a landscape, there exist 'patches', and dispersal is one major mechanism for connections among patches (Figure 1.2). It is increasingly clear that our understanding of many seemingly intractable ecological and evolutionary problems becomes improved when they are viewed in terms of patch dynamics. For example, it is now possible to incorporate spatial as well as temporal variations into population dynamics theory (Tilman, 1982). It is increasingly clear that a subject population (however delimited) cannot be understood unless both its history and neighbouring populations are taken into account. Such an approach is now being widely exploited in a variety of organisms (e.g. Wiens, 1976; Wolff, 1980) as well as by modellers (e.g. Gadgil, 1971; Vance, 1980; Hastings, 1983; Horn, 1983; Stenseth, 1986), and forms the basis for the newly formulated field of landscape ecology (Hansson, 1977; Forman and Godron, 1986; Lidicker, 1988b; Chapter 8). In our own experience, this perspective has contributed to the study of microtine multi-annual cycles (Hansson, 1977; Stenseth, 1977, 1980, 1986; Rosenzweig and Abramsky, 1980; Bowen, 1982; Cockburn and Lidicker, 1983; Lidicker, 1985, 1988a; Ostfeld and Klosterman, 1986; Ostfeld *et al.*, 1985).

1.6 THE BOOK

It seems clear to us that dispersal is a key element in the understanding of a great many population processes. Moreover, we claim that the study of dispersal is enjoying a renaissance of interdisciplinary activity. This has resulted from numerous conceptual and methodological advances, and has led to rapid progress in our knowledge and understanding. As evidence for this, we merely call attention to the large number of monographs and multi-authored volumes on the subject that have appeared since 1978 when three were published in the same year (Baker, 1978; Dingle, 1978; Schmidt-Koenig and Keeton, 1978). These events served to focus attention on a new era of intense activity in this field of investigation (Gauthreaux, 1980; Lidicker and Caldwell, 1982; Swingland and Greenwood, 1983; Rankin, 1985; Chepko-Sade and Halpin, 1987).

In view of all this enterprise, what then is the contribution that this book attempts to make? First of all it goes into depth in reviewing the extensive literature on small mammals, much of which is very recent and for which no such book-length review is available. Second, the commentaries covering insects to primates broaden the perspective of the book and thereby assure that dispersal is viewed in a much broader context.

(a)

Figure 1.2 Patchy habitats. (a) A heterogenous landscape in Østerdalen of eastern Norway (photo from Rolf Anker Ims). In the foreground, a scaled-down experimental study-area for studies on landscape ecology and dispersal; in the background, a real-scale forested and patchy environment. (b) (p. 16) Habitat heterogeneity on a spatial scale meaningful to small rodents (after Ostfeld and Klosterman, 1986). A grassland habitat in north coastal California; patches are outlined according to dominant plant species. Black dots represent trap sites.

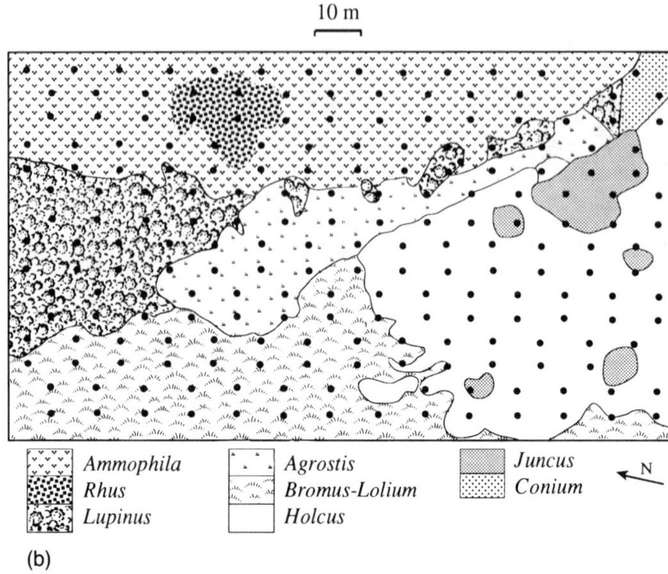

Figure 1.2 (b).

Along with the conceptual advances, there have been numerous and major improvements in the methodology of studying dispersal. In fact, it is probably true that dispersal was virtually ignored for so long partly because it is so difficult to study. In Appendix 1 we review the 'state of the art' with respect to methodologies of studying dispersal and analysing data on dispersal. Appendices 2 and 3 give detailed accounts of two important new techniques (i.e. radiotelemetry and radioisotopes).

We hope that this book taken as a whole will not only introduce our subject conceptually and methodologically, but will make meaningful connections among the diversity of researchers utilizing various organisms in their work and employing diverse technologies. The subject of organism movements is emphatically not monolithic, but we all can gain from increased communication.

REFERENCES

Abramsky, Z. and Tracy, C.R. (1979) Population biology of a 'noncycling' population of prairie voles and a hypothesis on the role of migration in regulating microtine cycles. *Ecology*, **60**, 349–61.

Alerstam, T. and Enckell, P.H. (1979) Unpredictable habitats and evolution of bird migration. *Oikos*, **33**, 228–32.

Anderson, P.K. (1970) Ecological structure and gene flow in small mammals. *Symposium of the Zoological Society of London*, no. 26, 299–325.

Anderson, P.K. (1980) Evolutionary implications of microtine behavioral systems on the ecological stage. *The Biologist*, **62**, 70–88.

Bailey, E.D. (1969) Immigration and emigration as contributory regulators of populations through social disruption. *Canadian Journal of Zoology*, **47**, 1213–5.

Baker, R.R. (1978) *The Evolutionary Ecology of Animal Migration*, Holmes and Meier, New York, 1024 pp.

Bengtsson, B.O. (1978) Avoiding inbreeding: at what cost? *Journal of Theoretical Biology*, **73**, 439–44.

Bourlière, F. (1975) Mammals, small and large; the ecological implications of size, in *Small Mammals: their productivity and population dynamics* (eds F.B. Golley, K. Petrusewicz and L. Ryszkowski), Cambridge University Press, London, pp. 1–8.

Bowen, B.S. (1982) Temporal dynamics of microgeographic structure of genetic variation in *Microtus californicus. Journal of Mammalogy*, **63**, 625–38.

Brown, K.L. (1985) Demographic and genetic characteristics of dispersal in the mosquitofish, *Gambusia affinis* (Pisces: Poeciliidae). *Copeia*, **1985**, 597–612.

Bull, J.J., Thompson, C., Ng, D. and Moore, R. (1987). A model for natural selection of genetic migration. *American Naturalist*, **129**, 143–57.

Charnov, E.L. and Finerty, J.P. (1980) Vole population cycles; a case for kin-selection? *Oecologia*, **45**, 1–2.

Chepko-Sade, S.D. and Halpin, Z.T. (eds) (1987) *Mammalian Dispersal Patterns: the effects of social structure on population genetics*, University of Chicago Press, Chicago, 342 pp.

Cockburn, A. and Lidicker, W.Z. Jr (1983) Microhabitat heterogeneity and population ecology of an herbivorous rodent, *Microtus californicus. Oecologia*, **59**, 167–77.

Crow, J.F. and Kimura, M. (1970) *An Introduction to Population Genetics Theory*, Harper and Row, New York, 591 pp.

DeLong, K.T. (1967) Population ecology of feral house mice. *Ecology*, **48**, 611–34.

Dingle, H. (1972) Migration strategies of insects. *Science*, **175**, 1327–35.

Dingle, H. (ed.) (1978) *Evolution of Insect Migration and Diapause*, Springer-Verlag, New York, 284 pp.

Dingle, H. (1980) Ecology and evolution of migration, in *Animal Migration, Orientation, and Navigation* (ed. S.A. Gauthreaux, Jr), Academic Press, New York, pp. 1–101.

Forman, R.T.T. and Godron, M. (1986) *Landscape Ecology*, Wiley, New York, 619 pp.

Gadgil, M. (1971) Dispersal: population consequences and evolution. *Ecology*, **52**, 253–61.

Gauthreaux, S.A., Jr (ed.) (1980) *Animal Migration, Orientation, and Navigation*, Academic Press, New York, 387 pp.

Hamilton, W.D. (1964) The genetical evolution of social behaviour I and II. *Journal of Theoretical Biology*, **7**, 1–52.

Hansson, L. (1977) Spatial dynamics of field voles Microtus agrestis in heterogeneous landscapes. *Oikos*, **29**, 539–44.
Hastings, A. (1983) Can spatial variation alone lead to selection for dispersal? *Theoretical Population Biology*, **24**, 244–51.
Hestbeck, J.B. (1982) Population regulation of cyclic mammals: the social fence hypothesis. *Oikos*, **39**, 157–63.
Horn, H.S. (1983) Some theories about dispersal, in *The Ecology of Animal Movement* (eds I.R. Swingland and P.J. Greenwood), Clarendon Press, Oxford, pp. 54–62.
Kennedy, J.S. (1961) A turning point in the study of insect migration. *Nature*, **189**, 785–91.
Krebs, C.J. (1979) Dispersal, spacing behaviour, and genetics in relation to population fluctuation in the vole *Microtus townsendii. Fortschritte der Zoölogie*, **25**, 61–77.
Levins, R. (1968) Evolution in changing environments, some theoretical explorations, *Monographs in Population Biology*, No. 2, Princeton University Press, Princeton, New Jersey, 120 pp.
Levins, R. (1970) Extinction, in *Some Mathematical Questions in Biology* (ed. J.D. Cowan), Lectures on mathematics in the life sciences, *Am. Math. Soc.*, Vol. 2, pp. 75–107.
Lidicker, W.Z. Jr (1975) The role of dispersal in the demography of small mammals, in *Small Mammals: their production and population dynamics* (eds F.B. Golley, K. Petrusewicz and L. Ryszkowski), Cambridge University Press, London, pp. 103–28.
Lidicker, W.Z. Jr (1985) Population structuring as a factor in understanding microtine cycles. *Acta Zoologica Fennica*, **173**, 23–7.
Lidicker, W.Z. Jr (1988a) Solving the enigma of microtine 'cycles'. *Journal of Mammalogy*, **69**, 225–35.
Lidicker, W.Z. Jr (1988b) The synergistic effects of reductionist and holistic approaches in animal ecology. *Oikos*, **53**, 278–81.
Lidicker, W.Z. Jr and Caldwell, R.L. (1982) Dispersal and migration. *Benchmark Papers in Ecology* 11, Hutchinson Ross, Stroudsburg, Pennsylvania, 311 pp.
Lidicker, W.Z. Jr and Patton, J.L. (1987) Patterns of dispersal and genetic structure in populations of small rodents, in *Mammalian Dispersal Patterns: the effects of social structure on population genetics* (eds B.D. Chepko-Sade and Z.T. Halpin), University of Chicago Press, Chicago, pp. 144–61.
MacArthur, R.H. (1972) *Geographical Ecology: patterns in the distribution of species*, Harper and Row, New York, 269 pp.
Maynard Smith, J. (1976) Group selection. *Quarterly Review of Biology*, **51**, 277–83.
Maynard Smith, J. (1982) *Evolution and the Theory of Games*, Cambridge University Press, Cambridge, 224 pp.
Maynard Smith, J. (1989) *Evolutionary Genetics*, Oxford University Press, Oxford, 325 pp.
Odum, E.T. (1953) *Fundamentals of Ecology*, W.B. Saunders, Philadelphia, 384 pp.
Ostfeld, R.S. and Klosterman, L.L. (1986) Demographic substructure in a Califor-

nia vole population inhabiting a patchy environment. *Journal of Mammalogy*, **67**, 693–704.

Ostfeld, R.S., Lidicker, W.Z. Jr and Heske, E.J. (1985) The relationship between habitat heterogeneity, space use, and demography in a population of California voles. *Oikos*, **45**, 433–42.

Petrusewicz, K. (1957) Investigation of experimentally induced population growth. *Ekologia Polska, Seria A*, **5**, 281–301.

Petrusewicz, K. (1963) Population growth induced by disturbance in the ecological structure of the population. *Ekologia Polska, Seria A*, **11**, 87–125.

Pianka, E.R. (1988) *Evolutionary Ecology*. Harper and Row, New York, 468 pp.

Rankin, M.A. (ed.) (1985) *Migration: mechanisms and adaptive significance*, Contributions in Marine Science, Suppl. 27, 868 pp.

Rosenzweig, M.L. and Abramsky, Z. (1980) Microtine cycles: the role of habitat heterogeneity. *Oikos*, **34**, 141–6.

Schmidt-Koenig, K. and Keeton, W.T. (eds) (1978) *Animal Migration, Navigation, and Homing*. Springer-Verlag, New York, 480 pp.

Smith, M.H., Garten, C.T. Jr and Ramsey, P.R. (1975) Genic heterozygosity and population dynamics in small mammals, in *Isozymes*, IV. *Genetics and evolution* (ed. C.L. Markert), Academic Press, New York, pp. 85–102.

Solbreck, C. (1978) Migration, diapause and direct development as alternative life histories in a seed bug, *Neacoryphus bicrucis*, in *Evolution of Insect Migration and Diapause* (ed. H. Dingle), Springer-Verlag, New York, pp. 195–217.

Southwood, T.R.E. (1977) Habitat, the templet for ecological strategies. *Journal of Animal Ecology*, **46**, 337–65.

Southwood, T.R.E. (1978) *Ecological Methods with Particular Reference to the Study of Insect Populations*, 2nd edn, Chapman & Hall, London, 524 pp.

Stenseth, N.C. (1977) On the importance of spatio-temporal heterogeneity for the population dynamics of rodents: towards a theoretical foundation of rodent control. *Oikos*, **29**, 545–52.

Stenseth, N.C. (1980) Spatial heterogeneity and population stability: some evolutionary consequences. *Oikos*, **35**, 165–84.

Stenseth, N.C. (1983) Causes and consequences of dispersal in small mammals, in *The Ecology of Animal Movement* (eds I. Swingland and P.J. Greenwood), Oxford University Press, Oxford, pp. 63–101.

Stenseth, N.C. (1986) On the interactions between stabilizing social factors and destabilizing trophic factors in small rodent populations. *Theoretical Population Biology*, **29**, 365–85.

Stenseth, N.C. (1988) The social fence hypothesis: a critique. *Oikos*, **52**, 169–77.

Swingland, I.R. and Greenwood, P.J. (eds) (1983) *The Ecology of Animal Movement*, Oxford University Press, Oxford, 311 pp.

Taylor, L.R. (1986) Synoptic dynamics, migration and the Rothamsted insect survey. *Journal of Animal Ecology*, **55**, 1–38.

Taylor, L.R. and Taylor, A.J. (1977) Aggregation, migration and population mechanics. *Nature*, **265**, 415–21.

Tilman, D. (1982) *Resource Competition and Community Structure*. Princeton University Press, Princeton, New Jersey, 296 pp.

Turner, N. (1960) The effect of inbreeding and cross-breeding on numbers of insects. *Annals of the Entomological Society of America*, **53**, 686–8.

Vance, R.R. (1980) The effect of dispersal on population size in a temporally varying environment. *Theoretical Population Biology*, **18**, 343–62.

Van Valen, L. (1971) Group selection and the evolution of dispersal. *Evolution*, **25**, 591–8.

Wade, M.J. and Breden, F.J. (1987) Kin selection in complex groups: mating structure, migration structure, and the evolution of social behaviors, in *Mammalian Dispersal Patterns: the effects of social structure on population genetics* (eds B.D. Chepko-Sade and Z.T. Halpin), University of Chicago Press, Chicago, pp. 273–83.

Wade, M.J. and McCauley, D.E. (1980) Group selection: the phenotypic and genotypic differentiation of small populations. *Evolution*, **34**, 844–55.

Warkowska-Dratnal, H. and Stenseth, N.C. (1985) Dispersal and the microtine cycle: comparison of two hypotheses. *Oecologia*, **65**, 468–77.

Wiens, J.A. (1976) Population responses to patchy environments. *Annual Review of Ecology and Systematics*, **7**, 81–120.

Wilson, D.S. (1977) Structured demes and the evolution of group-advantageous traits. *American Naturalist*, **111**, 157–85.

Wilson, D.S. (1980) *The Natural Selection of Populations and Communities*, Benjamin/Cummings, Menlo Park, California, 186 pp.

Wolff, J.O. (1980) The role of habitat patchiness in the population dynamics of snowshoe hares. *Ecological Monographs*, 50, 11–130.

– 2

To disperse or not to disperse: who does it and why?

W.Z. LIDICKER JR. AND N.C. STENSETH

2.1 INTRODUCTION

Dispersal is an integrative and synthetic phenomenon bridging the disciplines of ecology, genetics, behaviour, and evolution. It is therefore a key element in the understanding of a great many population processes (Chapter 1). Moreover, it becomes critical to our understanding of these processes to know which individuals disperse and what factors cause them to initiate this behaviour.

Motivations can usefully be viewed both from an ultimate (evolutionary) or from a proximate (triggering) perspective. In this chapter we provide a framework for asking questions about two vitally important issues, namely, who are the dispersers and why do they disperse. This analysis then leads us to propose a classification matrix for dispersal behaviour. Hopefully, this will provide a heuristic framework for organizing our thinking and our knowledge about dispersal, and for designing research protocols. Our classification is based on the observation that dispersal can sometimes be motivated by forces outside of the subject individual (extrinsic) and at other times by properties of the individuals themselves (intrinsic). In most cases, however, we will argue that both intrinsic and extrinsic factors will be involved, and distinguishing between them may become arbitrary and even misleading.

2.2 WHO ARE THE DISPERSERS?

In an earlier era (e.g. Elton, 1927, 1930), dispersers were thought to be of only one type. These were those hapless individuals faced with a home environment that was economically or socially inhospitable. Thus environmentally triggered, such individuals left home out of desperation. Their chances for future survival or reproductive success were thought to be minimal. Later, it was recognized that there were individuals genetically programmed to disperse (e.g. Andrewartha and Birch, 1954; Howard,

1960; Lidicker, 1962); they would do so at a certain age or season regardless of conditions at home. Odds for their future success were thought to be very much more favourable. This dichotomy of genetic imperative or environmental necessity was a very useful beginning, but, at least for small mammals, it turned out to be incomplete. Accumulating evidence now strongly suggests that dispersal is generally a heterogeneous phenomenon, defying attempts to characterize dispersers or their motivations in any simple fashion.

The demographic, behavioural, and genetic attributes of dispersers are reviewed, among others, by Gaines and McClenaghan (1980) for small mammals generally and for *Microtus* by Lidicker (1985a). These, plus other reviews, support the conclusions that often all post-weaning ages and both sexes can be involved, and that even the same sex/age cohort may disperse at different times for different reasons. Also, the same individual may disperse more than once and for possibly different reasons. These and related issues are discussed by others in this book (particularly in Chapters 5 and 9).

Because of this heterogeneity, there is a natural tendency to classify the various kinds of dispersal so as to better understand it and communicate about it. While it is generally helpful to simplify phenomena in order to understand them better, we should keep in mind that if a complex phenomenon is studied only under the assumption that it is simple, understanding will likely be incomplete. It is important to remember that classification systems by their nature tend to simplify reality and hence to caricature it. On the other hand, classification systems are essential for helping us to organize our data, our thinking, and our research protocols. At the same time we must be wary of accepting them as substitutes for reality. The Linnaean taxonomic system for classifying biotic diversity is a good case in point.

The first step in constructing a classification of dispersal is to clarify what is dispersal and what is not (Chapter 1). Here we use 'dispersal' to mean one-way movements of individuals away from their home ranges (sites). Although this is an objective definition, the boundaries of the phenomenon are not as clear as we would like. One source of uncertainty is the temporal scale over which movements occur. As observers we must be careful to think on a scale appropriate to the life history of our subject organisms. For small mammals, home ranges are usually conceptualized on a time scale of days, or weeks. If a larger time scale is specified, what may be defined as dispersal may seem to be philopatry. Or, on a scale of an hour, movements may appear nomadic even though they would be defined as philopatric within the usual definition of home range. The second source of confusion occurs because there are several ambiguous kinds of closely related movements. We refer specifically to nomadism,

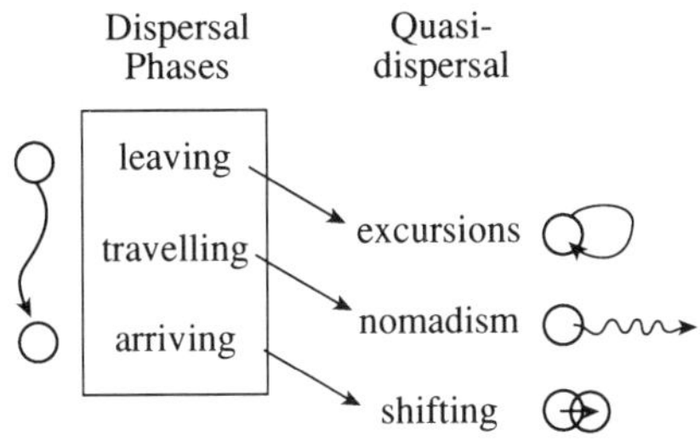

Figure 2.1 The three phases of normal dispersal, each of which generates a related form of quasi-dispersal. The four behaviours are also shown symbolically.

exploration, and shifting. All three are extra-home range movements which are generally excluded from dispersal.

Nomadism is a behaviour which can be viewed as a failure to establish a home range anywhere, and hence is like being in a chronic dispersal mode. In other cases, individuals become nomads (floaters) when they are unsuccessful in establishing a new home range. Nomads can be recognized by their failure to achieve a consistent centre of activity over a short time period, such as a few days. They tend to cover a larger area than do the residents, and move randomly with respect to past movements or anticipated known home range behaviour. An example of this latter situation in adult male *Mus musculus* is described by Lidicker (1976). Obviously, it is difficult to distinguish operationally a disperser that is in the travelling phase from a nomad. Another possibly ambiguous circumstance will occur when an individual moves slowly, that is in a time period exceeding a few days, around a very large home range. Such an individual will appear to be nomadic on a short time scale but on a large scaler is actually philopatric.

Explorers are those who make short-term excursions outside of their home ranges. They may be searching for mates, some scarce resource, or better living conditions generally. Excursions may be preludes to actual dispersal, and occur commonly among species of small mammals (reviewed in Lidicker, 1985b). If an explorer fails to return home, it becomes indistinguishable from a disperser.

Shifters are those who manage to move their home ranges by gradually adding bits of real estate to one edge while simultaneously subtracting it from another. Thus they never actually abandon completely their current home range. This behaviour has been described for *Microtus pennsylvanicus* by Madison (1980a,b; see also Appendix 2).

These three behaviours are clearly related closely to dispersal, but are distinguishable in various ways (Figure 2.1). Collectively, they can be called 'quasi-dispersal'. True dispersal has three phases (Figure 2.1): leav-

Table 2.1 Possible bases for classifications of dispersal behaviour. The utility of any particular scheme depends on the purposes for which it is intended. The first viewpoint relates to 'why' questions while the remaining concern 'how' questions (*sensu* Mayr, 1961)

Perspective	*Question*	*Criteria*
Evolutionary	Is it adaptive?	Fitness contribution
Demographic	Is it related to resource availability?	Relation to carrying capacity or density
	Is it density responsive?	Responsiveness?
Genetic	Is it genetically directed?	Heritabilities
Ethological	What environmental circumstances induce it?	Environmental cues or behavioural stimuli
Physiological	What internal states motivate it?	Internal drive

ing, travelling, and arriving. The new home range must be disjunct from the abandoned one, although it is possible that an individual may at some future time disperse back to the original home range. This round trip style of dispersal, especially if it is organized on a seasonal basis, is called 'migration' by vertebrate biologists (Lidicker and Caldwell, 1982; Lidicker, 1985b; see Table 1.1). The arrival phase can be omitted if death of the disperser aborts the process (see Figure 1.1). Nomads are deviants of the travelling phase of dispersal. Explorers leave their home range, sometimes repeatedly, but avoid becoming dispersers by returning home. Shifters are deviants of the arrival phase by continually establishing new parts to the home range without incorporating the other dispersal phases, except in the form of short excursions (Figure 2.1). While remaining cognizant of the somewhat fuzzy boundary to dispersal behaviour caused by quasi-dispersal, the following discussion will focus primarily on dispersal in the restricted sense. That is, our discussion will concern one-way movements away from an individual's home range (site).

In considering further subdivisions within the dispersal category, it became apparent to us that the utility and inherent logical structure of any scheme depended strongly on either the purposes for which it was intended or on the perspectives of the investigator (Table 2.1). Some will be inclined to ask 'how?' a phenomenon works, while others will ask 'why?' a phenomenon occurs (Mayr, 1961; Pianka, 1988). A physiologist, for example, might want to classify dispersal on the basis of appetitive states or physiological condition, while an ethologist will be most interested in a system based on conspecific behavioural stimuli. Ecologists and evolutionary biologists are most comfortable with criteria concerning effects of dispersal on survival and reproduction (and hence fitness), and how dispersal relates to critical resources and other species in the community. Each perspective could generate a unique and useful classification system.

In our view, the criteria that may at this stage of our understanding be meaningful across the widest spectrum of viewpoints are the motivations which generate or trigger dispersal behaviour. This puts the emphasis on the 'how' rather than the 'why' approach. Moreover, motivations help to identify and characterize dispersers in a context which is a necessary first step to enhanced understanding at other levels. The fact that motivations are often difficult to determine will, of course, reduce the operational utility of any classification based upon them. However, alternative criteria, such as those predicated on a 'why' perspective, will be even more difficult to establish empirically. We intend to devise our scheme of motivations so as to make them as amenable as possible to operational determination. We feel that a classification is immensely more satisfying if the criteria on which it is based are unambiguously defined in operational terms. Therefore, before devising our classification, we turn to an examination of the factors motivating dispersal. To the extent possible, we will suggest criteria for identifying these factors in realistic situations.

2.3 WHAT MOTIVATES DISPERSAL?

It is apparent that just as dispersers are a heterogeneous assemblage, so too are the motivations that drive them (Lidicker, 1975, 1985a,b; Lidicker and Caldwell, 1982: Part II; Stenseth, 1983; Chapters 4, 5 and 9). The immediate stimulus, however, for dispersal may be quite different and functionally disconnected from the evolutionary issues that transcend it. Hence, we must keep the distinction between proximate and ultimate factors in mind. The situation is analogous to the well-known phenomenon in birds in which current clutch size may be triggered by some factor that signals future food supply (Lack, 1968).

Ultimate factors are the selective forces that shape the evolution of the behaviour. These act via the fitness traits of survival and reproduction. If dispersal enhances these functions, it will be selected for independently of whatever proximate factors may serve to trigger it. Another evolutionary issue is the fitness bonanza that often follows successful colonization of empty habitat or the discovery of new habitat beyond the species' current range. Also of potential concern is the possibility of inbreeding (Falconer, 1960) or outbreeding depression (Shields, 1982). The importance of these last factors is controversial, and at least seems to vary with different species. In particular, species whose life histories encourage inbreeding seem not to experience inbreeding depression (May, 1980; Ralls, *et al.*,

1986; Templeton, 1987). Outbreeding depression, however, might be a realistic risk for such species. Such depression occurs when mating occurs between individuals sufficiently different genetically that their offspring are less fit than either parental type. Typically outbred species, on the other hand, characteristically experience inbreeding depression (Ralls *et al.*, 1979; Ballou and Ralls, 1982; Ralls and Ballou, 1982; Templeton and Read, 1983). A final evolutionary issue concerns the maintenance of an appropriate level of genetic variability in a population (Haldane, 1937; Brown, 1958; Brues, 1964; Dobzhansky, 1965; Kolata, 1974; Nevo, 1978; Cooper and Kaplan, 1982). This is often considered a population level process involving the long-term probability of demic survival and reproduction. From an individual's perspective, however, dispersal may improve its chances of living in a population with better than average long-term prospects for success.

In some cases, proximate mechanisms are closely allied to ultimate factors. This is the case, for example, when the choice is simply dispersal or death. Starvation directly impacts fitness, and if it leads to dispersal the proximate and ultimate motivations are virtually the same. In such instances, the issue is clearly one of increasing chances for survival even though the proximate force may be only indirectly related. In other cases, it is extremely difficult to determine the evolutionary explanation for a proximate mechanism, such as the genetic programming of dispersal by one sex at weaning.

Historically, the most significant initial division between dispersal acts, at the level of the individual organism, is the separation of those that can be attributed to *intrinsic* factors and those explained by *extrinsic* ones. This division corresponds to the innate versus environmental dichotomy proposed 30 years ago by Howard (1960) and supported by other authors. Ultimately, it is of course a false dichotomy because a complex behaviour such as dispersal is always the result of both intrinsic and extrinsic factors. It remains a useful distinction, nevertheless, because often motivations can be attributed primarily to internal or to external factors, even while recognizing that both forces are operating.

A second major organizing axis for motivations concerns the volition of the behaviour, i.e. whether or not the act is voluntary. We do not imply any conscious decision making on the part of voluntary dispersers, but only that their act is not forced by intolerable circumstances. This dichotomy can be made to correspond approximately to the distinction proposed by Stenseth (1983) for adaptive versus non-adaptive dispersal. As we will explain, the recognition of these two major modes of dispersal will depend on: (a) the circumstances under which it occurs, and (b) the availability of evidence for the fitness consequences of the act.

In Chapter 1 a model is described to show how potential dispersers could, through natural selection, make adaptively correct decisions about dispersal if they had correct information about the costs to fitness of alternative behaviours (Table 1.2). This model allows us to objectively define non-adaptive dispersal as that which violates the fitness-based advice provided by equations such as (1.2). Involuntary dispersal will mostly be of this type. Voluntary dispersal, on the other hand, will generally equate to adaptive dispersal as defined by the fitness inequalities. A complication for the investigator, as well as possibly for the potential disperser, is that the alternative fitness measures may both be so low that either choice appears to be both maladaptive and involuntary, even though one is marginally better. This is like the familiar problem in science of incorrectly not rejecting a null hypothesis when in fact two quantities are different but close ('type 2 error').

Involuntary dispersers can now be categorized as those that leave because their future fitness as a resident is rapidly approaching zero (physical environment has become intolerable or resources are inadequate for survival), or they are literally forced to leave by others (either conspecifics or some other species). In the former cases, dispersal may not really be non-adaptive as the fitness value of dispersal is probably greater than zero, although perhaps not much greater. In the latter cases, dispersal is likely to be truly non-adaptive. The involuntary category thus includes saturation dispersers (Lidicker, 1975) who leave because population density exceeds carrying capacity (economic reasons), and social subordinates who leave as a consequence of territorial behaviour, unsuccessful intra-sexual competition, or parental aggression (social reasons). All age and sex groups may be represented among involuntary dispersers, but we would expect that they would be most commonly found among socially subordinate individuals, and those living at high densities, or in poor or unstable habitats. Such circumstances would certainly suggest involuntary dispersal, although ideally we would like to be able to establish that an individual's physical, biotic, or social environment had become so intolerable that to stay home would reduce fitness to zero.

Voluntary dispersers are those that leave home even when their home environment is not physically, economically, or socially desperate. For such behaviour to evolve, there must on the average be benefits received in excess of the disadvantages typically associated with venturing forth into the unknown (Chapter 1). This is therefore an intriguing class of dispersal, deserving of considerable attention. To recognize a disperser as voluntary, one must have evidence that the fitness of that individual would not be expected to drop drastically in the near future if it did not leave home. For example, under some experimental regimes, it may be possible to place a disperser back in its home range and thereby test whether or not this environment has zero survival capacity for the individual. In other situations, direct observations of individuals just before they disperse could provide clues as to whether or not it is a voluntary act.

2.3.1 Intrinsic factors

Intrinsic control often implies a relatively strong genetic influence over dispersal behaviour [see Lidicker and Caldwell (1982: Part I) for a summary of the genetic basis of dispersal]. We include as intrinsic many cases in which dispersal is strongly sex biased, narrowly restricted to a given age or developmental stage, or associated with reproduction or other physiological condition.

Reproduction is a motivating factor that can clearly be ultimate as well as proximate. It is common for dispersal to serve a reproductive function. This can take the form of (1) achieving earlier reproductive maturation; (2) finding suitable mates (both quantity and quality may be at stake); or (3) enhancing the survival of weaned offspring. All of these motivations have been described for various species of *Microtus* (Lidicker, 1985a), and probably occur in most other groups: for example, there is evidence that dispersing voles reach sexual maturity at a younger age than residents, adult males often disperse at the beginning of the breeding season, and in five species females have been recorded abandoning their home ranges to their weaned litters (see also Cockburn, 1988).

Strong genetic control is also implied in those species in which dispersal is strongly sex biased or occurs regularly at a particular age or developmental stage (e.g. Chapter 5). In such cases, proximate motivations may be largely intrinsic, i.e. they are built into the neuroendocrine system. How such controls evolved remains an important question, though for sex biases in dispersal, it has been suggested that inbreeding avoidance is involved (Packer, 1979, 1985; Greenwood, 1980; Brandt, 1985; Chapter 5). Others feel that it has to do with mate competition between parents and their like-sexed offspring (Anderson, 1980, 1989; Dobson, 1982; Moore and Ali, 1984). Still another possibility is that the dispersive sex corresponds to the territorial sex (Chapter 9). The argument is that territorial spacing behaviour encourages dispersal in individuals of the territorial sex who lack territories. This is not necessarily a sufficient explanation, however, as in the genus *Microtus* a slight male bias in dispersal commonly occurs even though some species have male territoriality, some female territoriality, and still others show group territoriality by mated pairs (Lidicker, 1985a; Ostfeld, 1985).

Age or developmentally related dispersal has been traditionally viewed as favouring those with maximal survival capabilities. Moreover, losses to the population could be minimized if only individuals with relatively low reproductive value disperse. This would have the corollary effect of greatly improving the situation for the non-dispersers as well. However, it is not at all clear how this pattern could evolve under individual-level selection, unless dispersal was strictly involuntary (see below). In contrast, Dingle (1972) has argued that this type of dispersal (ontogenetic) should favour individuals with maximal reproductive value so as to maximize colonizing ability.

Physiological condition may comprise another important category of intrinsic motivations for dispersal. Health or levels of stored energy could logically be used by organisms as cues for dispersal.

2.3.2 Extrinsic factors

The first category of extrinsic motivating factors to be mentioned are physical (abiotic) factors. An individual's home range may deteriorate because of floods, fire, erosion, loss of suitable shelter, unfavourable temperatures, etc. Although survival and even reproduction may still be possible, voluntary dispersal to another, better location may improve conditions. Such changes may occur regularly on a seasonal basis triggering dispersal episodes that are typical features of a species' life history. For example, *Microtus xanthognathus* undergoes two seasonally connected dispersal episodes per year (Wolff and Lidicker, 1980), as do lemmings (*Lemmus lemmus*; Kalela, 1961).

Economic factors form a large group of extrinsic motivating forces. These factors comprise the basic resources of food, water, and shelter, those components generally considered to determine a habitat's carrying capacity. Again, it is obvious that inadequate resource levels can provoke involuntary dispersal. But, even when population density is below carrying capacity, economic issues can motivate voluntary dispersal. Individuals may be able to predict future shortages of resources while they are still adequate, or, more commonly, they may be sensitive to the possibility of moving to a better home range. This last circumstance can be called 'the greener pasture syndrome', and is particularly likely to occur following exploratory excursions. Lastly, there may be individual variation in the ability to extract or sequester resources from the habitat, producing an uneven distribution of the potentially available resources among individuals (Łomnicki, 1978, 1980). Because of this, some individuals may face shortages even though the total level of resources in an area could support more individuals. Resource levels can also act to inhibit dispersal. If resources are inadequate in the area surrounding an individual's home range, it may remain at home even if motivated for other reasons to leave. Such motivational conflicts could lead to frustrated dispersal among individuals so affected (Lidicker, 1975).

Social mechanisms, besides acting to force involuntary dispersal, may also lead to voluntary dispersal through changing intensity or quality of interactions. It has also been suggested that social interactions may inhibit dispersal ('social fence') when aggressive encounters are more intense outside of the home social group as compared to within the group (Hestbeck, 1982). More data on this latter scenario are especially needed.

The last category of extrinsic factors is that of interspecific co-actions. We refer here to interspecific competition, predation, and parasitism (interference). Although little studied, it seems likely that such effects may be widespread. Fulk (1972), for example, has reported that movements of *Microtus pennsylvanicus* can be responsive to the presence of the predatory shrew *Blarina brevicauda*; and Power *et al.* (1985) show how predatory bass (*Micropterus*) stimulate dispersal in a species of minnow (*Campostoma*) in small Oklahoma streams.

2.4 A CLASSIFICATION OF DISPERSAL

This overview of who disperses and the factors motivating this behaviour emphasizes the need for a classification scheme in which the heterogeneity and complexity of the phenomenon can be portrayed. We propose a system based on proximate motivating factors, acknowledging that these are often difficult to determine. Other possible criteria of interest, such as ultimate motivating factors or relation to population carrying capacity, are even more difficult to assess. Therefore, proximate factors may offer the best compromise of insightfulness and operational utility. After all, like most things that are interesting, almost everything about dispersal is difficult to study, so we should not be deterred by the seeming elusiveness of determining proximate mechanisms. We expect that imaginative research protocols will make it feasible to distinguish among the possibilities. Lastly, if dispersal is frustrated by barriers or the lack of a dispersal sink (Lidicker, 1975; Tamarin, 1977, 1980), it is the proximate motivations that will likely determine the kind of behavioural or physiological pathologies that may develop.

Having focused on proximate motivations, it becomes immediately apparent that they do not lend themselves to a simple dichotomous classification. This is because intrinsic and extrinsic factors are often inextricably intertwined, and because even within one of these categories multiple and even conflicting motivations often exist. For example, resource factors may encourage remaining a resident while social factors may be promoting dispersal. The results of such conflicts are sometimes difficult to predict. In view of these complexities we suggest a matrix configuration portrayed as a three-dimensional volume (Figure 2.2). The primary axes for this dispersal polygon are the intrinsic and extrinsic factors potentially motivating dispersal. To this is added the dichotomy of enforced (or involuntary)

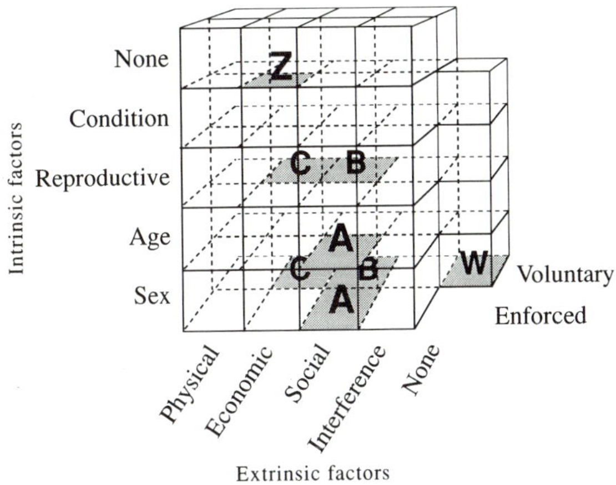

Figure 2.2 A proposed three-dimensional dispersal polygon to be used for classifying dispersal phenomena in small mammals based primarily on proximate motivating factors. Letters in the matrix cells refer only to examples of specific dispersal behaviour explained in the text.

versus voluntary because this is interesting from an evolutionary perspective. 'Enforced' dispersal simply implies that a contrary decision would reduce future fitness to zero. This array yields a dispersal volume with 50 cells. Some cells, however, will be more frequently occupied than others, and six cells are null or nonsense combinations and can be omitted (Figure 2.2). One of the exciting future prospects for the remaining 44 cells is to determine if there are some which are filled frequently and hence represent common types of dispersal as well as cells that remain empty. It would be interesting to discover if dispersal syndromes cluster together much like taxa do. Cells are also provided for instances of dispersal that can be attributed largely to intrinsic or extrinsic factors alone. For example, male dispersal that occurs independently of environmental or social triggers is simply 'sex-biased voluntary dispersal' (cell 'W' in Figure 2.2). Similarly, as a second example, dispersal forced by exhaustion of food resources is 'enforced economic dispersal' (cell 'Z' in Figure 2.2).

In most cases, however, some combination of extrinsic and intrinsic factors will be involved. Moreover, many cases of dispersal will involve more than one cell of the matrix. Three examples involving two cells are: (1) juvenile male *Antechinus* that disperse through maternal aggression (Cockburn *et al.*, 1985; Chapter 4) would be represented by cells 'A'; (2) post-reproductive females that abandon their home range to their weaned litter by cells 'B'; (3) if a pregnant female moves its home range, having found on an excursion a better place to raise her young, this would be characterized by 'C'. Other examples could be provided which involve more than two cells of the polygon. The important points to be derived from the three-dimensional approach are that dispersal is often multiply motivated and that intrinsic and extrinsic factors tend to be intertwined.

Earlier attempts at dichotomous classifications have been meritorious, but necessarily apply to a more limited perspective (Table 2.1). The genetic versus environmental distinction (Howard, 1960) has already been mentioned. It can now be seen that this classification omits physiological motivating factors and, more critically, forces us into what is often an unrealistic 'nature versus nurture' dichotomy. The distinction between saturation and presaturation dispersal (Lidicker, 1975) was important, we feel, in emphasizing that not all dispersal is forced by intolerable economic or social conditions (see also Chapter 9), but is limited to relating dispersal to various demographic issues such as a population's relation to its carrying capacity and density-regulating mechanisms. Stenseth's (1983) suggestion of adaptive versus non-adaptive dispersal places the subject in an evolutionary context. It directs us to ask whether or not a dispersal act is likely to increase an individual's fitness relative to what it would have been if the decision was made not to disperse (see also Chapter 1). Another more complex classification proposed by Lidicker (1985b) helped to emphasize the heterogeneity of dispersal, but it is based on a mixture of criteria that define only the general circumstances of dispersal and de-emphasize multiple causations. These earlier and more limited classification schemes have all been helpful, but now we feel that the subject is ready for a more comprehensive classificatory device.

Our proposed classification scheme relates 12 basic categories of motivating factors in a 44 cell polygon, and reality demands that any serious attempt to understand dispersal must take this potential complexity into account. Otherwise, we will harvest only confusion when we ask who disperses and why.

2.5 CONCLUSIONS

1. Dispersal is a key element in the understanding of many population processes. Its interdisciplinary nature bridges the fields of ecology, genetics, behaviour, and evolution in addition to several applied fields (Chapter 1).
2. Dispersers are a heterogeneous assemblage potentially including all sex and age groups in a population. In fact, dispersal occurs at a variety of times and under diverse circumstances.
3. The diversity of dispersal behaviour prompts the development of classification systems so as to organize our knowledge and to facilitate

communication. At the same time we must keep in mind that classifications simplify and hence risk distorting reality.

4. The distinction between ultimate (that is, evolutionary) motivating factors and proximate ones must be maintained. Even though an understanding of proximal factors is necessary, full understanding of dispersal will require placing it in an evolutionary context.
5. A classification scheme is proposed as a framework for discussion which is based primarily on the diversity of proximate motivating factors. The primary axes of the classification matrix are intrinsic (innate and physiological) and extrinsic (environmental) factors. Each of the resulting cells of the matrix is further divided into involuntary (enforced) and voluntary categories to give 44 cells.
6. We argue that reality demands that any serious attempt to understand who disperses and why must take this potential diversity and complexity into account.

ACKNOWLEDGMENTS

For helpful comments on earlier drafts of this manuscript we warmly thank A. Cockburn and R.S. Ostfeld.

REFERENCES

Anderson, P.K. (1980) Evolutionary implications of microtine behavioral systems on the ecological stage. *The Biologist*, **62**, 70–88.

Anderson, P.K (1989) *Dispersal in Rodents: a resident fitness hypothesis*, Special Publication No. 9, American Society of Mammalogists, 141 pp.

Andrewartha, H.G. and Birch, L.C. (1954) *The Distribution and Abundance of Animals*. University of Chicago Press, Chicago, 782 pp.

Ballou, J. and Ralls, K. (1982) Inbreeding and juvenile mortality in small populations of ungulates: a detailed analysis. *Biological Conservation*, **24**, 239–72.

Brandt, C.A. (1985) The evolution of sexual differences in natal dispersal: a test of Greenwood's hypothesis, in *Migration: mechanisms and adaptive significance* (ed. M.A. Rankin), Contributions in Marine Science, Suppl. 27, pp. 386–96.

Brown, W.L. Jr (1958) General adaptation and evolution. *Systematic Zoology*, 7, 157–68.

Brues, A.M. (1964) The cost of evolution versus the cost of non-evolving. *Evolution*, **18**, 379–83.

Cockburn, A. (1988) *Social Behaviour in Fluctuating Populations*, Croom Helm, Sydney, 239 pp.

Cockburn, A., Scott, M.P. and Scotts, D.J. (1985) Inbreeding avoidance and sex-biased dispersal in *Antechinus* spp. (Marsupialia, Dasyuridae). *Animal Behaviour*, **33**, 908–15.

Cooper, W.S. and Kaplan, R.H. (1982) Adaptive 'coin-flipping': a decision-theoretic examination of natural selection for random individual variation. *Journal of Theoretical Biology*, **94**, 135–51.

Dingle, H. (1972) Migration strategies of insects. *Science*, **175**, 1327–35.

Dobson, F.S. (1982) Competition for mates and predominant juvenile male dispersal in mammals. *Animal Behaviour*, **30**, 1183–92.

Dobzhansky, Th. (1965) Genetic diversity and fitness. *Genetics Today*, **3**, 541–52.

Elton, C.S. (1927) *Animal Ecology*, MacMillan, New York, 207 pp.

Elton, C.S. (1930) *Animal Ecology and Evolution*, Oxford University Press, Oxford, 96 pp.

Falconer, D.S. (1960) *Introduction to Quantitative Genetics*, Ronald Press, New York, 365 pp.

Fulk, G.W. (1972) The effect of shrews on the space utilization of voles. *Journal of Mammalogy*, **53**, 461–78.

Gaines, M.S. and McClenaghan, L.R. Jr (1980) Dispersal in small mammals. *Annual Review of Ecology and Systematics*, **11**, 163–96.

Greenwood, P.J. (1980) Mating systems, philopatry and dispersal in birds and mammals. *Animal Behaviour*, **28**, 1140–62.

Haldane, J.B.S. (1937) The effect of variation on fitness. *American Naturalist*, 71, 337–49.

Hestbeck, J.B. (1982) Population regulation of cyclic mammals: the social fence hypothesis. *Oikos*, **39**, 147–63.

Howard, W.E. (1960) Innate and environmental dispersal in individual vertebrates. *American Midland Naturalist*, **63**, 152–61.

Kalela, O. (1961) Seasonal change of habitat in the Norwegian lemming, *Lemmus lemmus* (L.). *Annals of the Academy of Sciences Fennici 4A* (*Biol.*), **55**, 1–72.

Kolata, G.B. (1974) Population genetics: reevaluation of genetic variation. *Science*, **184**, 452–4.

Lack, D. (1968) *Ecological Adaptations for Breeding in Birds*, Metheun, London, 409 pp.

Lidicker, W.Z. Jr (1962) Emigration as a possible mechanism permitting the regulation of population density below carrying capacity. *American Naturalist*, **96**, 29–33.

Lidicker, W.Z. Jr (1975) The role of dispersal in the demography of small mammals, in *Small Mammals: their production and population dynamics* (eds F.B. Golley, K. Petrusewicz and L. Ryszkowski), Cambridge University Press, London, pp. 103–28.

Lidicker, W.Z. Jr (1976) Social behaviour and density regulation in house mice living in large enclosures. *Journal of Animal Ecology*, **45**, 677–97.

Lidicker, W.Z. Jr (1985a) Dispersal, in *Biology of New World* Microtus (ed. R.H. Tamarin), American Society of Mammalogists, Special Publication No. 8, pp. 420–54.

Lidicker, W.Z. Jr (1985b) An overview of dispersal in non-volant small mammals, in *Migration: mechanisms and adaptive significance* (ed. M.A. Rankin), Contributions in Marine Science, Suppl. **27**, pp. 359–75.

Lidicker, W.Z. Jr and Caldwell, R.L. (1982) Dispersal and Migration, Benchmark Papers in Ecology 11, Hutchinson Ross, Stroudsburg, Pennsylvania, 311 pp.

Łomnicki, A. (1978) Individual differences between animals and the natural regulation of their numbers. *Journal of Animal Ecology*, **47**, 461–75.

Łomnicki, A. (1980) Regulation of population density due to individual differences and patchy environment. *Oikos*, **35**, 185–93.

Madison, D.M. (1980a) Space use and social structure in meadow voles, *Microtus pennsylvanicus*. *Behavioural Ecology and Sociobiology*, **7**, 65–71.

Madison, D.M. (1980b) Movement types and weather correlates in free-ranging meadow voles, in *Proceedings of the Fourth Eastern Pine and Meadow Vole Symposium* (ed. R.E. Byers), Henderson, North Carolina, pp. 34–42.

May, R.M. (1980) Inbreeding among zoo animals. *Nature*, **283**, 430–1.

Mayr, E. (1961) Cause and effect in biology. *Science*, **134**, 1501–6.

Moore, J. and Ali, R. (1984) Are dispersal and inbreeding avoidance related? *Animal Behaviour*, **32**, 94–112.

Nevo, E. (1978) Genetic variation in natural populations: patterns and theory. *Theoretical Population Biology*, **13**, 121–77.

Ostfeld, R.S. (1985) Limiting resources and territoriality in microtine rodents. *American Naturalist*, **126**, 1–15.

Packer, C. (1979) Inter-troop transfer and inbreeding avoidance in *Papio anubis*. *Animal Behaviour*, **27**, 1–36.

Packer, C. (1985) Dispersal and inbreeding avoidance. *Animal Behaviour*, **33**, 676–8.

Pianka, E.R. (1988) *Evolutionary Ecology*, Harper and Row, New York, 468 pp.

Power, M.E., Matthews, W.J. and Stewart, A.J. (1985) Grazing minnows, piscivorous bass, and stream algae: dynamics of a strong interaction. *Ecology*, **66**, 1448–56.

Ralls, K. and Ballou, J. (1982) Effects of inbreeding on infant mortality in captive primates. *International Journal of Primatology*, **3**, 491–505.

Ralls, K., Brugger, K. and Ballou, J. (1979) Inbreeding and juvenile mortality in small populations of ungulates. *Science*, **206**, 1101–3.

Ralls, K., Harvey, P.H. and Lyles, A.M. (1986) Inbreeding in natural populations of birds and mammals, in *Conservation Biology: the science of scarcity and diversity* (ed. M.E. Soulé), Sinauer, Sunderland, Massachusetts, pp. 35–56.

Shields, W.M. (1982) *Philopatry, Inbreeding, and the Evolution of Sex*, State University of New York Press, Albany, New York, 245 pp.

Stenseth, N.C. (1983) Causes and consequences of dispersal in small mammals, in *The Ecology of Animal Movement* (eds I. Swingland and P.J. Greenwood), Oxford University Press, Oxford, pp. 63–101.

Tamarin, R.H. (1977) Dispersal in island and mainland voles. *Ecology*, **58**, 1044–54.

Tamarin, R.H. (1980) Dispersal and population regulation in rodents, in *Biosocial Mechanisms of Population Regulation* (eds M.N. Cohen, R.S. Malpass and H.G. Klein), Yale University Press, New Haven, pp. 117–33.

Templeton, A.R. (1987) Inferences on natural population structure from genetic studies on captive mammalian populations, in *Mammalian Dispersal Patterns: the effects of social structure on population genetics* (eds B.D. Chepko-Sade and Z.T. Halpin), University of Chicago Press, Chicago, pp. 257–72.

Templeton, A.R. and Read, B. (1983) The elimination of inbreeding depression in a captive herd of Speke's Gazelle, in *Genetics and Conservation* (eds C.M. Schonewald-Cox, S.M. Chambers, B. MacBryde and W.L. Thomas), Benjamin/Cummings, London, pp. 241–61.

Wolff, J.O. and Lidicker, W.Z. Jr (1980) Population ecology of the taiga vole, *Microtus xanthognathus*, in interior Alaska. *Canadian Journal of Zoology*, 58, 1800–12.

3

The genetic consequences of dispersal

N.H. BARTON

3.1 INTRODUCTION

What are the genetic consequences of dispersal? This is a question which includes many controversial components. Does environmental heterogeneity maintain genetic variation, and if so, how does dispersal interact with this heterogeneity to affect levels of variability? Does gene flow bind a species together into a homogeneous unit, or can adaptation and divergence occur despite free intermingling? Do genetic factors, such as levels of inbreeding, or of variability, determine the evolution of dispersal itself? Several chapters in this book (e.g. Chapters 1, 4, 5 and 9) consider this question. This chapter therefore concentrates on the effects of environmental heterogeneity and gene flow on patterns of genetic variation, rather than on the evolution of dispersal itself; examples are drawn from a wide variety of taxa.

The effects of dispersal on variation within and between populations are interesting in their own right. However, if these effects can be understood, we may also be able to invert our arguments, and use patterns of genetic variation to make inferences about population structure. Such indirect methods of measuring population structure have a long history, beginning in Wright's collaboration with Dobzhansky (Provine, 1981). They can give estimates which are more robust, and more widely applicable, than uncertain and laborious direct measurements. It may be that the most useful contribution of theory is, at present, to the application of these methods to particular cases, rather than to the elusive resolution of broader evolutionary questions. The second part of this chapter is, therefore, concerned with showing how population genetical methods can be used to estimate population structure in nature.

3.2 DISPERSAL, ENVIRONMENTAL HETEROGENEITY, AND GENETIC VARIATION

In the small mammal literature, 'dispersal' refers to 'cases in which individuals leave their home ranges (or home sites) and do not return' (Chapters 1 and 2). I am concerned here with the movement of genes: this may occur without dispersal (e.g. through gradual 'shifting' of the home range (Chapter 2)), and conversely, dispersal may not lead to effective gene movement if dispersers do not reproduce. However, it seems likely that the most important cause of gene movement is in fact 'dispersal', and so it is possible to concentrate on the 'genetic consequences of dispersal' without being misled.

The genetic effects of dispersal can be divided into those which arise through the movements of individuals within one generation of one random-mating population, and those which arise through the movements of genes over longer times, and longer distances. Individual movements within a generation may have many important biological functions and consequences. However, they can only affect the composition of the population by exposing the individuals to different physical, biological, or social environments, and hence to different selection pressures. A pervasive idea in population genetics has been that environmental heterogeneity may, in some way, maintain genetic heterogeneity (Hedrick *et al.*, 1976; Felsenstein, 1976). If this is so, then movements within one generation may, by affecting the degree of heterogeneity perceived by the individuals, alter the variability maintained in the population. I will argue here that we should not expect any simple relation between movement and environmental heterogeneity, on the one hand, and the level of polymorphism on the other.

3.2.1 Theory

The idea that environmental heterogeneity maintains genetic polymorphism was first quantified by Levene (1953). He imagined that, at the beginning of each generation, juveniles are scattered at random into a number of patches: there is no spatial subdivision of the population at reproduction, and no selection of particular patches by particular genotypes. Individuals spend their whole lives on a single patch. Different genotypes may have different chances of surviving on different patches. However, if the numbers produced by each patch are directly proportional to the number of individuals which enter the patch (i.e. if selection is 'hard'; Wallace, 1975), this environmental heterogeneity in fitness will not by itself maintain any genetic heterogeneity. If, on the other hand, the numbers surviving on each patch are reduced by crowding of that patch (i.e. if selection is 'soft'), polymorphism may be maintained. The reason is that a rare genotype will find that the patch on which it survives best is less crowded, simply because that genotype is rare, and so it will gain an advantage. This negative frequency–dependent selection causes the rarest genotype to increase, and so may maintain a balanced polymorphism. In the extreme case, where the numbers surviving on each patch are independent of those entering it, the condition for polymorphism is that the arithmetic mean fitness of each genotype across patches should be greater than one, but the harmonic mean should be less than one.

Under this model, polymorphism can only be maintained if individuals spend most of their lives in a single patch, or in other words, if the environment is 'coarse grained' (rather than 'fine grained') with respect to movements within a generation. Together with similar models of temporal heterogeneity (Haldane and Jayakar, 1963), this model has been greatly elaborated (Hedrick *et al.*, 1976; Felsenstein, 1976; Karlin, 1982). It has inspired many attempts to correlate levels of variation – in particular, of electrophoretic variation – with measures of environmental 'grain', including measures of dispersal (e.g. Selander and Kaufman, 1973; Nevo, 1978; Nevo *et al.*, 1984). It has also formed the basis of models of sympatric speciation (Maynard Smith, 1966; Felsenstein, 1981). Here, assortative mating evolves so as to match genotypes more closely to the patches in which they live.

However, there are serious objections to the view that this form of environmental heterogeneity can account for a significant amount of genetic variation. Maynard Smith and Hoekstra (1980) point out that the conditions for polymorphism are extremely restrictive: unless selection pressures are extremely strong, the fitnesses of the various genotypes must be nicely matched to the carrying capacities of the patches; slight changes in conditions destroy the polymorphism (Figure 3.1a; see also Hoekstra *et al.*, 1985). This mechanism cannot, therefore, maintain polymorphism at more than a few loci, and certainly cannot account for much electrophoretic variation. Even if the genetic load were reduced by epistatic interactions (e.g. by truncation selection; Wills, 1978; Wallace, 1970, 1975), selection cannot act strongly ($s \approx 1$) at more than a handful of loci without causing an unacceptably high variance in fitness.

The simple Levene model can be strengthened slightly by supposing that individuals tend to return to the type of patch on which their mother was reared. However, such modifications do not make the conditions for polymorphism substantially less restrictive when selection is weak (Figure 3.1a). The model may also be strengthened by allowing some spatial subdivision of the mating population (Karlin, 1982). Again, this is of little help: in general, although spatial subdivision can allow substantial variation between populations, it cannot maintain much variation within populations unless the scale of environmental heterogeneity is closely matched to the scale of spatial differentiation (Barton and Clark, 1990). A robust polymorphism can only be maintained if different genotypes tend to begin in different patches, a situation termed 'habitat selection'. We can see, from Figure 3.1b, that polymorphism is now possible for a wide range

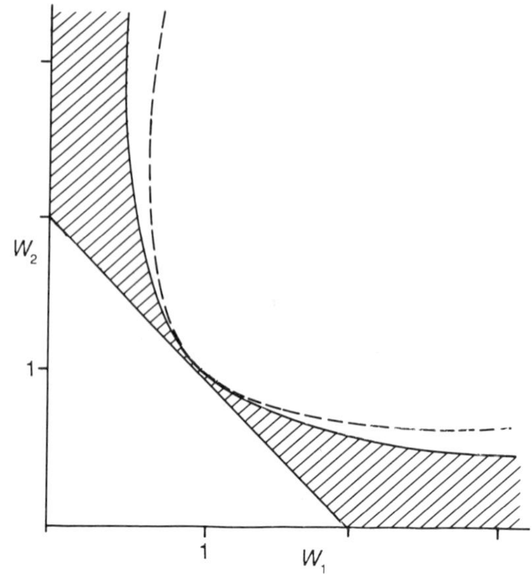

(a)

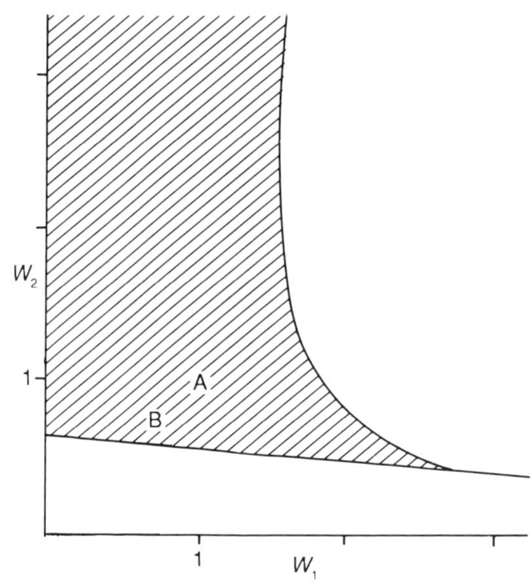

(b)

Figure 3.1(a) The conditions for stable polymorphism in a heterogeneous environment. Two patches of equal carrying capacity produce fixed and equal numbers of individuals. Two genotypes (in the haploid case), or two phenotypes (in the diploid case, with complete dominance), have relative viabilities $1{:}W_1$, $1{:}W_2$ in the two patches. The shaded region shows the range of viabilities for which a stable polymorphism is maintained ($(W_1 + W_2)/2 > 1$, $W_1W_2 < 1$). The dotted line shows the conditions when individuals return to the type of patch on which their mother was raised. (b) The conditions with habitat selection; type 1 has 75% probability of moving to patch 1, and type 2 has 75% probability of moving to patch 2.

of fitnesses. Indeed, balanced polymorphism can be maintained even if all genotypes have the same chance of surviving in each patch (A in Figure 3.1b), or even if individuals foolishly choose to live in the patches on which they have lower viability (B in Figure 3.1b). 'Habitat selection' has introduced a new and more powerful cause of polymorphism: if genotypes tend to live in patches whose densities are separately regulated, then rarer types will necessarily increase, regardless of whether they survive well in these patches. Indeed, most models of negative frequency-dependent selection can be seen to depend on this process. For example, polymorphic patterns are maintained by Batesian mimicry because these patterns exploit different, and separately regulated resources; in this case, the resources consist of the predators which have learned to avoid the various different patterns (Turner, 1981).

3.2.2 Evidence

These arguments suggest that genetic variation is more likely to be maintained not by environmental heterogeneity *per se*, as envisaged by Levene (1953), but rather by a tendency for different genotypes to exploit different, and separately regulated, resources; these 'resources' may be interpreted very generally, and need not be in different places. This implies that, even if frequency-dependent selection is an important force maintaining polymorphism, there will be no simple relation between dispersal, and levels of variation. For example, Pérez-Tomé and Toro (1982) have shown that competing *Drosophila* larvae survive ≈5% better if they are half-siblings than if they are full siblings (see also Martin *et al.*, 1988). This indicates that different genotypes do exploit different limiting resources; however, there is no suggestion that these resources are spatially segregated within the medium; they might, for example, consist of different, evenly distributed, chemicals. Analogous effects are well known in agriculture, where mixtures or species or varieties often perform better when grown together than would be expected from their individual yields (Barrett, 1980). Such synergistic effects may be caused by variation in the use of separate resources, and would be expected to maintain variation. To take a mammalian example, Hafner *et al.* (1983) have shown that in pocket gophers there is a polymorphism in pelage colour over a region ≈200 km long. Although this could be interpreted as being maintained by the neutral admixture of two divergent races, strong linkage disequilibria suggest that the polymorphism is maintained by balancing selection within this region (section 3.3). Here again, even if the polymorphism is maintained by adaptation to heterogeneous soil colours, the broad spatial extent of the polymorphism shows that dispersal *per se* is unlikely to affect it: exchange of individuals between populations separated even by many kilometers would not alter gene frequencies significantly. Only genotype-dependent movements, in which particular genotypes choose particular microhabitats, could affect polymorphisms of this sort.

These are only suggestive examples; what evidence do we have on the general relation between environmental heterogeneity and polymorphism? There are few, if any, examples where differential survival on different habitats, as in Levene's (1953) model, has been shown to maintain polymorphism (Hedrick *et al.*, 1976; Futuyma and Mayer, 1980; Futuyma *et al.*, 1981). Correlations between enzyme heterozygosity and measures of environmental heterogeneity have been reported; for example, from an extensive survey of heterozygosity in 1111 species, including 184 mammals, Nevo *et al.* (1984) shows that the levels of heterozygosity and polymorphism are better correlated with ecological characteristics (such as degree of specialization and habitat complexity) than with demographic factors (such as population size and species distribution). Nevo argues that this supports the direct maintenance of the enzyme polymorphisms by environmental heterogeneity. However, such correlations are hard to interpret. If enzyme variation is largely neutral, then heterozygosity is determined by the product of the effective population size, and the mutation rate to effectively neutral alleles. Unfortunately, this product bears no simple relation to any actual, observable, parameters of the present population. For example, if most mutations are slightly deleterious, rather than being precisely neutral, the rate of production of effectively neutral mutations is a decreasing function of population size (Kimura, 1983). This is because in a small population, a greater proportion of mutations will have effects so small as to be dominated by drift. The effective population size may be greatly altered by past bottlenecks (Nei *et al.*, 1975), and may also depend on apparently 'ecological' characteristics. For example, in Nevo *et al.*'s (1984) survey, geographic range is significantly and positively correlated with heterozygosity within the mammals. Nevo classes geographic range as an 'ecological' characteristic, acting through its effect on environmental diversity; however, increased spatial distribution would also be expected to increase the long-term effective population number, and hence to increase heterozygosity under the neutral theory. More subtly, the rate of drift in a population, and hence the effective population size, depends on the variance of fitness between individuals (Crow and Kimura, 1970). The genetic component of this variance may substantially alter the rate of drift, through hitch-hiking with linked loci (Nei and Murato, 1966; Maynard Smith and Haigh, 1974). A likely example of this effect is described by Nadean and Collins (1983) and Nadean *et al.* (1982), who show that enzymes linked to the highly polymorphic H-2 locus in mice are themselves significantly more polymorphic. Conversely, overall levels of heterozygosity in asexual bacterial populations is greatly reduced by the occasional 'periodic selection' of particular clones (Hartl and Dykhuizen, 1981; Dykhuizen, 1990).

Thus, a general correlation between heterozygosity and ecological characteristics may be better explained through differences in effective population size (perhaps caused by selection on other loci), than by direct selection on the observed genes. In that case, the relation between dispersal and the evolutionarily significant selected loci could not easily be inferred from studies on electrophoretic markers.

3.3 GENE FLOW AND SPATIAL DIFFERENTIATION

3.3.1 Definitions

The most important genetic consequence of dispersal is that it usually results in gene flow. The term 'gene flow' has been used in a variety of senses. This variation has caused unnecessary confusion in debates over the definition of the biological species. Gene flow has been defined as 'movement of genes, and their subsequent incorporation into new gene pools' (Endler, 1977; Gaines and McClenaghan, 1980). This definition can be ambiguous, since it is not obvious when a gene has been incorporated into a new gene pool. Clearly, dispersing individuals must breed at their new site in order for gene flow to occur. However, a much more stringent requirement has often been applied: that an allele should not only enter the new population, but should also not be eliminated from it by natural selection. Thus, hybrid zones have been taken to delimit separate biological species, on the grounds that because they involve stable genetic differentiation, alleles cannot successfully introgress across them (Dobzhansky, 1940, p. 354; Wright, 1978; p. 5; Mayr, 1982, p. 285). This interpretation is unsatisfactory, since it relies on the behaviour of particular alleles, rather than on any property affecting the whole genome: though some genes may remain sharply differentiated at hybrid zones, other genes may flow freely between the interacting populations (Barton and Hewitt, 1985). I will therefore take gene flow to mean the entry of selectively neutral alleles into a new gene pool. Gene flow, in this sense, is affected by selection if the genomes migrating into the new population are associated with lower fitness in the new environment or in the new genetic background (Barton, 1979; Spirito *et al.*, 1983; Bengtsson, 1985; Barton and Bengtsson, 1986): neutral alleles will be carried into the new population in association with alleles which cause lower fitness, and so will be reduced in frequency until they recombine onto the new genetic background. (This is a geographic analogue of the hitch-hiking effect discussed above.) However, this effect is due to the cumulative action of selection at many linked loci, rather than on properties peculiar to particular alleles.

A second source of ambiguity depends on whether 'gene flow' refers to the proportion of genes which move, or to the number of genes which move. Slatkin (1982, 1985a) refers to gene flow as the number of individuals moving between adjacent demes. This is the product of population size and dispersal rate (Nm). It corresponds to Wright's 'neighbourhood size' which applies in a continuous two-dimensional population, and is proportional to the product of density and dispersal rate. Nm and neighbourhood size determine the relative strengths of gene flow and sampling drift (see below). However, here 'gene flow' is used to refer to the proportion of genes which move; it is this parameter which is important in determining the strength of dispersal relative to selection.

3.3.2 Theory

Does gene flow cause genetic homogeneity? Mayr (1982, pp. 411–7, 562–6) has traced this question back to a disagreement between Darwin and Wagner. The latter held that 'the formation of a genuine variety which Mr. Darwin considers to be an incipient species, will succeed in nature only where a few individuals transgress the limited borders of their range, and segregate themselves spatially for a long period from other members of their species' (1841, pp. 199–200). On the other hand, Darwin 'did not doubt that many species have been formed at different points of an absolutely continuous area, of which the physical conditions graduate from one point to another in the most insensible manner' (1968, p. 266) (though he also stressed the importance of isolation in his early writings). The belief that gene flow prevents genetic divergence, which originated with Wagner, has been used to justify the definition of biological species as populations bounded by genetic barriers to gene exchange (Mayr, 1942). More recently, the debate has centred on theoretical models of clinal differentiation and parapatric speciation (Fisher, 1930; Haldane, 1948; White, 1968; Endler, 1977) in which divergence occurs within a continuous habitat, and on claims that measured rates of gene flow are too low to be of evolutionary significance (Ehrlich and Raven, 1969).

Theoretically, it is clear that gene flow is not particularly effective in preventing divergence by selection. Provided that dispersal is not pathologically leptokurtic, i.e. provided that no individuals move extremely long distances, and provided that selection is not extremely strong, the effects of gene flow on allele frequencies in a continuously distributed species can be approximated by diffusion (Fisher, 1937; Nagylaki, 1978). Spatial patterns will be smoothed out at a rate which depends only on the variance of distance between parent and offspring (σ^2). This parameter has the dimensions of (distance^2 generation^{-1}); since selection pressures (s) have the dimensions of (generation^{-1}), the importance of selection relative to gene flow is determined by a 'characteristic scale', $l = \sigma/\sqrt{s}$ (Slatkin, 1973). In a wide range of models, differential selection can be shown to maintain stable differentiation, despite gene flow, provided that the selective differences are sustained over an area larger than some critical value, of the same order as l^2 (Nagylaki, 1978; for a review see Felsenstein, 1976).

Although gene flow is, theoretically, more effective in smoothing out the effects of sampling drift, it still cannot prevent stochastic divergence in populations with low neighbourhood size. Consider, first, a balanced polymorphism, maintained by a selection pressure s; in the case of a neutral polymorphism, maintained by recurrent mutation, s should be replaced by the mutation rate, μ. In an island model, where a single deme, consisting of N diploid individuals, exchanges a proportion m each generation with some large external pool of migrants, the standardized variance of allele frequency is $F_{ST} = 1/(1 + 4Nm)$ (assuming $m >> s$; Wright, 1943). In a two-dimensional, continuous population, with density x, the effects of drift depend on the 'neighbourhood size', $Nb = 4\pi x\sigma^2$; $F_{ST} = 1/(1 + Nb/\log(1/2s))$ (Wright, 1943; Malecot, 1948). Thus, spatial differentiation can be produced by drift, provided that the product of density and rate of gene flow ($4Nm$, or $Nb = 4\pi x\sigma^2$) is small. The pattern of differentiation will reach equilibrium quite rapidly over small areas, but will take much longer to equilibrate over larger areas ($T \approx A/\sigma^2$). Provided that $Nb > 1$, Kimura and Maruyama (1971) show that the heterozygosity of neutral alleles should be essentially independent of neigbourhood size. The great majority of species are indeed likely to have moderate to large neigbourhood sizes over most of their range (Wright, 1978; Slatkin, 1982, 1985b; see next section).

Of course, random fluctuations in a neutral character, or around a single stable equilibrium, are of little evolutionary importance. However, the same parameters determine the effect of gene flow on shifts between different adaptive peaks; such shifts are essential to both speciation, and to adaptation under Wright's (1931) 'shifting balance' model (Barton and Charlesworth, 1984). Lande (1979) has shown that a new chromosome arrangement can become established in a deme, and can then spread via random extinction and recolonization, provided that $Nm < \approx 2$; in a two-dimensional, continuous population, peak shifts can occur if neighbourhood size (Nb) is $< \approx 10$ (Rouhani and Barton, 1987). Thus, stochastic divergence is possible even when gene flow occurs through either steady dispersal or occasional extinction and recolonization, provided that the product of density and dispersal is not too high.

3.3.3 Evidence

Observations on local differentiation support the argument that gene flow does not prevent divergence. To take a classic example, Haldane (1948) showed that a cline in *Peromyscus* coat colour, 35 km across (Sumner,

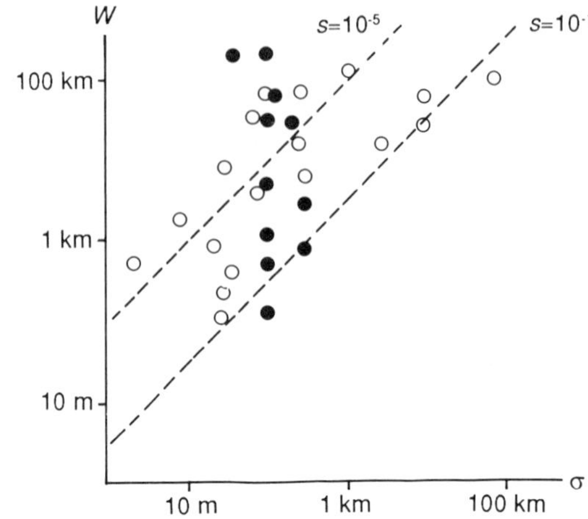

Figure 3.2 The relation between width and estimated dispersal for clines in 25 species. Some species are represented by several clines; filled circles denote mammals. For detailed references, see Barton and Hewitt (1985, Fig. 3).

1929), could be maintained by local differences in selection pressure of only 0.001, despite gene flow at an assumed rate of $\sigma \approx 1$ km generation$^{-1/2}$. Even in the bat, *Uroderma bilobatum*, where dispersal can be inferred from genetic data to be much greater ($\sigma \approx 11$ km; section 3.4), karyotypic clines ≈50 km wide could be maintained by selection against heterozygotes of $s \approx 37\%$ (Baker, 1981; Barton, 1982). In mammals, such sharp clines seem widespread: they are found in karyotype, enzymes, coat colour, and morphology (for reviews, see Endler, 1977; White, 1978; Barton and Hewitt, 1985). Admittedly, these clines often seem much wider than would be expected from direct measurements of dispersal (Figure 3.2), implying that gene flow is more important than these measures indicate (section 3.4). Nevertheless, these clines are usually much narrower than the species' range, and give direct evidence that gene flow does not prevent selection from producing local adaptation and divergence.

The effects of drift are harder to demonstrate: though many analyses implicitly assume that spatial differentiation is caused by drift, it is hard to distinguish the action of drift from that of divergent selection. However, in a few cases, spatial differentiation has been shown to be greater in regions of low neighbourhood size (Cavalli Sforza, 1969; Selander, 1970; Patton and Feder, 1978), indicating that it has been produced by sampling drift. More generally, the apparent correlation across taxa between the rate of chromosomal evolution, lack of enzyme heterozygosity, and the rate of speciation (Wilson *et al.*, 1975; Bush *et al.*, 1977; Coyne, 1984; Figure

3.3) is most easily explained by the common influence of population structure. If this correlation is real, and if it is indeed due to differences in population structure, it implies that both chromosomal changes, and those changes responsible for reproductive isolation, are largely caused by drift.

Though both theory and observation show that existing levels of gene flow are not sufficient to prevent local differentiation, we must still explain an apparently conflicting pattern: over larger scales, morphological and electrophoretic differences are often correlated with geographic distance (e.g. Patton and Yang, 1977; Grant, 1980; Pounds and Jackson, 1981; Dillon, 1984); such correlations are more obviously explained by the homogenizing effect of gene flow than by any increase in environmental differences with physical distance. More directly, Pounds and Jackson (1981) have shown that populations of the fence lizard, *Sceloporus grammicus*, which are separated by rivers differ more in morphology than populations a similar distance apart, but not separated by major barriers to dispersal. They take this as direct evidence that gene flow reduces divergence.

This apparent conflict may be resolved if the similarities between populations are caused not be contemporary gene exchange (which, as argued above, should be negligibly weak in comparison with selection pressures), but instead by common descent from a single group of colonizing individuals. Over distances greater than ≈ 100 dispersal ranges, diffusive gene flow would not have had time to cause much homogenization since the last glaciation ($\approx 10^4 - 10^5$ generations ago). It is true that quite weak selection ($s > 10^{-4}$) should allow local adaptation over these scales, and so would have obscured any patterns formed by ancient range changes. However, if selection allows populations to evolve towards many alternative 'adaptive peaks', different colonizing groups may reach different equilibria, and may be held there by selection. In this case, historical patterns may be the dominant factor determining geographic patterns, even though selection pressures would seem able to outweigh present levels of gene flow (Grant, 1980; Barton and Hewitt, 1985). If extinction and recolonization is indeed more important than contemporary gene flow, we are faced with the difficult problem of measuring their rate; given the problems involved in measuring present dispersal, this is a daunting prospect.

3.4 INFERENCE FROM GENETIC PATTERNS

If we have a good enough understanding of the effects of dispersal on genetic differentiation, then we should be able to use this genetic information to infer rates of gene flow, population densities, and selection pressures. Such indirect methods may be much less laborious, and more

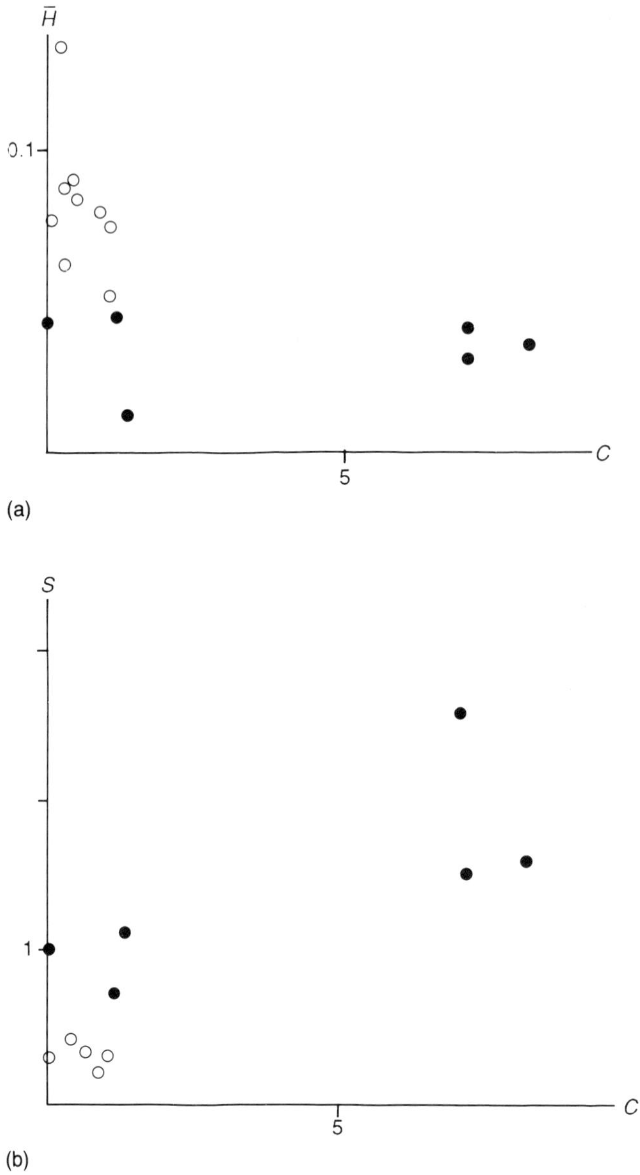

Figure 3.3(a) The correlation between rate of chromosomal evolution (change in number per 10^8 years) and the mean electrophoretic heterozygosity; data from Coyne (1984) and Wilson *et al.* (1975). Filled circles denote mammalian groups; the single outlier is The Carnivora. (b) The correlation between rate of chromosomal evolution (as above), and rate of speciation (corrected rate, *S*, per 10^6 years; Bush *et al.*, 1977).

generally applicable, than methods using direct observations of marked animals. They may therefore give information about a wider range of populations, and of species. Moreover, they complement direct measurements, since they give information about rates of gene flow, rather than of dispersal, and about effective population density, rather than actual density.

The clearest finding to emerge from theoretical work on the interaction between selection and gene flow is that clines maintained by a balance between dispersal and selection should have a width of the same order as the characteristic scale, $l = \sigma/\sqrt{s}$. For example, if we define cline width as the inverse of the maximum gradient in allele frequency, then selection against heterozygotes s should maintain a cline of width $\sqrt{8}l$ (Bazykin, 1969). Definitions of l and s, and the precise relation between them, will vary from model to model; however, the width will always be of the same order of magnitude as l. This suggests that if we know the cline width and the selection pressure, we can infer the rate of gene flow, σ. In practice, we rarely have any direct information about selection strength, and so cannot say more than that σ must be less than the cline width. The relation has instead been applied to infer selection pressure from estimates of dispersal and cline width (e.g. Endler, 1977; Hafner, 1982; Hafner *et al.*, 1983). When cline width is plotted against dispersal rates, for those species where published estimates are available (Figure 3.2), it can be seen that clines are often much wider than would be expected if even weak selection is involved. This could be interpreted to mean that these clines are neutral, and that they reflect transient mixing after secondary contact (Hafner, 1982; Hafner *et al.*, 1983). However, this seems unlikely for the morphological and chromosomal differences involved in the majority of these cases; in many instances, there is direct evidence of selection (Barton and Hewitt, 1985). Three other explanations seem more plausible. First, the clines plotted in Figure 3.2 often involve many coincident differences; thus, although the total selection pressure may be strong, the selection at any individual locus may be much weaker. Second, positive assortative mating may reduce the frequency of hybrids, and so broaden the clines. Finally, and more important, estimates of dispersal distances are notoriously difficult to obtain; if even a small proportion of long distance migrants go undetected, the rate of gene flow (σ) may be greatly underestimated. Occasional extinctions and recolonizations may also be important, and yet go unnoticed. The values used in Figure 3.2 are often crude, and refer to populations or even species different from those involved in the cline. For these reasons, it is rarely practicable to use cline widths to estimate either selection or dispersal.

More robust estimates can be obtained if we have data on two or more coincident clines at different loci. Then, the continual influx of parental combinations of alleles will generate linkage disequilibrium at the centre of the set of clines (Li and Nei, 1974; Slatkin, 1975). This will be broken down by recombination. Hence, the level of disequilibrium at the centre will be proportional to the ratio between the rate of gene flow, and the product of the rate of recombination and the square of the cline width ($D = \sigma^2/rw^2$, where w is the cline width σ^2 is the variance in parent–offspring distance, and r is the recombination rate; Barton, 1982). Since the disequilibrium, the cline width, and the recombination rate can all be measured, σ can be estimated. (This method is not restricted to linked loci. The term 'linkage disequilibrium' refers to a tendency for alleles at different loci to be associated with each other within a population of gametes; it does *not* imply either linkage, or lack of equilibrium.) The rate of gene flow may be overestimated if some linkage disequilibrium is generated by epistatic selection; however, gene flow is likely to be the dominant process when many loci are involved (Barton, 1983), In *Uroderma bilobatum*, clines for three Robertsonian fusions coincide (Baker, 1981). The genotype frequencies in a sample of 25 bats from the centre of these clines produce an estimate, using the method described above, of $\sigma = 11$ km. Since linkage disequilibrium is a common feature of coincident clines (Barton and Hewitt, 1985, 1989; Szymura and Barton, 1986), this method may provide a general way of estimating gene flow and selection.

We saw above that in a spatially distributed population, sampling drift will produce fluctuations in allele frequency which have a standardized variance inversely proportional to the product of effective population density and rate of gene flow. Selection cannot affect the pattern of differentiation over distances shorter than the characteristic scale, l; moreover, in two dimensions, drift produces fluctuations with a total variance, summed over all scales of measurement, which depends only weakly on the selection pressure maintaining the polymorphism (Malecot, 1948; Kimura and Weiss, 1964). The variance of local fluctuations can therefore be used to estimate neighbourhood size, whilst the distance at which the correlation between nearby populations begins to decline gives an upper limit to the rate of gene flow, σ. (In a demic population, the corresponding parameters are the product of deme size and migration rate, Nm, and the deme spacing.)

Although Wright's F coefficients have been used to describe spatial patterns in a wide range of organisms, and although Wright himself has used them to estimate population structure (Wright, 1978; Provine, 1981), they have rarely been combined with the relevant theory to make quantitative estimates of population structure (for exceptions, see Wright *et al.*, 1942; Yokoyama, 1979; Sokal and Wartenberg, 1983). Not all populations are suitable for such analysis: ideally, the population should have been reasonably stable for several hundred generations, and should be more or less continuously distributed. If local demes have instead descended through bottlenecks from one or more ancestral populations, then Wright's hierarchic analysis may be more appropriate (Wright, 1978; Neel, 1978), and one may not be able to make any quantitative estimates at all. However, if the population structure is suitable, and if enough samples are available (>50, say), this approach offers a general method for estimating neighbourhood size and gene flow. Moreover, comparisons between loci can be used to test whether the observed patterns could plausibly have been produced by drift: if patterns at different loci are significantly discordant, then some other process must be acting (Lewontin and Krakauer, 1973; Gaines and Whittam, 1980). For example, in the alpine grasshopper *Podisma pedestris*, four polymorphic enzyme loci, three almost monomorphic loci, and fluctuations in a chromosomal cline, all show concordant patterns; one polymorphic locus (PGM) is much more spatially differentiated, and its pattern cannot be explained by population structure alone (Figure 3.4a). The remaining enzyme loci give an estimate of neighbourhood size of 335 (95% confidence limits 265–440). Correlations in frequency fall away over distances of less than 30 m, so that the dispersal range must be less than this value (Figure 3.4b,c). These estimates are consistent with direct measurements (Barton and Hewitt, 1981; Barton *et al.*, in preparation).

Slatkin (1982) has recently proposed a simple method of estimating gene flow which seems at first sight quite different from Wright's approach, but is in fact closely related. In an electrophoretic survey, each allele can be characterized by the fraction of samples in which it is found (P), and by its average frequency in those samples ($\bar{p}$) (Figure 3.5a,b). Slatkin finds that, in simulations of both island and two-dimensional demic populations, the graph of the conditional average frequency, $\bar{p}$, against P, depends strongly on Nm, but only weakly on other parameters; in particular, on selection pressures. This graph can therefore be used to estimate Nm. A useful estimator is the frequency of 'private' alleles (i.e. alleles which are found in only one sample) in the sample in which they are present. This declines as gene flow and population density increase (Slatkin, 1985a; Figure 3.5c). Graphs for *Thomomys bottae* (Patton and Yang, 1977), and *Peromyscus californicus* (Smith, 1979) are shown in Figure 3.5b; they give estimates of $Nm = 0.86$ and 2.2, respectively (Slatkin, 1985a).

The reason for this relation between the frequency of private alleles and Nm is that, if gene flow is low, and the population sparse, then drift will cause large fluctuations in frequency from place to place. Thus, even if an allele is so rare that it is only found in one sample, it may nevertheless have a high frequency in that sample. Conversely, if gene flow and population density are high, then allele frequencies will be uniform, and so alleles which are so rare that they are found in only one sample, are likely to be

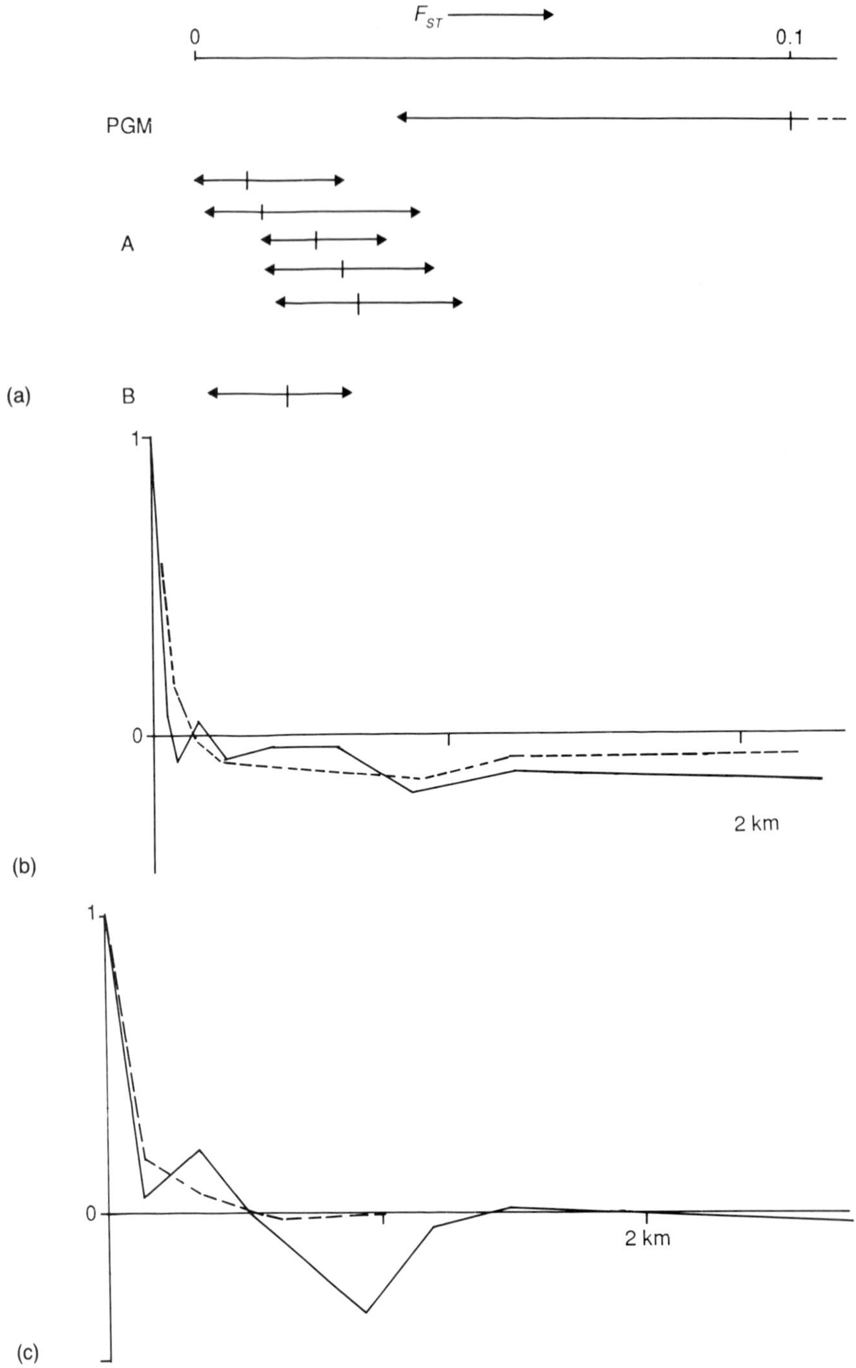

Figure 3.4(a) The estimated F_{ST}, and confidence limits, for five polymorphic enzyme loci (38 samples; A), fluctuations in a chromosomal cline (118 samples; B), and one aberrant locus (38 samples; PGM). Samples were taken from an area about 1 km^2. (b,c) The correlation between fluctuations at points different distances apart. (b) Five polymorphic enzymes (54 samples); (c) fluctuations in a chromosomal cline (118 samples). (Barton *et al.*, in preparation.) Dotted lines indicate theoretical expectations.

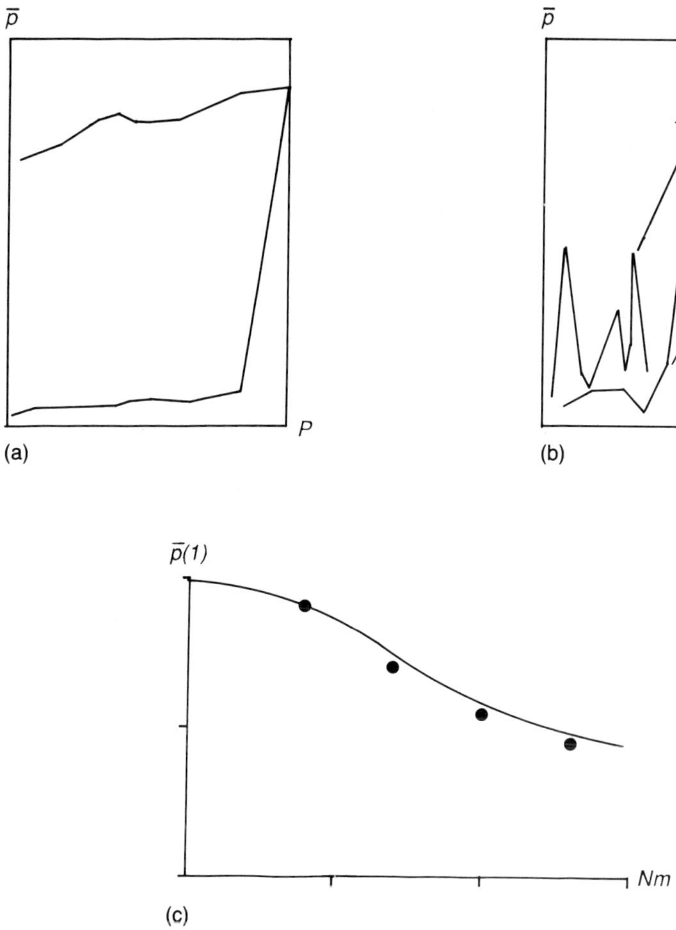

Figure 3.5(a) The simulated relation between the proportion of demes in which an allele is found (P), and its average frequency in those demes ($\bar{p}$) (Island model; $\mu = 10^{-4}$, 10 demes, $N = 25$; $Nm = 0.025$ (upper curve), 2.5 (lower curve); Slatkin, 1982). The frequency of private alleles is given by the left-hand intercept. (b) The observed relation in *Thomomys bottae* (T; Patton and Yang, 1977) and *Peromyscus californicus* (P; Smith, 1979). (c) The theoretical relation between Nm, and the frequency of private alleles, compared with simulated values (crosses).

rare in that sample. Slatkin's method thus gives a measure of the variability in allele frequency, which is analogous to Wright's F_{ST}, but much easier to calculate. Indeed, the theoretical relation between Slatkin's measure, and Nm, can be derived from Wright's (1943) island model, and fits well with the corresponding simulations (Figure 3.5c; Barton and Slatkin, 1986). The method is strengthened by the fact that it uses only rare alleles: these will be affected relatively more strongly by drift than by selection (Kimura, 1983), and so will give a more robust measure of population structure.

3.5 CONCLUSIONS

Although small-scale movements of individuals within a generation will alter the range of selective environments which they experience, neither theory nor observation suggest any clear relation between dispersal, and the amount of genetic variation held within populations. In contrast, the effect of dispersal over larger scales is clear: by causing the diffusive flow of genes from place to place, it will tend to impede local adaptation, and reduce stochastic divergence. However, except in the most vagile organisms, gene flow will not be powerful enough to prevent these processes occurring: the frequency of narrow clines, and evidence of the long-term importance of drift in producing molecular and chromosomal evolution support this view. We cannot, therefore, invoke dispersal as the cause of either morphological stasis, or widespread geographic uniformity (Charlesworth *et al.*, 1982).

The theory of the interaction between gene flow, selection, and drift is by now well understood, and well supported by field studies. It can therefore be used to estimate parameters such as rate of gene flow, population density, and selection pressure; this indirect approach promises to be a valuable complement to more traditional methods of measuring population structure. It is possible that inferences from genetic patterns will also allow us to understand the longer term processes of extinction and recolonization which may dominate over local dispersal in determining broad geographic and evolutionary patterns.

ACKNOWLEDGEMENTS

I would like to thank B.O. Bengtsson, W.Z. Lidicker and N.C. Stenseth for their helpful comments on the manuscript, and B.O. Bengtsson for his valuable improvements to my English. This work was supported by grants from the SERC (GR/C/91529), and the Royal Society.

REFERENCES

Baker, R.J. (1981) Chromosome flow between chromosomally characterized taxa of a volant mammal, *Uroderma bilobatum* (Chiroptera: Phyllostomatidae). *Evolution*, **35**, 296–305.

Barrett, J.A. (1980) Pathogen evolution in multilines and variety mixtures. *Zeitschrft fur Pflanzenkrankheiten und Pflanzenschutz*, **87**, 383–96.

Barton, N.H. (1979) Gene flow past a cline. *Heredity*, **43**, 333–9.

Barton, N.H. (1982) The structure of the hybrid zone in *Uroderma bilobatum. Evolution*, **36**, 863–6.

Barton, N.H. (1983) Multilocus clines. *Evolution*, **37**, 454–71.

Barton, N.H. and Bengtsson, B.O. (1986) The barrier to genetic exchange between hybridizing populations. *Heredity*, **57**, 357–76.

Barton, N.H. and Charlesworth, B. (1984) Genetic revolutions, founder effects, and speciation. *Annual Review of Ecology and Systematics*, **15**, 133–64.

Barton, N.H. and Clark, A. (1990) Population structure and process in evolution, in *Population Biology* (eds K. Wohromann and S.K. Jain), Springer-Verlag, Berlin, pp. 115–74.

Barton, N.H., Halliday, R.B. and Hewitt, G.M. Spatial patterns and population structure. III. Electrophoretic variation in the grasshopper *Podisma pedestris*. (in preparation).

Barton, N.H. and Hewitt, G.M. (1981) A chromosomal cline in the grasshopper *Podisma pedestris. Evolution*, **35**, 1008–18.

Barton, N.H. and Hewitt, G.M. (1985) Analysis of hybrid zones. *Annual Review of Ecology and Systematics*, **16**, 113–48.

Barton, N.H. and Hewitt, G.M. (1989) Adaptation, speciation, and hybrid zones. *Nature*, **341**, 497–503.

Barton, N.H. and Slatkin, M. (1986) A quasi-equilibrium theory of the distribution of rare alleles in subdivided populations. *Heredity*, **56**, 409–16.

Bazykin, A.D. (1969) Hypothetical mechanism of speciation. *Evolution*, **23**, 685–7.

Bengtsson, B.O. (1985) The flow of genes through a genetic barrier, *Evolution: essays in honour of J. Maynard Smith* (eds P. Harvey, M. Slatkin and J. Greenwood), Cambridge University Press, Cambridge, pp. 31–42.

Bush, G.L., Case, S.M., Wilson, A.C. and Patton, J.L. (1977) Rapid speciation and chromosomal evolution in mammals. *Proceedings of the National Academy of Sciences, USA*, **74**, 3942–6.

Cavalli Sforza, L.L. (1969) Genetic drift in an Italian population: with biographical sketch. *Scientific American*, **221**, 30–7.

Charlesworth, B., Lande, R. and Slatkin, M. (1982) A neo-Darwinian commentary on macroevolution. *Evolution*, **36**, 474–98.

Coyne, J.A. (1984) Correlation between heterozygosity and rate of chromosome evolution in animals. *American Naturalist*, **123**, 725–9.

Crow, J.F. and Kimura, M. (1970) *An Introduction to Population Genetics Theory*, Harper and Row, New York.

Darwin, C. (1968) *The Origin of Species by Means of Natural Selection*, 1st edn (reprint), Penguin, London. (First published in 1859, J. Murray, London.)

Dillon, R.T. (1984) Geographic distance, environmental difference, and divergence between isolated populations. *Systematic Zoology*, **33**, 69–82.

Dobzhansky, T. (1940) Speciation as a stage in evolutionary divergence. *American Naturalist*, **74**, 312–21.

Dykhuizen, D.E. (1990) Experimental studies of natural selection in bacteria. *Annual Review of Ecology and Systematics*, **21**, 373–98.

Ehrlich, P.R. and Raven, P.H. (1969) Differentiation of populations. *Science*, **165**, 1228–32.

Endler, J.A. (1977) *Geographic Variation, Speciation, and Clines*, Princeton University Press, Princeton, New Jersey.

Felsenstein, J. (1976) The theoretical population genetics of variable migration and selection. *Annual Review of Genetics*, **10**, 253–80.

Felsenstein, J. (1981) Skepticism towards Santa Rosalia, or why are there so few kinds of animals? *Evolution*, **35**, 124–38.

Fisher, R.A. (1930) *The Genetical Theory of Natural Selection*, Oliver and Boyd, London.

Fisher, R.A. (1937) The wave of advance of advantageous genes. *Annals of Eugenics*, **7**, 355–69.

Futuyma, D.J. and Mayer, G.C. (1980) Non-allopatric speciation in animals. *Systematic Zoology*, **29**, 254–71.

Futuyma, D.J., Leipertz, S.L. and Mitter, C. (1981) Selective factors affecting clonal variation in the fall cankerworm *Alsophila pometaria*, (Lepidoptera: Geometridae). *Heredity*, **47**, 161–72.

Gaines, M.S. and McClenaghan, L.R. (1980) Dispersal in small mammals. *Annual Review of Ecology and Systematics*, **11**, 163–96.

Gaines, M.S. and Whittam, T. (1980) Genetic changes in fluctuating vole populations. *Genetics*, **96**, 767–78.

Grant, V. (1980) Gene flow and the homogeneity of species populations. *Biol. Zbl.* **99**, 157–69.

Hafner, J.C. (1982) Genetic interaction at a contact zone of *Uroderma bilobatum* (Chiroptera: Phyllostomatidae). *Evolution*, **36**, 852–66.

Hafner, J.C., Hafner, D.J., Patton, J.L. and Smith, M.F. (1983) Contact zones and the genetics of differentiation in the pocket gopher *Thomomys bottae* (Rodentia: Geomyidae). *Systematic Zoology*, **32**, 1–20.

Haldane, J.B.S. (1948) The theory of a cline. *Journal of Genetics*, **48**, 277–84.

Haldane, J.B.S. and Jayakar, S.D. (1963) Polymorphism due to selection of varying direction. *Journal of Genetics*, **58**, 237–42.

Hartl, D.A. and Dykhuizen, D. (1981) Potential for selection among nearly neutral allozymes of 6 phosphoglucomutase dehydrogenase in *Escherischia coli*. *Proceedings of the National Academy of Sciences, USA*, **78**, 6344–88.

Hedrick, P.W., Ginevan, M.E. and Ewing, E.P. (1976) Genetic polymorphism in heterogeneous environments. *Annual Review of Ecology and Systematics*, **7**, 1–32.

Hockstra, R.F., Bijlsma, R. and Dolman, A.J. (1985) Polymorphism from environmental heterogeneity: models are only robust if the heterozygote is close in fitness to the favoured homozygote in each environment. *Genetical Research*, **45**, 299–314.

Karlin, S. (1982) Classification of selection-migration structures and conditions for a protected polymorphism. *Evolutionary Biology*, **14**, 61–203.

Kimura, M. (1983).*The Neutral Theory of Molecular Evolution*, Cambridge University Press, Cambridge.

Kimura, M. and Maruyama, T. (1971) Pattern of neutral polymorphism in a geographically structured population. *Genetic Reseach*, **18**, 125–31.

Kimura M. and Weiss, G.H. (1964) The stepping stone model of population structure and the decrease of genetic correlation with distance. *Genetics*, **49**, 561–76.

Lande, R. (1979) Effective deme sizes during long-term evolution estimated from rates of chromosomal rearrangement. *Evolution* **33**, 234–51.

Levene, H. (1953) Genetic equilibrium when more than one ecological niche is available. *American Naturalist*, **87**, 331–3.

Lewontin, R.C. and Krakauer, J. (1973) Distribution of gene frequency as a test of the theory of the selective neutrality of polymorphisms. *Genetics*, **74**, 175:195.

Li, W.H. and Nei, M. (1974) Stable linkage disequilibrium without epistasis in subdivided populations. *Theoretical Population Biology*, **6**, 173–83.

Malecot, G. (1948) *La Mathematique de l'Heredite*, Masson et Cie, Paris.

Martin, M.J., Pérez-Tomé, J.M. and Toro, M.A. (1988) Competition and genotypic variability in *Drosophila melanogaster*. *Heredity*, **60**, 119–24.

Maynard Smith, J. (1966) Sympatric speciation. *American Naturalist*, **100**, 637–50.

Maynard Smith, J. and Haigh, J. (1974) The hitch-hiking effect of a favourable gene. *Genetical Research*, **18**, 125–31.

Maynard Smith, J. and Hoekstra, R. (1980) Polymorphism in a varied environment: how robust are the models? *Genetical Research*, **35**, 45–57.

Mayr, E. (1942) *Systematics and the Origin of Species*, Columbia University Press, New York.

Mayr, E. (1982) *The Growth of Biological Thought: diversity, evolution and inheritance*, Belknap Press, Cambridge, Massachusetts.

Nadeau, J.H. and Collins, R.L. (1983) Does associative overdominance account for the extensive polymorphism of H-2 linked loci? *Genetics*, **105**, 241–4.

Nadeau, J.H., Collins, R.L. and Klein, J. (1982) Organization and evolution of the mammalian genome. I. Polymorphism of the H–2 linked loci. *Genetics*, **102**, 583–98.

Nagylaki, T. (1978) Conditions for the existence of clines. *Genetics*, **80**, 595.

Neel, J.V. (1978) The population structure of an Amerindian tribe, the Yanomama. *Annual Review of Genetics*, **12**, 365–413.

Nei, M., Maruyama, T. and Chakraborty, R. (1975) The bottleneck effect and genetic variability in populations. *Evolution*, **29**, 1–10.

Nei, M. and Murato, M. (1966) Effective population size when fertility is inherited. *Genetic Research*, **8**, 257–60.

Nevo, E. (1978) Genetic variation in natural populations: patterns and theory. *Theoretical Population Biology*, **13**, 121–77.

Nevo, E., Beiles, A. and Ben-Shlomo, R. (1984) The evolutionary significance of genetic diversity, ecological, demographic and life history correlates. *Lecture Notes in Biomathematics*, Vol. 53, Springer–Verlag, Berlin, pp. 13–213.

Patton, J.L. and Feder, J.H. (1978) Genetic divergence between populations of the pocket gopher, *Thomomys umbrinus* (Richardson). *Zeitschrift Saugertierkind*, **43**, 17–30.

Patton, J.L. and Yang, S.Y. (1977) Genetic variation in *Thomomys bottae* pocket gophers: macrogeographic patterns. *Evolution*, **31**, 697–720.

Pérez-Tomé, J.M. and Toro, M. (1982) Competition of similar and non-similar genotypes. *Nature*, **299**, 153–4.

Pounds, J.A. and Jackson, J.F. (1981) Riverine barriers to gene flow and the differentiation of fence lizard populations. *Evolution*, **35**, 516–28.

Provine, W.B. (1981) Origins of 'The Genetics of Natural Populations' series, in *Dobzhansky's Genetics of Natural Populations.* (eds R.C. Lewontin, J.A. Moore, W.B. Provine and B. Wallace), Columbia University Press, New York, pp. 1–85.

Rouhani, S. and Barton, N.H. (1987) Speciation and the 'shifting balance' in a continuous population. *Theoretical Population Biology*, **31**, 465–92.

Selander, R.K. (1970) Behaviour and genetic variation in natural populations. *American Zoologist*, **10**, 53–66.

Selander, R.K. and Kaufman, D.W. (1973) Genic variability and strategies of adaptation in animals. *Proceedings of the National Academy of Sciences, USA*, **70**, 1875–7.

Slatkin, M. (1973) Gene flow and selection in a cline. *Genetics*, **75**, 733–56.

Slatkin, M. (1975) Gene flow and selection in a two locus system. *Genetics*, **81**, 209–22.

Slatkin, M. (1982) Estimating levels of gene flow in natural populations. *Genetics*, **99**, 323–35.

Slatkin, M. (1985a) Rare alleles as indicators of gene flow. *Evolution*, **39**, 53–65.

Slatkin, M. (1985b) Gene flow in natural populations. *Annual Review of Ecology and Systematics*, **16**, 393–430.

Smith, M.F. (1979) Geographic variation in genic and morphological characters in *Peromyscus californicus. Journal of Mammalogy*, **60**, 705–22.

Sokal, R.R. and Wartenberg, D.E. (1983) A test of spatial autocorrelation analysis using an isolation by distance model. *Genetics*, **105**, 219–37.

Spirito, F., Rossi, C. and Rizzoni, M. (1983) Reduction of gene flow due to the partial sterility of heterozygotes for a chromosome mutation. I. Studies on a 'neutral' gene not linked to the chromosome mutation. *Evolution*, **37**, 785–97.

Szymura, J. and Barton, N. (1986) Genetic analysis of a hybrid zone between the fire-bellied toad (*Bombina bombina*) and the yellow-bellied toad (*B. Variegata*) in Southern Poland. *Evolution*, **40**, 1141–59.

Sumner, F.B. (1929) The analysis of a concrete case of intergrades between two subspecies. *Proceedings of the National Academy of Sciences, USA*, 15, 110.

Turner, J.R.G. (1981) Adaptation and evolution in *Heliconius*, a defence of neo-Darwinism. *Annual Review of Ecology and Systematics*, **12**, 99–122.

Wagner, M. (1841) *Reisen im der Regentschaft Algier dem Jahren 1836, 1837, und 1838*, Leopold Voss, Leipzig.

Wallace, B. (1970) *Genetic Load: its biological and conceptual aspects*, Prentice-Hall, New York.

Wallace, B. (1975) Hard and soft selection revisited. *Evolution*, **29**, 465–73.

White, M.J.D. (1968) Models of speciation. *Science*, **158**, 1068–70.

White, M.J.D. (1978) *Modes of Speciation*, W.H. Freeman, San Francisco.

Wills, C. (1978) Rank-order selection is capable of maintaining all genetic polymorphism. *Genetics*, **89**, 403–17.

Wilson, A.C., Bush, G.L., Case, S.M. and King, M. (1975) Social structuring of mammalian populations and the rate of chromosomal evolution. *Proceedings of the National Academy of Sciences, USA*, **72**, 5061–5.

Wright, S. (1931) Evolution in Mendelian populations. *Genetics*, **16**, 97–159.

Wright, S. (1943) Isolation by distance. *Genetics*, **28**, 114–38.

Wright, S. (1946) Isolation by distance under diverse systems of mating. *Genetics*, **31**, 39–59.

Wright, S. (1978) *Evolution and the Genetics of Populations*. IV. *Variability within and among Natural Populations*, University of Chicago Press, Chicago.

Wright, S., Dobzhansky, T. and Hovanitz, W.H. (1942) The allelism of lethals in the third chromosome of *Drosophila pseudoobscura*. *Genetics*, **27**, 363–94.

Yokoyama, S. (1979) The rate of allelism of lethal genes in a geographically structured population. *Genetics*, **93**, 245–62.

PART TWO

The process of dispersal

Dispersal involves a confusing complex of potential circumstances and motivations (Chapter 2), but central to most of them are issues of habitat patchiness and/or social behaviour. In this section, we have included two reviews, each focusing on one of these two aspects.

Chapter 4 by Andrew Cockburn focuses on habitat patchiness while Chapter 5 by Charles Brandt reviews the role of social behaviour. It seems significant that each invokes the importance of the other, thus emphasizing the inextricable linkages between patchiness (extrinsic) and social (intrinsic) influences. Both are critical of many recently proposed interpretations of dispersal behaviour, and both warn against overly enthusiastic extrapolations from merely local data bases or perspectives.

Cockburn has worked extensively on small mammal populations in Australia, both of rodents and marsupials. He has also researched California vole (*Microtus californicus*) populations in coastal California, including the observation of a peak population reaching well in excess of 1000 voles per hectare. We are indeed fortunate to have his critical, but carefully reasoned synthesis of both theory and empirical data. Brandt has done a major study on the social behaviour of pikas (*Ochotona princeps*). His chapter is a scholarly review of a massive and complex literature relating behaviour to dispersal.

Collectively, these two chapters not only let us know where we are empirically with respect to the processes of dispersal, but guide us to productive approaches in the future.

– 4

Habitat heterogeneity and dispersal: environmental and genetic patchiness

A. COCKBURN

4.1 INTRODUCTION

Dispersal may enhance the fitness of an animal in two ways: by changing either the habitat in which it lives, or the animals with which it associates. Adaptive interpretation of dispersal therefore requires an understanding of heterogeneity in the physical, genetical and social environments to which animals are exposed. This chapter is concerned with the physical and genetical environment, and introduces theory and relevant data which suggests that the importance of these factors has been underestimated. Chapter 5 deals with the social environment. Here I am concerned with the adaptive consequences of dispersal and transfer, and deal only cursorily with enforced movements by subordinate animals. I assume that these enforced movements are made by individuals with low potential fecundity (i.e. are probably doomed), and have few, if any, adaptive consequences.

In my discussion of dispersal, I include reference to transfer, or movement between social groups without complete change of habitat (Greenwood and Harvey, 1982). I also follow other authors in recognizing two important dichotomies in dispersal. The first is the distinction between presaturation dispersal and saturation dispersal (*sensu*, Lidicker, 1975), and the second is the distinction between natal and breeding dispersal (*sensu* Greenwood and Harvey, 1982).

The methodological difficulties in measuring dispersal by small mammals within these classifications have proved formidable, and are discussed in detail in the appendices to this book. Because some of these difficulties relate directly to the effect of habitat heterogeneity, they warrant brief discussion here. Indices which are used to identify dispersers from residents are often inadequate (Dobson, 1981; Holekamp, 1984a; Tamarin, 1984; Cockburn *et al.*, 1985; Lidicker 1985a), and therefore experimental intervention is often necessary to identify movements. The usual technique is to measure movement on to plots from which all animals have been removed. Immigrants are then assumed to represent a random sample of dispersers. This technique relies on the artificial creation of habitat heterogeneity, or a 'dispersal sink'. All recent studies which compared the results of a removal plot study with actual intrapopulation movements suggest that results obtained need to be treated with some caution (Krohne *et al.*, 1984; Tamarin *et al.*, 1984; Bondrup-Nielsen, 1985; Krohne and Miner, 1985; Boutin *et al.*, 1985). The most damaging results were provided by Dobson (1981), who showed that movement by the ground squirrel *Spermophilus beecheyi* on to a removal plot was female-biased, while immigration on to unperturbed plots was male-biased. Information concerning which animals move into vacant habitat, or dispersal sinks, may be the chief benefit of this approach.

Also, in assessing sex differences or other trends in dispersal it is necessary to use the source population as a reference point. For example, the sex ratio of microtine populations is often strongly female-biased, so an excess of females among dispersers should not be taken to indicate female-biased dispersal (Pokki, 1981). Unfortunately, dispersal may itself contribute to the biases in the sex ratio of the source population, so unless dispersal and mortality can be distinguished circularity may ensue. The problem of choice of reference point is particularly severe when attempting to relate dispersal behaviour to some aspect of population growth or density (e.g. Stenseth, 1983), particularly when these parameters are themselves habitat-dependent (Cockburn and Lidicker, 1983). If we are interested, for example, in changes in natal dispersal through time, it is necessary to assess what proportion of young individuals disperse, and what proportion are philopatric. It is hardly of interest to say that natal dispersal is unimportant in declining populations if no young are born during this period. The best reference point is probably the proportion of weanlings which disperse, but appropriate data are unavailable for most small mammal populations.

A more intractable difficulty is illustrated by a study of dispersal in great tits. Greenwood and Harvey (1982) point out that a measure of distance moved has little biological meaning unless it is related to the number of home ranges traversed, and that the size of home ranges will depend on both population density, characteristics of the animal including body size, diet and phylogeny, and the geometry of the habitat in which the animal lives (Ostfeld, 1985). The density dependence of home range size will obviously confound the distinction between saturation and presaturation dispersal where home range fluctuates strongly with density, or where dispersal movements are small relative to home range diameter.

Despite these difficulties, the study of dispersal occupies a central role in the analysis of the demography of small mammals, because its consequences for population and evolutionary dynamics are profound.

4.1.1 Demographic significance

Confined populations grow more rapidly than populations from which dispersal is permitted (Lidicker, 1975; Krebs, 1979; Tamarin *et al.*, 1984), and demography in island populations where dispersal is restricted differs greatly from nearby mainland populations (Lidicker, 1973; Tamarin, 1977, 1978; Gliwicz, 1980), clearly demonstrating the demographic significance of dispersal. The prevalence of dispersal in rapidly expanding populations is held to be the best evidence for presaturation dispersal.

Because dispersal reduces the growth rate of source populations, it is generally believed that emigration is not balanced by immigration, and that mortality of emigrants occurs as a result of movement into a 'sink' of unfavourable habitat. If such dispersal is age- or sex-biased, the demography of the population is markedly affected, as a consequence of differences in mortality in the dispersive sex or age class. Habitat heterogeneity consequently underlies this interpretation of dispersal and its demographic consequences, although the spatial variability of environments is rarely assessed in dispersal studies.

4.1.2 Evolutionary significance

Effective dispersal can have a profound influence on the genetic structure of populations, as it determines the rate at which genes may flow through a population, and thus sets a limit to how much genetic divergence can occur between subpopulations in response to selection or random genetic drift (Chapter 3). Subdivision of populations into locally differentiated demes which undergo selection forms the basis of Wright's (1978) 'shifting balance' theory of evolution, and is the subject of massive theoretical speculation (Futuyma, 1986). Fusion of populations through migration or expansion of the area colonized also reduces the frequency of homozygous genotypes (Wahlund, 1928), which may influence the genetic consequences of different mating systems. The number of migrants necessary to reduce local genetic differentiation appears to be small (but see Chapter 3), and the countermanding effects of selection or drift are only likely to be significant where selection is strong, or where effective population size is quite small, and populations are isolated by some distance. It is noteworthy that the spatial structure of small mammal populations is again of fundamental significance in interpreting the genetic consequences of dispersal.

If dispersal tendency is influenced by genotype, persistent emigration which is not balanced by immigration may profoundly affect the genotypic composition of populations. This observation, coupled with the demographic significance of dispersal, has led to optimism among some authors that the Chitty Polymorphic Behavioural Hypothesis could be operating through variability in the propensity to disperse (Krebs, 1978; Chapter 7; but see the refutation by Stenseth 1981; Warkowska-Dratnal and Stenseth, 1985).

4.2 THE ADAPTIVE BASIS OF DISPERSAL

Thus far we have been concerned with the consequences of dispersal, and seen that analysis of these consequences at the demographic and population genetical level has led to speculation that the propensity to disperse may be influenced by genotype, and may be polymorphic within microtine populations. It has proved much less tractable to analyse why individual animals should depart from their home range, particularly as the imbalance of emigration and immigration suggests that dispersal often leads to mortality. Dispersers may be more susceptible to predation, may carry more parasites, and may often be in poor condition (reviewed by Lidicker, 1975). Whether poor condition leads to weakened animals being expelled, or is a consequence of the risk of movement, will normally be difficult to disentangle.

This chapter examines the theory and empirical information pertaining to three forms of dispersal and transfer frequently reported from small mammals: abandonment of the nest by the mother (breeding dispersal), male-biased natal dispersal, and transfer between winter nesting aggregations. The widespread but non-ubiquitous occurrence of these patterns makes them particularly amenable to adaptive interpretation. I conclude that they all may be interpreted in terms of the local interactions between family members, and show that this possibility is not considered in current theory and experimental design used in research on small mammals.

4.2.1 Breeding dispersal by females

One of the most perplexing observations to emerge from descriptive studies of microtine dispersal is a high frequency of breeding dispersal by females during the breeding season. Females from several species abandon their breeding home range, or perhaps the natal nest, to their offspring (Table 4.1). This behaviour has never been investigated experimentally or properly quantified, but it appears to be most common when populations are increasing rapidly (Stenseth, 1978, 1983). Here I attempt to show that such behaviour is related to variation in the durational stability of the habitats in which these species live.

There is abundant evidence that all environments are perceived by small mammals as patchy [see Cockburn and Lidicker (1983) for an extreme example]. All field workers are familiar with consistent trap success at

Table 4.1 Species for which there is evidence that mothers abandon the brood nest (breeding dispersal) as a form of parental investment in their offspring

Microtinae*†
Lemmus lemmus
Microtus pennsylvanicus
M. ochrogaster
M. agrestis
M. oeconomus
Clethrionomys glareolus
C. rutilis
Arvicola terrestris
Dicrostonyx groenlandicus
Hydromyinae‡
Pseudomys shortridgei
P. novaehollandiae
Sciuridae§
Spermophilus columbianus

* Jannett (1980); † Stenseth (1978); ‡ Cockburn *et al.* (1981); § Harris and Murie (1984).

certain sites in contrast to those only a short distance away, and the preferential use of 'hotspots' by researchers concerned with maximizing sample size is a major factor confounding the analysis of both dispersal and habitat preference. It is important to recognize that an environment which is patchy in space but stable in time may provide much the same opportunities for dispersal as one which is continuous in space and stable in time. There are theoretical reasons for believing that animals in temporally stable environments should distribute themselves so that their fitnesses in different habitat types will be approximately equal (Fretwell and Lucas, 1970; Fretwell, 1972; Rausher, 1984). A number of empirical observations support this concept of an evolutionarily stable 'ideal free' distribution (e.g. Whitham, 1980; Power, 1984), despite the formidable difficulties in designing appropriate tests. The importance of this theory for dispersal is that an animal need not increase its fitness by successful dispersal from a low quality patch to a high quality patch, though of course it does not preclude this outcome.

When temporal changes in habitat quality are superimposed on a pattern of spatial heterogeneity, the advantages of dispersal change markedly. Gadgil (1971) suggested that several aspects of habitat heterogeneity would promote the evolution of dispersal tendency. These include low durational stability of habitat patches relative to the species' generation time, close proximity of highly suitable patches, and catastrophic mortality creating vacant habitat patches. Variance in the ability of individuals to control local resources will exaggerate this effect, producing very low evolutionary thresholds (Łomnicki, 1978, 1980; Uchmanski, 1983). Extrapolations of this model to microtine life histories are discussed at some length by Stenseth (1983; see also Chapter 9).

The best model for habitat use by small mammals in temporally fluctuating environments had its genesis in Russian research in the middle of the century (e.g. Elton, 1942), and is normally attributed to Naumov (1954, 1964; but see Lidicker, 1985b). Wider availability of translations of this literature would be invaluable. The model has been reviewed on a number of recent occasions (e.g. Anderson, 1970, 1980; Hansson, 1977; Smith *et al.*, 1978), and distinguishes habitats according to the period they remain suitable for both occupancy and reproduction. Because there is little consistency in the use of terms to describe the different habitats recognized in the Naumov model, I have tried to draw together the varying terms under the general habitat templet proposed by Southwood (1977), as modified by Begon and Mortimer (1981) (Table 4.2; Figure 4.1).

There seems little doubt that populations which fluctuate in abundance are confined for much of the time to small, and often isolated, patches of habitat (Newsome, 1969a,b; Cockburn and Lidicker, 1983), which provide conditions suitable for year-round occupancy (constancy), and occasional (normally seasonal) reproduction. At least some of the population growth during fluctuations appears to be associated with an increase in the range of habitats occupied, particularly during periods adverse for reproduction (Cockburn and Lidicker, 1983; see however Batzli *et al.*, 1983), or an increase in the period over which reproduction occurs (e.g. Krebs, 1964; Myllymäki, 1977; Hansson, 1984; Kaikusalo and Tast, 1984).

Dispersal of animals into habitat where reproduction is unlikely or impossible (colonization:reception habitat, invasion habitat, and failure to traverse hostile areas between two habitable patches ($U > R_d$; Table 4.2), leads to the dispersal sink described by Lidicker (1975). Although these movements may have considerable economic significance (e.g. Chapter 8), and demographic significance (Lidicker, 1975), their adaptive importance is negligible (Anderson, 1970).

An animal which lives in a patchy environment and disperses from a constant habitat to colonization or invasion habitat, will only benefit if the new patch remains available for sufficiently long to permit breeding (colonization: production habitat; Table 4.2). Such movements will only be of evolutionary significance if reproduction is sufficiently prolific to ensure that at least some offspring recolonize patches constantly suitable for occupancy. Anderson (1970, 1980) has argued that the probability of recolonization is extremely low, and that the attention of evolutionary biologists should consequently focus on those individuals which live in

Table 4.2 A classification of habitat suitability for small mammals, modified from Southwood (1977) and Begon and Mortimer (1981), with synonyms provided by previous authors. Note the confusing use of the term survival habitat, and the greater precision in the classifications by Hansson (1977) and Myllymäki (1977). Note the use of R_d for migratory range to avoid the use of R_m by Begon and Mortimer (1981), with its connotations of similarity to the Malthusian parameter r_m

Suggested terms	*Constant*	*Colonization: production*	*Colonization: reception*	*Invasion*	*Traversable*
Anderson (1970, 1980)	Survival (reservation)	Colonization	Colonization	Colonization	Traversable
Smith et al. (1980)	Primary	Secondary	Secondary	Secondary	–
Myllymäki (1977)	Primary production	Temporary production	Survival	Invasion	Transition
Hansson (1977)	Donor	Induced donor	Reception	Invasion	Transition
Definitions following Southwood (1977); see also Begon and Mortimer (1981)					
Temporal axis					
Non-breeding	Constant	Unpredictable or seasonal	Unpredictable		?
Breeding	Seasonal	Unpredictable	Ephemeral or unpredictable	Ephemeral	
Spatial axis					
	Patchy or continuous	Patchy	Patchy	Continuous	
Algebraic formulation					
	$F_{ann}+L_{ann} \approx$ one year	$t < F$	$t > F$	$t > F$	$U < R_d$
	V is low	V is high	V is high	V is extreme	

t, generation time; F, period a habitat remains suitable for breeding; F_{ann}, breeding period within one year; L, period a habitat remains suitable for occupancy (non-breeding); L_{ann} non-breeding period within one year; $H = F + L$; R_f, the foraging or 'trivial' range of an organism; R_d, the migratory range, or dispersal capacity, of an organism; U, the distance between habitat patches, where $F = L = 0$, but movement is not frustrated; $V_{F,L}$, variance in the period of habitat suitability.

constant habitat, even if they represent a small fraction of the individuals present during peaks of abundance.

Data pertaining to this hypothesis are few. Cockburn (1981) demonstrated that *Pseudomys fumeus* which lives in habitat providing food through a late winter nutritional crisis survived to breed (often in two seasons), while the majority of the population died in early spring (Figure 4.2). Cockburn and Lidicker (1983) showed that female *Microtus californicus* in Californian valley grassland preferred habitat dominated by native

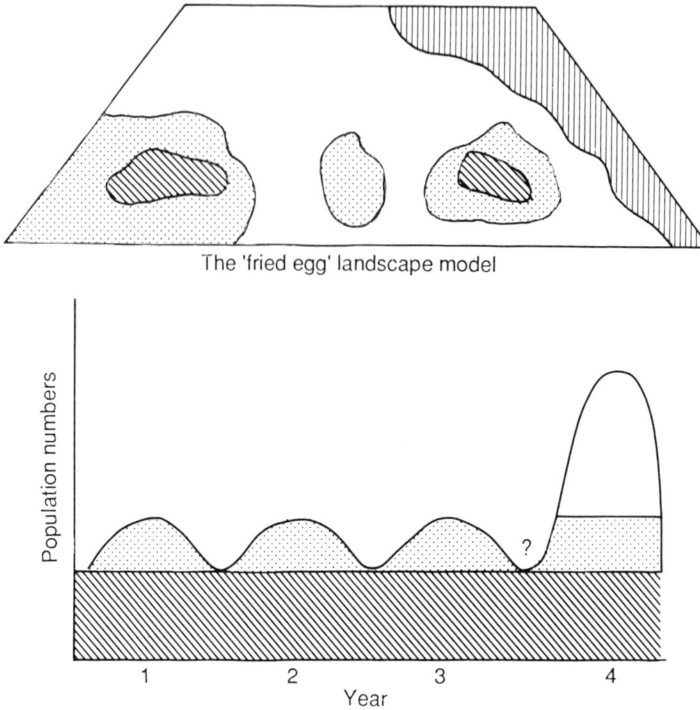

Figure 4.1 The 'fried-egg' landscape model. Some habitats are only temporarily available for both occupancy and breeding. Others are continually suitable for occupancy, but not for breeding (constant habitat; sloped lines). Seasonal fluctuations are the most important contributors to increased habitat availability. Expansion into new habitats (colonization habitat; dotted patches) contributes to the extent of population fluctuations. Under unusual circumstances vast extra areas may be colonized (invasion habitat; unshaded areas). Some areas are never suitable for breeding, and form a sink which restricts the growth of populations (vertical lines).

perennial plant species, even though these species were likely to provide poor nutrition in contrast to adjacent habitat during the period of reproduction. It appears that the food and protection from predation afforded by perennial grassland during the dry summer means that such habitat is to be preferred, despite any costs during the reproductive season (e.g. Krohne, 1980).

However, animals which live in the seasonally ephemeral habitat may approach the problem of recolonization in two ways. If reproduction is extremely rapid, and a large proportion of a patch of ephemeral habitat is colonized by the descendants of a single female, the prospects of success-

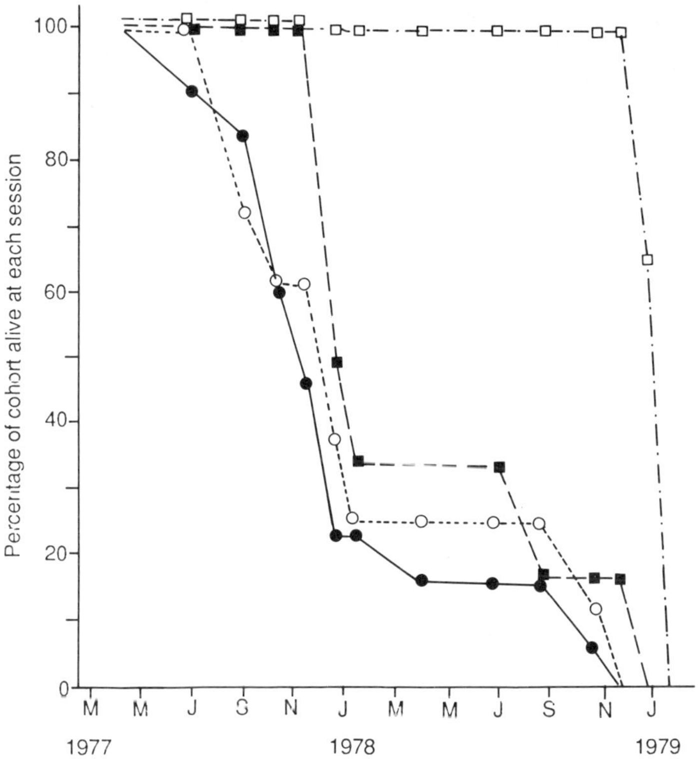

Figure 4.2 Percentage survival of the 1976–77 cohort of *Pseudomys fumeus* at Mt. William, Victoria. □, females, preferred habitat; ○, females, non-preferred habitat; ■, males, preferred habitat; ●, males, non preferred habitat. Individuals living in non-preferred habitat disappear by November, before they have had an opportunity to breed. From Cockburn (1981).

fully infiltrating at least a few descendants back into the favourable habitat are good if the area of ephemeral habitat is sufficiently large. Saturation of a large patch by the descendants of a single female is only likely if the female colonizes the patch early in the period during which it is available. Under these circumstances successful saturation may ensue if the female invests little time in each litter, abandoning them soon after weaning, and moving to a new site to breed again. It is worth mentioning that females should have little difficulty procuring mates given the high flux of immigrant males at this time (see below). Although the mother provides little post-weaning care, she may invest through the provision of a breeding site for her daughters, leading to eventual saturation of the patch.

As the prospects of successful movement by adult females diminishes,

Table 4.3 Breeding density in the preceding year, and proportion of immigrants among *Antechinus stuartii* females breeding on 7.5 ha study grids where breeding densities had fluctuated because of experimental manipulation, or as a consequence of extreme drought. Breeding densities are usually very stable

	Breeding density	*No. of immigrants*	*No. grid born*	*Prop. of immigrants*
Sherbrooke Forest				
1982	5	10	6	0.63
1983	16	0	22	0
Monga Forest				
1984	11	5	13	0.27
1985	18	1	34	0.03

and the need to relocate into a constant habitat increases, the optimum strategy will be to enforce dispersal of all young, to enhance the probability that at least some will successfully relocate. The prospect of at least one young surviving is greatly increased if this dispersal can be spread through time (e.g. Giesel, 1976). Reproduction should be extended, particularly where there is no prospect of surviving the season of greatest stringency. Cockburn and Lidicker (1983) showed that *M. californicus* living in habitat which had dried completely bred further into the summer than those which lived in areas in which perennial green vegetation persisted, apparently because the prospects of surviving the summer were negligible in the absence of some green vegetation. There is clearly potential for integration of this theoretical perspective into design of experiments on microtine dispersal.

Behaviour of this sort is not unique to microtines. Although populations are typically extremely stable, dispersal by female *Antechinus* appears to be most common in populations recovering from drought, or experimental manipulation (Table 4.3). Harris and Murie (1984) demonstrated that female Columbian ground squirrels (*Spermophilus columbianus*) often abandoned their nest site to a daughter, but retained their nest if yearling daughters were absent. This appears to be a form of parental investment, as mothers are dominant to their daughters and should be able to exclude them, and there is little evidence of benefit and some evidence for cost associated with abandonment. Although data were not available for populations which differed in density, Harris and Murie speculated that the low density at their study site contributed to the prevalence of nest abandonment.

Cockburn *et al.* (1981) showed that local survival of adult and juvenile *Pseudomys shortridgei* changed dramatically with time since fire, and suggested that this was a reflection of the relative frequency of breeding and natal dispersal. This species depends on highly productive heathland which has burnt recently, and disappears from mature habitat as productivity and seed production decline (Cockburn, 1978). In deteriorating habitat, adults are philopatric and long-lived, often breeding in each of three consecutive years. Young are forced to disperse from their natal home range. In recently colonized habitat, adults rarely survive beyond their first breeding season, but their offspring are philopatric and breed successfully in or near their natal home range. Cockburn *et al.* (1981) suggest that in recently burnt habitat adults should maximize juvenile survival, and may have a reasonable probability of successful dispersal to an uncolonized site nearby. In mature habitats, adults should enforce dispersal of young for as long as possible to minimize the probability that they will leave no young at all. This would occur if their home range was burnt, or no new habitat became available nearby, or if the productivity of seeds continued to decline, supporting an ever smaller resident population. Although the habitat in which *P. shortridgei* occurs may initially appear to have little in common with that occupied by species which fluctuate in a cyclic manner, evidence abounds that both microtines and snowshoe hares live in a spatially heterogeneous environment where the durational stability of patches is highly variable (Myllymäki, 1977; Hansson, 1977, 1979; Wolff, 1980).

These observations from unrelated rodents and lagomorphs suggest that temporal fluctuations in habitat suitability may influence the relative frequency of natal and breeding dispersal. In benign or improving conditions, breeding dispersal, particularly by females, may be more probable than when the population is crowded or conditions are deteriorating.

4.2.2 Sex differences

In common with many simple but illustrative models, the original mathematics used to derive the unsurprising conclusion that dispersal should be favoured in fluctuating environments were based on haploid (asexual) populations [see Stenseth (1983) for a review]. The plausibility of the results may be one reason why the effects of sex differences have only rarely been included in formal theory (e.g. Motro, 1982b), which therefore makes no predictions about which sex should be the predominant disperser. However, in the majority of bird and mammal populations one sex tends to move earlier and farther than the other (Greenwood, 1980, 1983; Table 4.4). Further, there is preliminary evidence that sex-biased dispersal confounds interpretation of genetic structure of populations, by creating local heterozygote excess when movements occur before mating (Prout, 1981; Kawata, 1985).

There are several suggestions why one sex should disperse more fre-

Table 4.4 The number of species and families of mammals and birds known to have male-biased, female-biased and no sex difference in natal dispersal (modified from Greenwood, 1980)

	Predominant dispersing sex					
	Mammals			*Birds*		
	Male	*Female*	*Both*	*Male*	*Female*	*Both*
Species	45	5	15	3	21	6
Families	23	4	7	1	11	5

quently than the other. Most of these arguments centre around the desirbility of choosing a genetically optimum mate, and differences between the sexes in the criteria by which they choose breeding sites. High rates of inbreeding are a direct consequence of philopatry of both sexes, and any philopatry will impose a spatial structure on a population which could be unrelated to the extrinsic characters of its habitat. Under these conditions relatedness will be proportional to geographic separation. Dispersal of one sex should help prevent close inbreeding, which tends to reduce fitness (Falconer, 1981), presumably because deleterious recessive alleles are more likely to be expressed, and beneficial interactions between different alleles at the same locus (heterosis) will be lost. There is convincing evidence that inbreeding is usually very rare in mammal populations (Ralls, *et al.*, 1986), despite claims to the contrary by Shields (1983).

Although inbreeding avoidance provides a possible theoretical explanation for sex-biased dispersal, it does not really predict which of the sexes should disperse, though the data in Table 4.4 suggest that this is reasonably predictable in both birds (females) and mammals (males). Greenwood (1980, 1983) pointed to a correlation between mating systems and the direction of the sex bias. In resource-defence mating systems, of which obligate monogamy in many species of birds provides a good example, the resource defender (usually the male) may have difficulty in establishing and maintaining an initial territory, and associated problems in subsequent dispersal to a new breeding site. Females, by contrast, will have a comparatively low investment in defence of resources but may benefit through dispersal by increased choice among male-defended resources. However, where mates are defended the relative investment in resources tends to shift in favour of the limiting sex. In most cases this tends to be females, as the reproductive success of males will be limited principally by the number of females to which they can gain access (Bateman 1948; Trivers, 1972), and the gametes of males will be comparatively abundant, though perhaps not to the extent that was originally believed (Dewsbury, 1982, 1985; Nakatsura and Kramer, 1982; Dewsbury and Sawrey, 1984). This will be particularly true in the polygynous mating systems prevalent among mammals, and will lead to selection for female philopatry, while males may benefit through increased access to females by undertaking dispersal.

4.2.3 Male-biased natal dispersal

Although small mammal dispersal has been reviewed on a number of occasions (Lidicker, 1975, 1985a; Gaines and McClenaghan, 1980; Tamarin, 1980; Stenseth, 1983), and there is general agreement that there are important demographic consequences of dispersal, there is no evidence of any real consensus over its causes or degree of concordance with theory. Indeed, there remain some authors who express scepticism as to whether dispersal is interpretable within an adaptive framework at all (Anderson, 1980; Shields, 1983). This is no doubt partly due to the methodological difficulties described above and the near impossibility of observing the proximate behaviour which underlies dispersal movements in these secretive animals. Despite these shortcomings, there is good and growing evidence that natal dispersal occurs more frequently than breeding dispersal, and that males are over-represented among natal dispersers in almost all species of microtines. This pattern is repeated throughout the mammals, and there are no good reasons to believe that its occurrence in microtines should be interpreted within the idiosyncratic selective milieu of cyclic or fluctuating populations.

In order to examine the hypothesis that social interactions in a source population led to expulsion of stressed voles, McDonald and Taitt (1982) measured the concentration of steroid hormones in blood plasma of *Microtus townsendii* moving on to a removal grid during a 'spring decline' in numbers. Perhaps to their surprise, they found that dispersing males, most of which were young and possibly reproductively inhibited, showed no signs of stress. The same was not true for females, which towards the end of the decline frequently had plasma total corticosterone concentrations in excess of their maximum corticosteroid binding capacity (Table 4.5). The significance of this parameter as a measure of stress is elaborated by Lee and McDonald (1985). These results suggest that the conditions underlying dispersal of juvenile males are unrelated to social pressure, and differ from the causes of female dispersal.

The ability to distinguish whether local competition for resources, competition for mates or inbreeding avoidance are principle contributors to sex-biased dispersal in higher vertebrates has been hampered by the absence of suitable study organisms. Dobson (1982) contends that all three factors are probably involved, but attributes particular importance to

Table 4.5 Total corticosterone and maximum corticosterone binding capacity (MCBC) (μmol L^{-1}) in the plasma of male and female *Microtus townsendii* moving on to a removal grid during a spring decline in numbers. Data from McDonald and Taitt (1982)

	Male		*Female*	
	Corticosterone	*MCBC*	*Corticosterone*	*MCBC*
16 April	0.72	1.12	5.9	7.2
29 April	0.84	1.27	7.6	6.2

intrasexual competition for mates. Holekamp (1984a,b) uses similar data to implicate inbreeding avoidance. Liberg and von Schantz (1985) suggest that attempts by subordinate males to copulate with females controlled by dominant males will be a major form of intrasexual conflict leading to male bias in dispersal. Moore and Ali (1984) suggest that intrasexual competition is a parsimonious explanation in all published instances where inbreeding avoidance has been implicated as the cause of sex-biased dispersal [see Packer (1985) for a refutation]. However, a recent study by Cockburn *et al.* (1985), using the unique life history of the marsupial genus *Antechinus*, suggests that this scepticism is unwarranted. Male generations of *Antechinus* are discrete, and punctuated by an abrupt post-mating mortality of all males. Mothers always enforce dispersal of all of their sons immediately after weaning, but tolerate their daughters, and nest communally with them for several months. There is no experimental evidence which distinguishes whether mothers recruit unrelated males to live in the female kin groups, or sons will only settle near or with family groups of females, but circumstantial evidence suggests that both effects are important. Competition for mates among males should not promote complete dispersal by males, particularly as there is no class of experienced males. There is no risk of kleptogamy by younger males in *Antechinus*. Any maternal advantage is unlikely to accrue from reduced competition for resources, as group size does not change significantly as a result of natal dispersal (Figure 4.3). Although I have no data confirming the deleterious effects of inbreeding (it has never occurred on my study sites), incest avoidance seems to be a parsimonious and sufficient cause for the evolution of male-biased dispersal in *Antechinus* (Cockburn *et al.*, 1985). Males do not move to nearby nests, but cross several home ranges before they are recruited, suggesting inbreeding with relatives other than siblings is also avoided.

Dispersal to avoid inbreeding seems to suggest that the costs proposed by Bengtsson (1978) are insufficient to counter the benefits obtained.

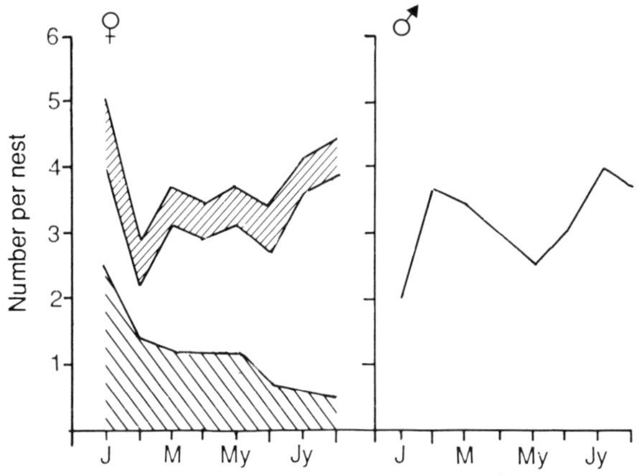

Figure 4.3 Mean number of male and female *Antechinus* sharing communal nests at Sherbrooke Forest, Victoria. Sample size for January is two nests; sample size thereafter varies from seven to ten nests. Upward hatching, adults; downward hatching, juveniles living in the nest in which they were born; unhatched areas, juveniles not born in the nest. Data from Scotts (1983) and Cockburn *et al.* (1985).

Bateson (1983) and Shields (1983) have argued that there are several costs associated with choosing a mate on the basis of genetic dissimilarity. In particular, local adaptation will break down when favourable gene complexes are scrambled. In addition, there may be other costs of outcrossing which counterbalance the costs of inbreeding. These are summarized in Table 4.6 (see also Bateson 1983; Shields 1983; Chapter 1). In particular, Bengtsson (1978) has suggested low population density and patchy distribution will lead to very high costs for dispersal to achieve optimum mate choice. Such conditions are characteristic of microtine rodents during much of the population cycle. These costs and benefits have lead Bateson to propose the concept of optimum outbreeding, for which some supportive evidence is available from both birds (Bateson *et al.*, 1980; Bateson, 1982, 1983; McGregor and Krebs, 1982; see however Millington and Price, 1985) and plants (Price and Waser, 1979; Waser and Price, 1983).

In order to probe this possibility further, I have recently examined dispersal near a hybrid zone for teat phenotypes of *Antechinus stuartii*. The study site is located where plateau-adapted animals with ten teats and coastal-adapted animals with eight teats interact and hybridize. Teat number in *Antechinus* is almost perfectly correlated with litter size, and varies along climatic gradients, but in most sites intergradation is very gradual (Cockburn *et al.*, 1983). Animals with nine teats have low but variable weaning success in comparison to eight and ten teat animals (Cockburn, in

Table 4.6 The hypothetical disadvantages of inbreeding and outbreeding (from Bateson, 1983). In any one species only some of these costs would be likely to operate. Note particularly Cost of outbreeding no. 5, and see also Bengtsson (1978)

Costs of inbreeding

1. Deleterious genes are more likely to be fully expressed.
2. Beneficial interactions between different alleles at the same genetic locus (heterozygous advantage or overdominance) are likely to be lost.
3. Offspring insufficiently variable for one or another to cope with a varying environment.
4. Offspring more like each other and so compete more intensely.

Costs of outbreeding

1. Genes required for adaptation to particular environment lost or suppressed.
2. Co-adapted gene complexes broken up by recombination.
3. In polygynous species advantage of having extra-closely related offspring lost and parental genes less well represented in next generation.
4. Infection from pathogens carried by mate more likely.
5. Dispersal may be costly and dangerous.
6. Acquired skills useful in one environment and not appropriate in another.
7. Mismatch of habits acquired by mates in different environments disrupts parenting.

preparation), and consequently have the characteristics of low fitness hybrids (Endler, 1977). Available data suggest that hybridization is rare except when population density is very low, and that effective dispersal by males across the hybrid zone at other times is negligible, although these are the movements most likely to be revealed by the configuration of the study plot. These very preliminary data indicate that optimization of mate choice according to genetic criteria may shape social evolution and dispersal behaviour in *Antechinus*, and support the contention of Bateson (1983) that optimum outcrossing may be of importance in social dynamics in a range of vertebrate taxa.

4.2.4 Transfer between nesting constellations

Transfer is of particular interest as movement of this type changes only the individuals with which an animal associates, rather than the habitat in which it lives. Widespread transfer between nesting groups has been documented at the start of winter in *Microtus xanthognathus* (Wolff and Lidicker, 1981) and during winter in *M. pennsylvanicus* (Webster and Brooks, 1981; Madison *et al.*, 1984). In both cases the effect of transfer is to reduce the level of relatedness within nests as family groups break up and individuals reaggregate more or less randomly, though female relatives are more likely to nest together than male relatives.

Thermoregulatory benefits of winter grouping are well documented. For example, Grodziński *et al.* (1977) compared *M. arvalis* living alone and in groups. The voles showed a monotonic decrease in energy requirements up to a group size of seven, after which there were no additional savings (or thermoregulatory costs). Madison *et al.* (1984) suggest that local depletion of food resources and increased risk of detection by predators may set an upper limit to winter group size, and for *M. pennsylvanicus* the optimum group size is between two and three, though their argument is confounded by the use of a 'what is must be optimal' argument, and is thus circular (Gould and Lewontin, 1979). There now appears to be consensus that optimum group sizes will rarely be attained in nature, because of conflicts of interest, and variance in costs and benefits associated with joining a group, or expelling an animal from a group (Pulliam and Caraco, 1984). Certainly in the case of *M. xanthognathus*, individuals in winter groups co-operate to cache food, which is likely to compensate for any costs of local food depletion (Wolff and Lidicker, 1981).

Such benefits shed little light on the reason for transfer between nesting constellations, which appears to offer benefits to unrelated 'cheaters' without any guarantee of reciprocity. Wolff and Lidicker (1981) hypothesize two chief benefits: (i) spreading the risk of predation and (ii) avoiding inbreeding. They argue that discovery of a nest by a predator (usually mustelids in the case of *M. xanthognathus*) could lead either to total destruction of a nest or reduction in group size below that which would facilitate energy saving through huddling. By dispersal of some family members, there may be some individuals which survive the winter. I believe that both these explanations are unlikely to provide a general explanation for the phenomenon of winter transfer. Substantial transfer between nests which appears analogous to that observed in microtines has been documented recently in *A. stuartii* (Scotts, 1983; Saxon, 1984; Cockburn, in preparation). In these dasyurid marsupials there is an abrupt long-distance dispersal of all males immediately after weaning, eliminating any possibility of nuclear family incest (Cockburn *et al.*, 1985; see above). However, there is substantial subsequent transfer between nests which gradually reduces the number of related females associated with a nest, without reducing group size (Figure 4.3), suggesting that inbreeding avoidance is unlikely to be relevant in this case. Nests are generally located in small hollows in stages, trees or fallen logs which are unlikely to be accessible to predators (mainly snakes and feral foxes and cats). Long-term studies of occupancy of these nests have failed to reveal a single instance of successful nest predation. Cessation of occupancy of a nest is chiefly caused by abandonment of the site when female groups break up at the time of mating, and mothers find a new nest in which they can rear their young (Cockburn, in preparation). The ability of mothers to locate new natal nests indicates that nests are not a limiting resource during winter.

I now develop theory which suggests that transfer may function to achieve a dispersion of daughters which reduces local competition among female relatives, but increases the risk of competition with neighbouring matrilines. Evidence is accumulating that as a consequence of female natal philopatry mothers compete with their daughters in many species of small mammals, leading to suggestions that where competition is prevalent, mothers should bias sex allocation towards the production of sons (Clark, 1978; Taylor, 1981; Silk, 1984; Cockburn *et al.*, 1985). Indeed, W.D. Hamilton (personal communication cited in Charnov, 1982) has argued that where there is strong local resource competition it can be evolutionarily stable for polytocous species to produce a single daughter, and that additional investment in daughters may reduce the fitness of their female relatives. One pertinent body of theory is the prediction by Hamilton and May (1978) that at least some dispersal also appears to be evolutionarily stable in homogeneous environments, where all available sites for effective dispersal are already colonized. This is because the territories of mothers which produce only non-dispersers are subject to invasion by the offspring of females which enforce dispersal of some of their offspring. The non-disperser phenotype can never replace a disperser phenotype, but the converse will occasionally be true (Hamilton and May, 1978; Comins *et al.*, 1980). This effect occurs in both conventional population genetical and ESS models containing a variety of assumptions, and even where there is more than one disperser phenotype in the population (Motro 1982a,b, 1983; Comins, 1982). Two specific cases are of particular interest. Firstly, Hamilton and May (1978) and Motro (1983) show that there may be parent-offspring conflict over the optimum dispersal fraction. A higher proportion of offspring should disperse, if dispersal frequency is controlled by genes in the mother rather than in her offspring. Secondly, the result holds even when dispersal is localized to nearby sites. Comins (1982) showed that although the probability of competing with an alternative allele is reduced by the gene frequency correlations caused by localized dispersal, any effect on the optimum dispersal frequency tends to be cancelled out by the necessity to increase dispersal to seek recolonization further afield. Unfortunately, empirical data pertaining to this theory will inevitably be counfounded by the ubiquity of spatial and temporal heterogeneity in the extrinsic characteristics of animal habitat. It is important to realise that these models pertain to an intrinsic heterogeneity, caused by the presence of different 'competing' alleles or genotypes within the population.

Advantages associated with limited dispersal of daughters may provide both an additional advantage of producing daughters in species which can control sex allocation (e.g. *Antechinus*), and a 'use' for daughters in species which have only limited capacity to influence the sex ratio (most microtines). Scattering of females into nearby nests may improve the capacity of mothers or their genotypes to 'capture' adjacent habitat, by improving the probability that they will succeed in competition with the kin clusters in nearby nests. In this context, it is relevant that mothers of both *A. stuartii* and *M. xanthognathus* sometimes participate in nest constellations throughout the winter, even though there is negligible probability of their surviving to breed in the following spring (Wolff and Lidicker, 1981; Cockburn, in preparation). The successful operation of this system does rely on a degree of reciprocity, or balance in exchange of individuals between kin clusters. This seems less probable in small mammal populations where aggregations exist principally for huddling, rather than the much closer knit groups which occur, for example, among the primates and social carnivores. It seems likely that expulsion from and recruitment into the group would take place in or near the nest. However, a kin group exporting daughters may suffer thermoregulatory costs unless it recruits at least some strange females, and/or juvenile males. Male-biased recruitment may help, and it is true that in *Antechinus* there is very high variance in the number of males which participate in winter nesting constellations (Cockburn *et al.*, 1985). However, living with a large number of males may also have costs. Female *Antechinus* are often hurt during the mating season, apparently as a result of attempts at copulation by males. The occurrence of fur loss and wounding is patchy within and between populations and it is possible that such damage is the result of a very high local operational sex ratio, with females subject to prolonged and repeated copulations (Scott and Tan, 1985).

In summary, it appears that there are both energetic reasons for winter aggregations in small mammals, and theoretical reasons for expecting a moderate level of female transfer into neighbouring nesting constellations. These influences may work in concert. The cost of a male-biased operational sex ratio at the time of mating may counterbalance the disadvantage of accepting unrelated females into nesting constellations. Unlike many aspects of dispersal behaviour, these ideas appear to be amenable to testing without any major methodological developments. Radiotelemetry has proved useful in revealing the composition of nesting constellations, and also can provide clues concerning the physiological condition and fate of individual animals. The fate of groups and females in populations with 'normal' and manipulated sex ratios is also testable. The data from *Antechinus* show that the spread and extinction of matrilines can be accurately quantified, provided kinship in natural populations can be described. The formal models necessary to integrate these observations are not particularly complex.

4.3 ARE DISPERSERS GENETICALLY DISTINCT?

The evidence presented in this chapter suggests that at least three common patterns in small mammal dispersal are amenable to adaptive interpretation. It is consequently useful to place these observations within the perspective of the literature on small mammals which has sought a genetic basis for dispersal [see Lidicker and Caldwell (1982) for a review]. This literature has emerged largely as a result of the optimism that dispersal may be the means through which the 'quality' of individuals changes through the course of population cycles in microtine rodents [see Krebs (1978) for a review].

Evidence for a genetic basis for dispersal comes from two sources: (i) quantitative genetic studies which estimate the heritability of tendency to disperse, and (ii) comparison of the frequency of allozyme morphs in samples of 'non-dispersers' (the source population), and 'dispersers' (animals appearing on removal grids).

4.3.1 Heritability

There are no completely convincing data which suggest that dispersal tendency in vertebrate populations is heritable. The most widely cited quantitative genetic analysis is a study of great tits by Greenwood *et al.*, (1979), where correlations between distance dispersed by parents and their offspring indicated that heritability was about 60%. van Noordwijk (1984) has recently shown that their original analysis was confounded by environmental heterogeneity, and by the restrictions to sampling created by a finite study area (see also Barrowclough, 1978). Correlations between dispersal tendency of microtine siblings have been reported more than once (Hilborn, 1975; Beacham, 1979), though these studies were based on very restricted samples within families, which may have confounded their results to an unknown extent. Unfortunately, the only study which has attempted formal measurements of heritability of dispersal tendency is found in an unpublished thesis (Anderson, 1975), though Krebs (1979) summarizes the chief results. *Microtus townsendii* offspring of known parentage were released on to both fenced and unfenced grids, and between-sib correlations were compared to estimate heritability (Falconer, 1981). If dispersal tendency was heritable Anderson anticipated and found a higher heritability for length of residency on the unfenced grid where dispersal was possible ($H^2 = 0.62$) than on the fenced grid ($H^2 = 0.37$).

Although these data are consistent with some genotypic influence on dispersal, the argument is very indirect, and Krebs (1979) rightly urges caution in their interpretation. Replication of this study with other species and population conditions should be a high priority for researchers.

Allozyme frequencies

Changes in allozyme frequencies as a result of dispersal have attracted considerable attention because such changes may underlie population fluctuations of this sort. Attempts to establish a difference between the frequency of electromorphs of enzymes in residents and dispersers have been simultaneously encouraging but equivocal. Evidence that dispersers are not a random sample of the population has been documented for *M. pennsylvanicus* (Myers and Krebs, 1971; Keith and Tamarin, 1981), *M. ochrogaster* (Myers and Krebs, 1971; Pickering *et al.*, 1974) and *M. townsendii* (Krebs *et al.*, 1976). Over-representation of genotypes among dispersers was sometimes restricted to certain phases of the population cycle. For example, Myers and Krebs (1971) found a preponderance of homozygotes for an allele of transferrin among dispersing males during the population peak and decline, but females dispersing during a period of population increase were more often heterozygous for transferrin than the source population. In contrast to these results, no differences between allozyme frequencies of residents and dispersers were detected in studies of *M. californicus* (Riggs, 1979; Bowen, 1982), *M. pennsylvanicus* and *M. ochrogaster* (Verner, 1979, cited by Gaines and McClenaghan, 1980), and *M. breweri* (Keith and Tamarin, 1981). Keith and Tamarin (1981) argued that the difference between *M. breweri*, an island dweller, and *M. pennsylvanicus*, a mainland form, reflect the restricted potential for dispersal in the former species.

Whether these results are helpful or obfuscatory remains uncertain, and even the most ardent advocates of the uses of electrophoresis urge caution in the interpretation of extremely indirect correlations of this sort. In particular, the distinction between cause and consequence is currently insoluble. Indeed, the interpretation of electrophoretic data from microtine populations is currently undergoing re-evaluation from a number of perspectives (Gaines and Whittam, 1980; Motro and Thomson, 1982; Gyllensten, 1985; Kawata, 1985). If, as I have claimed above, genealogical and genetic structure were to drive dispersal machinery, instead of vice versa, some of the conventional assumptions used in population genetics may be violated (Cockburn, 1985).

Thus, although there are data available which suggest dispersal tendency is influenced by genotype, we are some way from resolving this question satisfactorily. Furthermore, the search for genetic influences appears to be clouded by the dichotomy between saturation and presaturation dispersal.

Because dispersal is prevalent during periods of population increase when local conditions are benign, there appears to be consensus that dispersal is not 'enforced', and is hence the result of some individual predisposition. The data I have documented on both sex-biased dispersal and winter transfer suggest that such dispersal may be both adaptive and density independent, and enforced by interactions at the level of the family. Both these data and the associated theory suggest that it is often adaptive to enforce dispersal of part of a litter. Consequently, while dispersal may be subject to selection, its heritability is unlikely to be reflected in sibling correlations. While differences in allozyme frequencies may reflect between-family differences, the overwhelming importance of within-family effects may confound any use of this technique.

4.4 CONCLUSIONS

Previous studies of behaviour and life history evolution in small mammals have suffered from a tendency to focus exclusively on environmental exigencies of importance to local populations of a particular study species, and consequently ignore the broader constraints on the evolution of behaviour in all mammals. To quote a recent example, Pruitt (1984, p. 1) recently concluded that 'Snow is the all-pervasive factor that integrates various aspects of the winter ecology of small mammals'. I am unable to find a single significant aspect of the ecology of subnivean mammals which does not also occur in species in snow-free environments.

In this chapter I have suggested that at least three aspects of dispersal behaviour in small mammals are amenable to 'selectionist' interpretation (Charnov, 1982). These are male-biased natal dispersal, post-weaning breeding dispersal by females living in benign environments, and winter transfer between nests. The occurrence of these aspects of behaviour in the groups of small mammals referred to in this chapter are documented in Table 4.7, together with some important aspects of the ecology of these groups. Several patterns emerge. Firstly, there is no evidence that any type of dispersal is restricted to cyclic or other fluctuating populations. While I agree that dispersal patterns are an important contributor to the extent of population fluctuations (see other chapters in this book), I feel that the case that cyclic population fluctuations are an important selective influence on dispersal behaviour cannot be confirmed with available data. Dispersal and transfer do not appear to be influenced by diet, as has been implied by West and Dublin (1984), or degree of intergeneration overlap, as has been implied by Dobson (1982). Natal dispersal varies with environmental circumstances from nil to 100% in some species (e.g. *Pseudomys shortridgei*), and is restricted to one sex in other species (e.g. *Antechinus*). The chief influences on dispersal and transfer appear, as has long been recog-

Table 4.7 Summary of types of dispersal and important ecological features of four groups of small mammals (sources are given in the text)

	Microtines	*Ground squirrels*	*Pseudomys shortridgei*	*Antechinus stuartii*
Diet	Herbivores	Omnivores	Omnivore	Insectivore
Interannual variance in nos	High	Low	Moderate	Very Low
Periodicity	Yes	No	Decline with succession	No
Duration of life	<1 year	>1 year	1 year early in succession 3 years late in succession	1 year
Polyoestrous/ monoestrous	Poly	Poly	Poly	Mono
Mating system	Polygyny Monogamy Promiscuity	Polygyny Monogamy	Monogamy	Promiscuity
Habitat patchiness	Extreme	Moderate	High	Very low
Breeding dispersal by females	Yes	Sometimes	Yes	Yes
Male-biased natal dispersal	Common	Yes	No	Yes
Winter transfer	Yes	Hibernators	Unlikely	Yes
Prop. of litter which disperses or transfers	High	Most males	0–100%	High

nized, to be heterogeneity in the social and physical environment. However, genetic heterogeneity (the distribution of relatives, or of competing alleles) may be important in most mammalian populations, and may lead to an underlying pattern of dispersal which needs to be isolated before progress can be made in other areas.

These results suggest that many of the older methods used in studies of small mammal populations may be unsuited to the dissection of the evolutionarily and ecologically important patterns which underly social behaviour and dispersal. For example, it is unclear what aspects of dispersal are likely to be revealed by the removal plot technique. Cockburn (1985) has suggested that this method is most likely to reveal long-distance movements, and is quite unsuited to understanding the proximate social context of dispersal. In particular, the method does not distinguish between an increased frequency of dispersal, which is likely to be of considerable evolutionary significance, and increased distance dispersed, which will depend on number of home ranges traversed, genetic structure of the population, and habitat characteristics. Exciting developments in the study of dispersal are documented in Appendices 1–3.

Conspicuous species like ground squirrels and demographically tractable species like *Antechinus* currently offer the best opportunities for progress. However, now that a theory which relates genealogical and genetic structure is emerging (e.g. Cannings and Thompson, 1981), and advances in molecular biology seem likely to make genealogical structure accessible even in species with complex life histories or limited genetic variation (see also Tamarin *et al.*, 1983; Jeffreys *et al.*, 1985), I argue that the design of research into dispersal by small mammals warrants a fundamental reassessment. There seems little point in continuing to interpret dispersal against a background of population fluctuation, as the connections to both the proximate behavioural, and ultimate evolutionary, causes of dispersal are so diffuse as to be uninterpretable. While some aspects of dispersal will eventually be related to features unique to fluctuating populations, the most exciting progress seems likely to come from a synthetic approach which recognizes that dispersal patterns will both influence and be influenced by elements of social organization common to many mammalian species, and in particular by the interactions which occur within families.

ACKNOWLEDGEMENTS

This research was funded by the Australian Research Grants Scheme (#D18415745). Useful contributions to the ideas expressed were made by Hugh Comins, Charley Krebs, Bill Lidicker and Tony Lee during a number of postdoctoral assignments. Charles Brandt, Bill Lidicker and Nils Chr. Stenseth provided most helpful comments on the manuscript.

REFERENCES

Anderson, J.H. (1975) Phenotypic correlations among relatives and variability in reproductive performance in populations of the vole *Microtus townsendii*. Ph.D. Thesis, University of British Columbia.

Anderson, P.K. (1970) Ecological structure and gene flow in small mammals. *Symposium of the Zoological Society of London*, **26**, 299–325.

Anderson, P.K. (1980) Evolutionary implications of microtine behavioral systems on the ecological stage. *The Biologist*, **62**, 70–88.

Barrowclough, G.F. (1978) Sampling bias in dispersal studies based on finite area. *Bird Banding*, **49**, 333–41.

Bateman, A.J. (1948) Intra-sexual selection in *Drosophila*. *Heredity*, **2**, 349–68.

Bateson, P. (1982) Preference for cousins in Japanese quail. *Nature*, **295**, 236–7.

Bateson, P. (1983) Optimal outbreeding, in *Mate Choice* (ed. P. Bateson), Cambridge University Press, Cambridge, pp. 257–77.

Bateson, P., Lowick, W. and Scott, D.K. (1980) Similarities between the faces of parents and offspring in Bewick's swan and the differences between mates. *Journal of Zoology*, **191**, 61–74.

Batzli, G.O., Pitelka, F.A. and Cameron, G.N. (1983) Habitat use by lemmings near Barrow, Alaska. *Holarctic Ecology*, **6**, 255–62.

Beacham, T.D. (1979) Dispersal tendency and duration of life of littermates during population fluctuations of the vole *Microtus townsendii*. *Oecologia*, **42**, 11–22.

Begon, M. and Mortimer, M. (1981) *Population Ecology: a unified study of plants and animals*, Blackwell Scientific, Oxford.

Bengtsson, B.O. (1978) Avoid inbreeding: at what cost? *Journal of Theoretical Biology*, **73**, 439–44.

Bondrup-Nielsen, S. (1985) An evaluation of the effects of space use and habitat patterns on dispersal in small mammals. *Annales Zoologici Fennici*, **22**, 373–83.

Boutin, S., Gilbert, B.S., Krebs, C.J. *et al.* (1985) The role of dispersal in the population dynamics of snowshoe hares. *Canadian Journal of Zoology*, **63**, 106–15.

Bowen, B.S. (1982) Temporal dynamics of microgeographic structure of genetic variation in *Microtus californicus*. *Journal of Mammalogy*, **63**, 625–38.

Cannings, C. and Thompson, E.A. (1981) *Genealogical and Genetic Structure*, Cambridge University Press, Cambridge.

Charnov, E.L. (1982) *The Theory of Sex Allocation*, Princeton University Press, Princeton.

Clark, A.B. (1978) Sex ratio and local resource competition in a prosimian primate. *Science*, **201**, 163–5.

Cockburn, A. (1978) The distribution of *Pseudomys shortridgei* (Muridae: Rodentia) and its relevance to that of other heathland *Pseudomys*. *Australian Wildlife Research*, **5**, 213–9.

Cockburn, A. (1981) Population regulation and dispersion of the smoky mouse, *Pseudomys fumeus*. II. Spring decline, breeding success and habitat heterogeneity. *Australian Journal of Ecology*, **6**, 255–67.

Cockburn, A. (1985) Does dispersal increase as populations expand? *Oikos*, **44**, 367–8.

Cockburn, A., Braithwaite, R.W. and Lee, A.K. (1981) The response of the heath rat, *Pseudomys shortridgei*, to pyric succession: a temporally dynamic life history strategy. *Journal of Animal Ecology*, **50**, 649–66.

Cockburn, A., Lee, A.K. and Martin, R.W. (1983) Macrogeographic variation in litter size in *Antechinus* (Marsupialia: Dasyuridae) *Evolution*, **37**, 86–95.

Cockburn, A. and Lidicker, W.Z. (1983) Microhabitat heterogeneity and population ecology of an herbivorous rodent, *Microtus californicus*. *Oecologia*, **59**, 167–77.

Cockburn, A., Scott, M.P. and Dickman, C.R. (1985) Sex ratio and intrasexual kin competition in mammals. *Oecologia*, **66**, 427–9.

Cockburn, A., Scott, M.P. and Scotts, D.J. (1985) Inbreeding avoidance and sex-biased dispersal in *Antechinus* spp. (Marsupialia, Dasyuridae). *Animal Behaviour*, **33**, 908–15.

Comins, H.N. (1982) Evolutionarily stable strategies for localized dispersal in two dimensions. *Journal of theoretical Biology*, **94**, 579–606.

Comins, H.N., Hamilton, W.D. and May, R.M. (1980) Evolutionarily stable dispersal strategies. *Journal of Theoretical Biology*, **82**, 205–30.

Dewsbury, D.A. (1982) Ejaculate cost and male choice. *American Naturalist*, **119**, 601–10.

Dewsbury, D.A. (1985) Aggression, copulation, and differential reproduction of deer mice (*Peromyscus maniculatus*) in a semi-natural enclosure. *Behaviour*, **91**, 1–23.

Dewsbury, D.A. and Sawrey, D.K. (1984) Male capacity as related to sperm production, pregnancy initiation, and sperm competition in deer mice (*Peromyscus maniculatus*). *Behavioral Ecology and Sociobiology*, **16**, 37–47.

Dobson, F.S. (1981) An experimental examination of an artificial dispersal technique. *Journal of Mammalogy*, **62**, 74–81.

Dobson, F.S. (1982) Competition for mates and predominant juvenile male dispersal in mammals. *Animal Behaviour*, **30**, 1183–92.

Elton, C.S. (1942) *Voles, Mice and Lemmings: problems of population dynamics*, Oxford University Press, London.

Endler, J.A. (1977) *Geographic Variation, Speciation, and Clines*, Princeton University Press, Princeton.

Falconer, D.S. (1981) *Introduction to Quantitative Genetics*, 2nd edn, Longman, London.

Fretwell, S.D. (1972) *Populations in a Seasonal Environment*, Princeton University Press, Princeton, New Jersey.

Fretwell, S.D. and Lucas, H.L. (1970) On territorial behavior and other factors influencing habitat distribution in birds. I. Theoretical development. *Acta Biotheoretica*, **19**, 16–36.

Futuyma, D.J. (1986) *Evolutionary Biology*, 2nd edn, Sinauer, Sunderland, Massachusetts.

Gadgil, M. (1971) Dispersal: population consequences and evolution. *Ecology*, **52**, 253–61.

Gaines, M.S. and McClenaghan, L.R. (1980) Dispersal in small mammals. *Annual Review of Ecology and Systematics*, **11**, 163–96.

Gaines, M.S. and Whittam, T.S. (1980) Genetic changes in fluctuating vole populations: selective vs nonselective forces. *Genetics*, **96**, 767–78.

Giesel, J.T. (1976) Reproductive strategies as adaptations to life in temporally heterogeneous environments. *Annual Review of Ecology and Systematics*, **7**, 57–79.

Gliwicz, J. (1980) Island populations of rodents: their organization and functioning. *Biological Reviews of the Cambridge Philosophical Society*, **55**, 109–38.

Gould, S.J. and Lewontin, R.C. (1979) The spandrels of San Marco and the Panglossian paradigm: a critique of the adaptionist programme. *Proceedings of the Royal Society of London*, **B205**, 581–98.

Greenwood, P.J. (1980) Mating systems, philopatry and dispersal in birds and mammals. *Animal Behaviour*, **28**, 1140–62.

Greenwood, P.J. (1983) Mating systems and the evolutionary consequences of dispersal, in *The Ecology of Animal Movement* (eds I.R. Swingland and P.J. Greenwood), Clarendon Press, Oxford, pp. 116–31.

Greenwood, P.J. and Harvey, P.H. (1982) The natal and breeding dispersal of birds. *Annual Review of Ecology and Systematics*, **13**, 1–21.

Greenwood, P.J., Harvey, P.H. and Perrins, C.M. (1979) The role of dispersal in the great tit (*Parus major*): the causes, consequences and heritability of natal dispersal. *Journal of Animal Ecology*, **48**, 123–42.

Grodziński, W., Makomaska, M., Tertil, R. and Weiner, J. (1977) Bioenergetics and total impact of vole populations. *Oikos*, **29**, 494–510.

Gyllensten, U. (1985) Temporal allozyme frequency changes in density fluctuating populations of willow grouse (*Lagopus lagopus* L.). *Evolution*, **39**, 115–21.

Hamilton, W.D. and May, R.M. (1978) Dispersal in stable habitats. *Nature*, **269**, 578–81.

Hansson, L. (1977) Spatial dynamics of field voles *Microtus agrestis* in heterogeneous landscapes. *Oikos*, **29**, 539–44.

Hansson, L. (1979) On the importance of habitat heterogeneity in northern regions for the breeding population densities of homeotherms: a general hypothesis. *Oikos*, **33**, 182–9.

Hansson, L. (1984) Winter reproduction of small mammals in relation to food conditions and population dynamics, in *Winter Ecology of Small Mammals* (ed. J.F. Merritt), Carnegie Museum of Natural History, Pittsburgh, pp. 225–34.

Harris, M.A. and Muric, J.O. (1984) Inheritance of nest sites in female Columbian ground squirrels. *Behavioral Ecology and Sociobiology*, **15**, 97–102.

Hilborn, R. (1975) Similarities in dispersal tendency among siblings in four species of voles (*Microtus*). *Ecology*, **56**, 1221–5.

Holekamp, K.E. (1984a) Dispersal in ground-dwelling sciurids, in *The Biology of Ground-dwelling Squirrels* (eds J.O. Murie and G.R. Michener), University of Nebraska Press, London, pp. 297–320.

Holekamp, K.E. (1984b) Natal dispersal in Belding's ground squirrels (*Spermophilus beldingi*). *Behavioral Ecology and Sociobiology*, **16**, 21–30.

Jannett, F.J. (1980) Social dynamics of the montane vole, *Microtus montanus*, as a paradigm. *The Biologist*, **62**, 3–19.

Jeffreys, A.J., Wilson, V. and Thein, S.L. (1985) Hypervariable 'minisatellite' regions in human DNA. *Nature*, **314**, 67–73.

Kaikusalo, A. and Tast, J. (1984) Winter breeding of microtine rodents at Kilpisjärvi, Finnish Lapland, in *Winter Ecology of Small Mammals* (ed. J.F. Merritt), Carnegie Museum of Natural History, Pittsburgh, pp. 243–52.

Kawata, M. (1985) Sex differences in the spatial distribution of genotypes in the red-backed vole, *Clethrionomys rufocanus bedfordiae*. *Journal of Mammalogy*, **66**, 384–7.

Keith, T.P. and Tamarin, R.H. (1981) Genetic and demographic differences between dispersers and residents in cycling and noncycling vole populations. Journal of Mammalogy, **62**, 713–25.

Krebs, C.J. (1964) The lemming cycle at Baker Lake, Northwest Territories, during 1959–1967. *Technical Papers of the Arctic Institute of North America*, **15**, 1–104.

Krebs, C.J. (1978) A review of the Chitty hypothesis of population regulation. *Canadian Journal of Zoology*, **56**, 2463–80.

Krebs, C.J. (1979) Dispersal, spacing behaviour, and genetics in relation to population fluctuations in the vole *Microtus townsendii*. *Fortschritte der Zoologie*, **25**, 61–77.

Krebs, C.J., Wingate, I., LeDuc, J. *et al.* (1976) *Microtus* population biology: dispersal in fluctuating populations of *M. townsendii*. *Canadian Journal of Zoology*, **54**, 79–95.

Krohne, D.T. (1980) Intraspecific litter size variation in *Microtus californicus*. II. Variation between populations. *Evolution*, **34**, 1174–82.

Krohne, D.T., Dubbs, B.A. and Baccus, R. (1984) Analysis of dispersal in an unmanipulated population of *Peromyscus leucopus*. *American Midland Naturalist*, **112**, 146–56.

Krohne, D.T. and Miner, M.S. (1985) Removal trapping studies of dispersal in *Peromyscus leucopus*. *Canadian Journal of Zoology*, **63**, 71–5.

Lee, A.K. and Cockburn, A. (1985) *Evolutionary Ecology of Marsupials*, Cambridge University Press, Cambridge.

Lee, A.K. and McDonald, I.R. (1985) Stress and population regulation in small mammals. *Oxford Reviews in Reproductive Biology*, 7, 261–304.

Liberg, O. and von Schantz, T. (1985) Sex-biased philopatry and dispersal in birds and mammals: the Oedipus hypothesis. *American Naturalist*, **126**, 129–35.

Lidicker, W.Z. (1973) Regulation of numbers in an island population of the California vole, a problem in community dynamics. *Ecological Monographs*, 43, 271–302.

Lidicker, W.Z. (1975) The role of dispersal in the demography of small mammals, in *Small Mammals: productivity and dynamics of populations* (eds K. Petrusewicz, F.B. Golley and L. Ryszkowski), Cambridge University Press, Cambridge, pp. 103–28.

Lidicker, W.Z. (1985a) Dispersal, in *Biology of New World Microtus* (ed. R.H. Tamarin), Special Publication No. 8, American Society of Mammalogists, pp. 420–54.

Lidicker, W.Z. (1985b) Population structuring as a factor in understanding microtine cycles. *Acta Zoologica Fennica*, **173**, 23–7.

Lidicker, W.Z. and Caldwell, R.L. (1982) Editors comments on papers 1 through 10, in *Dispersal and Migration*, Hutchinson Ross, Stroudsburg, Pennsylvania, pp. 9–16.

Łomnicki, A. (1978) Individual differences between animals and natural regulation of their numbers. *Journal of Animal Ecology*, **47**, 461–75.

Łomnicki, A. (1980) Regulation of population density due to individual differences and patchy environment. *Oikos*, **35**, 185–93.

McDonald, I.R. and Taitt, M.J. (1982) Steroid hormones in the blood plasma of Townsend's vole (*Microtus townsendii*). *Canadian Journal of Zoology*, **60**, 2264–9.

McGregor, P.K. and Krebs, J.R. (1982) Mating and song types in the great tit. *Nature* **297**, 60–61.

Madison, D.M. (1984) Group nesting and its ecological and evolutionary significance in overwintering microtine rodents, in *Winter Ecology of Small Mam-*

mals (ed. J.F. Merritt), Carnegie Museum of Natural History, Pittsburgh, pp. 267–74.

Madison, D.M., Fitzgerald, R.W. and McShea, W.J. (1984) Dynamics of social nesting in overwintering meadow voles (*Microtus pennsylvanicus*): possible consequences for population cycling. *Behavioral Ecology and Sociobiology*, **15**, 9–17.

Maynard Smith, J. (1982) *Evolution and the Theory of Games*, Cambridge University Press, Cambridge.

Millington, S.J. and Price, T.D. (1985) Song inheritance and mating patterns in Darwin's finches. *Auk*, **102**, 342–6.

Moore, J. and Ali, R. (1984) Are dispersal and inbreeding avoidance related? *Animal Behaviour*, **32**, 94–112.

Motro, U. (1982a) Optimal rates of dispersal I. Haploid populations. *Theoretical Population Biology*, **21**, 394–411.

Motro, U. (1982b) Optimal rates of dispersal II. Diploid populations. *Theoretical Population Biology*, **21**, 412–29.

Motro, U. (1983) Optimal rates of dispersal III. Parent-offspring conflict. *Theoretical Population Biology*, **23**, 159–68.

Motro, U. and Thomson, G. (1982) On heterozygosity and the effective size of populations subject to size changes. *Evolution*, **36**, 1059–66.

Myers, J.H. and Krebs, C.J. (1971) Genetic, behavioural, and reproductive attributes of dispersing field voles *Microtus pennsylvanicus* and *Microtus ochrogaster*. *Ecological Monographs*, **41**, 53–78.

Myllymäki, A. (1977) Demographic mechanisms in the fluctuating populations of the field vole *Microtus agrestis*. *Oikos*, **29**, 468–93.

Nakatsura, K. and Kramer, D.L. (1982) Is sperm cheap? Limited male fertility and female choice in the Lemon Tetra (Pisces, Characidae). *Science*, **216**, 753–5.

Naumov, N.P. (1954) *The Ecology of Mammals*, Sov Nauka, Moscow (in Russian).

Naumov, N.P. (1964) Spatial features and mechanisms of the dynamics of numbers in terrestrial vertebrates, in *Contemporary Problems in the Study of the Dynamics of Numbers in Animal Populations*, Izd Akad Nauk SSR, Moscow (in Russian).

Newsome, A.E. (1969a) A population study of house mice temporarily inhabiting a South Australian wheatfield. *Journal of Animal Ecology*, **38**, 341–60.

Newsome, A.E. (1969b) A population study of house mice permanently inhabiting a reed bed in South Australia. *Journal of Animal Ecology*, **38**, 361–78.

van Noordwijk, A.R. (1984) Problems in the analysis of dispersal and a critique on its 'heritability' in the great tit. *Journal of Animal Ecology*, **53**, 533–44.

Ostfeld, R.S. (1985) Limiting resources and territoriality in microtine rodents. *American Naturalist*, **126**, 1–15.

Packer, C. (1985) Dispersal and inbreeding avoidance. *Animal Behaviour*, **33**, 676–8.

Partridge, L. (1983) Non-random mating and offspring fitness, in *Mate Choice* (ed. P. Bateson), Cambridge University Press, Cambridge, pp. 227–53.

Peterson, R.O., Page, R.E. and Dodge, K.M. (1984) Wolves, moose, and the allometry of population cycles. *Science*, **224**, 1350–2.

Pickering, J., Getz, L.L. and Whitt, G.S. (1974) An esterase phenotype correlated with dispersal in *Microtus. Transactions of the Illinois Academy of Sciences*, **67**, 471–5.

Pokki, J. (1981) Distribution, demography and dispersal of the field vole, *Microtus agrestis* (L.), in the Tvarminne archipelago, Finland. *Acta Zoologica Fennica*, **164**, 1–48.

Power, M.E. (1984) Habitat quality and the distribution of algae-grazing catfish in a Panamanian stream. *Journal of Animal Ecology*, **53**, 357–74.

Price, M.V. and Waser, N.M. (1979) Pollen dispersal and optimal out-crossing in *Delphinium nelsoni. Nature*, **280**, 389–91.

Prout, T. (1981) A note on the island model with sex dependent migration. *Theoretical and Applied Genetics*, **59**, 327–32.

Pruitt, W.O. (1984) Snow and small mammals, in *Winter Ecology of Small Mammals*, (ed. J.F. Merritt), Carnegie Museum of Natural History, Pittsburgh, pp. 1–8.

Pulliam, H.R. and Caraco, T. (1984) Living in groups: is there an optimal size? *Behavioural Ecology: an evolutionary approach*, 2nd edn (eds J.R. Krebs and N.B. Davies), Blackwell Scientific, Oxford, pp. 122–47.

Ralls, K., Harvey, P.H. and Lyles, A.M. (1986) Inbreeding in natural populations of mammals, in *Conservation Biology: the science of scarcity and diversity*, Sinauer, Sunderland, Massachusetts, pp. 35–56.

Rausher, M.D. (1984) The evolution of habitat preferences in subdivided populations. *Evolution*, **38**, 598–608.

Riggs, L.A. (1979) Experimental studies of dispersal in the California vole, *Microtus californicus*. Ph.D. Thesis, University of California at Berkeley.

Saxon, M. (1984) B.Sc. (Hons) thesis, Monash University, Victoria, Australia.

Scott, M.P. and Tan, T.N. (1985) A radiotracer technique for the determination of male mating success in natural populations. *Behavioral Ecology and Sociobiology*, **17**, 29–33.

Scotts, D.J. (1983) The social organization of *Antechinus stuartii* (Macleay) (Marsupialia, Dasyuridae) at Sherbrooke Forest, Victoria. B.Sc. (Hons) thesis, Monash University, Victoria, Australia.

Shields, W.M. (1983) Optimal inbreeding and the evolution of philopatry, in *The Ecology of Animal Movement* (eds I.R. Swingland and P.J. Greenwood), Clarendon Press, Oxford, pp. 132–59.

Silk, J.B. (1984) Local resource competition and the evolution of male-biased sex ratios. *Journal of Theoretical Biology*, **108**, 203–13.

Smith, M.H., Manlove, M.N. and Joule, J. (1978) Spatial and temporal dynamics of the genetic organisation of small mammal populations, in *Populations of Small Mammals Under Natural Conditions* (ed. D.P. Snyder), Pyamatuning Ecology Laboratory, Pittsburgh, pp. 99–113.

Southwood, T.R.E. (1977) Habitat, the templet for ecological strategies? *Journal of Animal Ecology*, **46**, 337–65.

Stenseth, N.C. (1978) Demographic strategies in fluctuating populations of small rodents. *Oecologia*, **33**, 149–72.

Stenseth, N.C. (1981) On Chitty's theory for fluctuating populations: the importance of genetic polymorphism in the generation of regular density cycles. *Journal of Theoretical Biology*, **90**, 9–36.

Stenseth, N.C. (1983) Causes and consequences of dispersal in small mammals, in *The Ecology of Animal Movement* (eds I.R. Swingland and P.J. Greenwood), Clarendon Press, Oxford, pp. 63–101.

Tamarin, R.H. (1977) Dispersal in island and mainland voles. *Ecology*, **58**, 1044–54.

Tamarin, R.H. (1978) Dispersal, population regulation, and K-selection in field mice. *American Naturalist*, **112**, 545–55.

Tamarin, R.H. (1980) Dispersal and population regulation in rodents, in *Biosocial Mechanisms of Population Regulation* (eds M.N. Cohen, R.S. Malpass and H.G. Klein), Yale University Press, New Haven, pp. 117–33.

Tamarin, R.H. (1984) Body mass as a criterion of dispersal in voles: a critique. *Journal of Mammalogy* **65**, 691–2.

Tamarin, R.H., Reich, L.M. and Moyer, C.A. (1984) Meadow vole cycles within fences. *Canadian Journal of Zoology*, **62**, 1796–804.

Tamarin, R.H., Sheridan, M. and Levy, C.K. (1983) Determining matrilineal kinship in natural populations of rodents using radionuclides. *Canadian Journal of Zoology*, **61**, 271–4.

Taylor, P.D. (1981) Intra-sex and inter-sex sibling interactions as sex ratio determinants. *Nature* **291**, 64–6.

Trivers, R.L. (1972) Parental investment and sexual selection, in *Sexual Selection and the Descent of Man* (ed. B. Campbell), Heinemann, London, pp. 136–79.

Uchmanski, J. (1983) The effect of emigration on population stability: a generalization of the model of regulation of animal numbers, based on individual differences. *Oikos*, **41**, 49–56.

Wahlund, S. (1928) Zuzammensetzung von Populationen und Korrelationserscheinungen vom Standpunkt der Verebungslehre aus betrachet. *Hereditas*, **11**, 65–106.

Warkowska-Dratnal, H. and Stenseth, N.C. (1985) Dispersal and the microtine cycle: comparison of two hypotheses. *Oecologia*, **65**, 468–77.

Waser, N.M. and Price, M.V. (1983) Optimal and actual outcrossing in plants, and the nature of plant-pollinator interactions, in *Handbook of Experimental Pollination Biology* (eds C.E. Jones and R.J. Little) Scientific and Academic Editions, New York, pp. 341–59.

Webster, A.B. and Brooks, R.J. (1981) Social behaviour of *Microtus pennsylvanicus* in relation to seasonal changes in demography. *Journal of Mammalogy*, **62**, 738–51.

West, S.D. and Dublin, H.T. (1984) Behavioral strategies of small mammals under winter conditions: solitary or social? in *Winter Ecology of Small Mammals* (ed. J.F. Merritt), Carnegie Museum of Natural History, Pittsburgh, pp. 293–99.

Whitham, T.G. (1980) The theory of habitat selection: examined and extended using *Pemphigus* aphids. *American Naturalist*, **115** 449–66.

Wolff, J.O. (1980) The role of patchiness in the population dynamics of snowshoe hares. *Ecological Monographs*, **50**, 111–30.

Wolff, J.O. and Lidicker, W.Z. (1981) Communal winter nesting and food sharing in taiga voles. *Behavioral Ecology and Sociobiology*, **9**, 237–40.

Wright, S. (1978) *Evolution and the Genetics of Populations, Vol. 4, Variability Within and Among Natural Populations*, University of Chicago Press, Chicago.

– 5

Social factors in immigration and emigration

C.A. BRANDT

5.1 INTRODUCTION

Social behaviour comprises behaviour directed at other individuals often for the purpose of affecting their actions, such as altering spacing, enhancing cohesiveness of group members, and establishing the order of access to resources. Because dispersal (emigration and immigration) of small mammals occurs in a social milieu, social behaviour has often been proposed as a cause of emigration or a limit to immigration. Likewise, social behaviour may in turn be affected by dispersal regimes; for example, emigration may decrease the incidence of aggressive behaviour as individuals become relatively more scarce. Social behaviour could also serve as an intervening variable correlated with but not directly a cause of particular dispersal patterns. Furthermore, social behaviour and dispersal may be connected by feedback loops such that a particular dispersal pattern may cause changes in social behaviour, which in turn may modify dispersal patterns. Types of social behaviour likely to be relevant to dispersal include aggressive behaviour (e.g. fighting, chasing, threatening postures), affiliative behaviour (e.g. naso-nasal contact, grooming), mating behaviour, and so-called altruistic behaviours (e.g. alarm calling, predator attraction through feigning injury, aiding in territory defence).

The first objective of this chapter is to describe and evaluate several of the more prominent theories regarding the relationships between social behaviour and both emigration and immigration. Emphasis is placed on the individual and how, through its own peculiar situation with regard to sex, age, ontogenetic status, and social surroundings, its movements may affect or be affected by others. The theories include proximal or mechanistic causes of emigration, such as the hypothesis that asocial behaviour by the prospective emigrant precedes its departure, as well as ultimate or evolutionary causes of emigration and immigration, such as to increase access to mates. Many theories fall between these extremes. It is probable that several causes may apply to the dispersal regimes of any single population.

Until recently, little was known about the relationships between social behaviour and the dynamics of dispersal of small mammals, in part due to logistic problems in observing interactions between individuals in the field. However, field and enclosure studies have now produced a great deal of data relating to social factors and dispersal. The second objective of this chapter is to examine recent data bearing on the validity of the various theories.

5.2 AGGRESSION, DOMINANCE, AND SPACING

What exactly do we mean by 'aggression' and 'dominance'? Aggression includes a multitude of behaviour patterns serving various functions, such as territorial defence, access to reproductive females, access to resources, discipline of juveniles, weaning, and predatory and antipredatory behaviour (Wilson, 1975; Benton, 1981). In an effort to avoid confusion, the term 'agonistic' was coined to refer only to those behaviours directly involved in fights or contests between individuals of the same species. Many researchers, however, continue to use 'aggressive' as equivalent to 'agonistic' behaviour. This approach is used in this chapter.

Dominance is sometimes used to describe a situation in which particular individuals have priority of access to contested resources (Clutton-Brock and Harvey, 1976); more often, dominance is inferred where one individual defeats or supplants another in agonistic encounters (e.g. Bronson, 1964; Rowel, 1974; Dewsbury, 1981). Some see these usages as equivalent: the ability to displace others in agonistic encounters is presumed to confer the ability to obtain more or better resources. However, defining dominance solely in terms of access to resources can lead to circularity unless the same individuals maintain priority of access to all types of resources under various contexts (Richards, 1974). Such congruity is rarely obtained. For example, Benton and Dalrymple-Alford (1981), examining a lab strain of *Mus musculus*, compared the number of wounds, priority of access to water, and priority of access to space in tubes and in open arenas as measures of dominance. No significant correlations were found between any of these measures. As Dewsbury (1981) has suggested, dominance is best defined on the basis of the outcomes of agonistic encounters. Whether agonistic dominance allows priority of access to resources becomes an empirical question open to experiment.

Aggression is intimately involved with dominance and spacing behaviour. However, aggressiveness *per se* does not generally confer dominance (Francis, 1988). Territorial boundaries are maintained principally through the use of aggressive behaviour, or the threat of its use as indicated by display calls, scent markings, scratching posts, etc. (Wilson, 1975). Dominance orders arise in part via aggressive interactions, which become less frequent over time (Christian, 1970). Dominance orders are expected to grade into territorial systems when critical resources are patchily distributed and defensible (Bradbury and Vehrencamp, 1977). Extremes of density may cause the breakdown of territorial behaviour into a system of overlapping home ranges with animals organized into a dominance hierarchy (Anderson, 1961; Rubenstein, 1981). The importance to dispersal of these two types of spacing behaviour is that, at equilibrium, territorial behaviour will usually keep population sizes well below the actual carrying capacity of the environment, whereas populations are expected to be closer to carrying capacity where individuals occupy overlapping home ranges (Fretwell and Lucas, 1969). Hence, emigration rates may be greater and immigration rates lower in territorial systems than in hierarchically organized systems with individuals in overlapping home ranges. Thus, when changes in resource quantity, quality, or dispersion cause a population to shift from dominance hierarchies to a territorial system, a wave of emigration of subordinate individuals may result (e.g. Carl, 1971; Myllymäki, 1977a; Fairbairn, 1978a).

5.2.1 Aggression and dominance as causes of emigration

Most hypotheses implicating aggression as a cause of emigration originated as fundamental elements of theories regarding the origin and maintenance of cyclicity in population density fluctuations. Chitty's (1967) hypothesis, as modified by Krebs (1978), assumes that populations consist of two genetically distinct types of individuals: aggressive and docile. Others suggest such behavioural types represent ends of a polymorphic spectrum (e.g. Charnov and Finerty, 1980; Stenseth, 1983). Still others suggest such phenotypes are not genetically distinct, but represent behavioural responses to such things as high population density or the absence of close kin (e.g. Wynne-Edwards, 1962; Chitty, 1987). In any case, all such theories rely upon the stated assumption that more aggressive individuals tend to philopatry and force the emigration of more docile types.

None of the hypotheses are explicit regarding the identities of the relevant participants (e.g. parents vs offspring, sibling vs sibling, neighbour vs neighbour) or their social system (territorial or overlapping home ranges). The Charnov–Finerty hypothesis and Stenseth's hypothesis imply that aggressive animals induce the emigration of less aggressive neighbours, but all hypotheses imply that, at densities near environmental carrying capacity, parents may drive off all their young. None of the hypotheses deal with aggression at low population densities.

Christian (1970) recognized aggressiveness as being an important factor in inducing emigration, though he also noted that, to be effective, such aggressiveness had to be performed by dominant individuals. Gauthreaux (1978) suggested much the same thing as Christian, but presented a scenario in which emigration of subordinates is a gradual process caused by the interactions of variously aggressive animals. Thus, Gauthreaux suggests, for example, that sexual differences in natal emigration are due to sexual differences in dominance.

The Chitty–Krebs hypothesis is unique in that aggressiveness is seen as the sole ultimate cause of emigration. In the Charnov–Finerty hypothesis and Stenseth's hypothesis, increased aggressiveness is a consequence of decreasing relatedness among neighbours, but it is aggressiveness itself that still drives dispersal. In the remainder of the hypotheses, aggression and dominance serve as variables connecting resource levels to population densities, and thus act as proximal causes of emigration.

These hypotheses have been examined in a number of studies conducted both in the field and the laboratory [see Lidicker (1985a) for a review of behaviour of dispersing *Microtus*]; the results from studies of 28 species of small mammals are outlined in Table 5.1.

The hypothesis that emigrants are generally less aggressive than residents is apparently supported for five species; conflicting results exist for eight other species. The evidence from an additional 12 species does not support the aggression hypothesis. No phylogenetic patterns are evident in these results, nor do the rules or exceptions conform to particular social or ecological situations.

The assumption that differences in rates of aggression, readiness to attack, or the use of more costly behavioural acts necessarily result in territory ownership or priority of access to resources is generally not supported. The ability of individuals to displace others in agonistic contests is seldom correlated with rates of aggressive behaviour or differences in attack latency (Busser *et al.*, 1974; Reinhardt and Reinhardt, 1975; Krebs and Perrin, 1978; Braithwaite, 1979; Dienske, 1979; Dewsbury, 1981; Mihok, 1981; Rushen, 1982). Neither are more aggressive individuals consistently able to displace others in contests over territories (Krebs and Perrin, 1978; Yasukawa, 1979; van Zegeren, 1980; Wiley and Wiley, 1980). The few studies of small mammals in which aggressiveness and dominance were positively correlated (Wolfe, 1966; Oakeshott, 1974; Fairbairn, 1978b) were each confounded by the positive correlations of both aggressiveness and dominance with body size. Hence the hypothesis that high aggressiveness produces dominance cannot be supported by these studies unless the effects of body size can be excluded experimentally or controlled statistically.

Because most emigrants are members of younger age classes, usually subadults or juveniles [except perhaps for *Cynomys ludovicianus* (King, 1955) and *Microtus breweri* (Keith and Tamarin, 1981)], it is generally

Table 5.1 Studies examining the relationship between social status, aggression, and emigration

Species	*Emigrant status vs resident* *subordinate?* (*age–sex class of emigrant, where known*)	*less aggressive?*	*Evidence*	*Reference*
Mus musculus	No	No	Attack latency of emigrants compared to residents in neutral arenas.	van Zegeren (1980)
	Yes (sibling)	No	Observations in lab enclosure prior to and during dispersal.	Busser *et al.* (1974)
Rattus norvegicus	Yes		Observations in enclosure prior to and during dispersal.	Calhoun (1962)
Peromyscus leucopus	Yes (female) No (male)		Weight comparisons of residents vs immigrants to removal grid.	Nadeau *et al.* (1981)
Peromyscus maniculatus		Yes (juvenile)	Emigration rates of juveniles released with aggressive or docile adults in woodlots.	Healey (1967)
		Yes (juvenile)	Emigration rates of juveniles with aggressive or docile parents.	Savidge (1974)
		No	Manipulation of enclosure population.	King (1983)
	Yes (female) No (male)		Weight comparisons of residents vs transients and immigrants.	Fairbairn (1977)
		Yes (male)	Neutral arena tests of residents vs transients and immigrants.	Fairbairn (1978b)
		No (juvenile)	Neutral arena tests of adult aggression levels vs time of juvenile emigration.	Fairbairn (1978a)
		No (sibling)	Neutral arena tests of siblings.	Halpin (1981)
Sigmodon hispidus	No	No	Dispersal tests in field enclosures; behaviour tests in neutral arenas	Gregory and Cameron (1988)
Clethrionomys gapperi		No (male)	Neutral arena tests of males of dispersal age.	Perrin (1981)

Clethrionomys rufocanus	Yes (male)	Yes (male)	Field indicators of aggression: trappability equated with social status.	Viitala (1977)
Microtus agrestis	Yes (male)	Yes (adult male) No (female) No (juve. male)	Field indicators of aggression and social status.	Myllymäki (1977a)
	Yes	Yes (male) No (female)	Field indicators of aggression: trappability equated with social status.	Viitala (1977)
Microtus breweri	No (male)		Weight comparisons of immigrant vs resident males.	Keith and Tamarin (1981)
	No (female)	No	Weight and neutral arena tests of immigrants vs residents.	Reich *et al.* (1982)
Microtus ochrogaster	No (male) Yes (female)	Yes (male)	Weight and neutral arena tests of immigrants vs residents.	Myers and Krebs (1971)
		No	Wounding levels vs density cycles.	Rose and Gaines (1976)
Microtus pennsylvanicus	No (male) Yes (female)	No (male)	Weight and neutral arena tests of immigrants vs residents.	Myers and Krebs (1971)
		Yes	Neutral arena tests of immigrants vs residents.	Turner and Iverson (1973)
		No (juvenile)	Introduction of juveniles into adult cages with direct observation of aggressive behaviour.	Boonstra (1984)
	Yes		Weight comparisons of immigrants vs source population.	Baird and Birney (1982a,b)
	No (male) Yes (female)		Weight comparisons of immigrants vs residents.	Keith and Tamarin (1981)
	No		Weight comparisons of immigrants vs residents.	Dueser *et al.* (1981)

Table 5.1 (cont'd)

Species	*Emigrant status vs resident subordinate? (age–sex class of emigrant, where known)*	*less aggressive?*	*Evidence*	*Reference*
Microtus townsendii	Yes		Neutral arena tests of immigrants vs residents.	Krebs (1979)
	Yes (male)		Neutral arena tests of immigrants vs residents.	Krebs *et al.* (1978)
		Yes	Comparisons of weight gain of docile, aggressive, and emigrant animals.	Beacham and Krebs (1980)
Spermophilus armatus		No (sibling)	Field observation of sibling behaviour prior to dispersal.	Slade and Balph (1974)
		No (juvenile)	Field observation of parent–offspring aggression prior to emigration.	Balph (1984)
Spermophilus beecheyi	No (sibling)	No (sibling) No (juvenile)	Field observations of sibling and parent–offspring aggression prior to emigration.	Dobson (1979, 1983)
Spermophilus beldingi		No (juve. male) Yes (juve. female)	Field observations of parent–offspring behaviour vs time of emigration.	Holekamp (1983)
	No (juve. male)		Weight comparisons of resident vs emigrant males before emigration.	Holekamp (1984a)
Spermophilus columbianus	No (sibling)	No (sibling) Yes (juve. male)	Field observations of parent–young and sibling–sibling aggression at time of emigration.	Festa-Bianchet and King (1984)
Spermophilus elegans		No (juve. male) Yes (juve. female) No (sibling)	Field observations of parent–young and sibling–sibling aggression vs time of emigration.	Pfeifer (1982)
Spermophilus parryii	Yes (male)	Yes (juvenile)	Field observations of aggression at time of dispersal.	Carl (1971)
Spermophilus richardsonii		No (juvenile)	Field observations of parent–offspring behaviour vs time of emigration.	Michener (1981), Yeaton (1972)

Spermophilus tereticaudus	No (sibling)	No (juvenile)	Field observations of dam–offspring, sibling–sibling behaviour, and weight vs emigrant status.	Dunford (1977a)
Marmota flaviventris		No (juve. male) Yes (juve. female)	Field observations of parent–offspring behaviour vs time of emigration.	Downhower and Armitage (1981)
Marmota monax		No (juve. male)	Field observations of parent–offspring behaviour vs time of emigration.	Bronson (1964)
Cynomys ludovicianus	No	No	Field observations of aggression and weight vs emigrant status.	King (1955), Garrett and Franklin (1988)
Lepus americanus	No (juve. male) Yes (juve. female) Yes (adult)		Weight comparisons of immigrant vs resident individuals.	Boutin *et al.* (1985)
Oryctolagus cuniculus		Yes (juve. male)	Field observations of parent–offspring behaviour vs time of emigration.	Daly (1981)
Ochotona princeps	Yes (sibling)	Yes (sibling)	Observations of aggression of lab-reared young.	Whitworth and Southwick (1984)
	No (sibling)	No (sibling)	Field observations of sibling aggression and dominance prior to emigration.	Brandt (1984)
Heterohyrax brucei	No (sibling)		Weight comparisons prior to emigration.	Hoeck (1982)
		No (juvenile)	Field observations of parent–offspring behaviour prior to emigration.	Hoeck *et al.* (1982)
Procavia johnstoni	No (sibling)		Weight comparisons prior to emigration.	Hoeck (1982)
		No (juvenile)	Field observations of parent–offspring behaviour prior to emigration.	Hoeck *et al.* (1982)

true that the average emigrant is subordinate to the average resident. On a more refined level, however, differences in sibling dominance do not appear to influence the emigration of juveniles (Table 5.1), though different studies of *Mus musculus* and *Ochotona princeps* give conflicting results. Juvenile male *Lepus americanus* emigrants may be dominant to their resident counterparts, while juvenile female emigrants may be subordinate to juvenile female residents. Studies of adult behaviour give less clear-cut results: peer comparisons of adult *Microtus ochrogaster* and *M. pennsylvanicus* generally indicate emigrant males to be dominant to resident males, but the opposite holds for emigrant females. The situation among adult *Peromyscus maniculatus* seems to be the exact opposite of *M. ochrogaster* and *M. pennsylvanicus*. Both sexes of emigrants seem to be subordinate to residents among adult *Peromyscus leucopus, Microtus breweri, M. agrestis*, and *Lepus americanus*.

Caution must be exercised in interpreting the studies reviewed in Table 5.1. Though suggestive, the studies described do not entirely support the conclusion that differences in aggressiveness or dominance induce emigration in some but not all species, though such a statement is probably true. Few of the studies reviewed in Table 5.1 provide adequate tests of the hypothesis that differences in dominance or aggression are correlated with the onset of emigration, and none test hypotheses of causality. For example, most of the studies of aggression and dominance among *Microtus, Peromyscus*, and *Clethrionomys* employed tests of field-caught animals in confrontational situations in laboratory enclosures (Myers and Krebs, 1971; Turner and Iverson, 1973; Fairbairn, 1978a, b; Krebs *et al.*, 1978; Krebs, 1979; Halpin, 1981; Mihok, 1981; Perrin, 1981; Reich *et al.*, 1982). Whether or not such arenas resemble the appropriate field conditions is unknown, but the researcher should establish the congruence of behaviour in arenas to behaviour in field situations likely to influence emigration. This has yet to be done for any of the studies mentioned.

The importance of the spatial context of aggressive behaviour cannot be stressed too highly. Female *Peromyscus leucopus* continue to increase in aggressiveness from proestrus to the second day of lactation when tested in their home cages, but show no aggressiveness until day five of lactation when tested in neutral arenas (Gleason *et al.*, 1980). Rates of aggression by male *Apodemus sylvaticus* are highest in winter and lowest in late summer and autumn when tested in neutral arenas, though the converse is found when tested in their home cages (Gurnell, 1978).

Other problems involve the identities of the groups being compared. Many studies have compared the behaviour of animals entering either areas emptied of conspecifics or other monitored regions with the behaviour of resident animals (e.g. Myers and Krebs, 1971; Turner and Iverson, 1973; Fairbairn, 1978a, b; Krebs *et al.*, 1978; Krebs, 1979; van

Zegeren, 1980; Reich *et al.*, 1982). Other researchers compared body weights rather than behaviour between the same groups of animals (Dueser *et al.*, 1981; Keith and Tamarin, 1981; Nadeau *et al.*, 1981; Baird and Birney, 1982a; Boutin *et al.*, 1985). Such experimental protocols have several failings, the most important being a failure to identify the appropriate sets of animals, i.e. potential emigrants and those likely to induce their emigration. Instead, potential immigrants were compared to residents, thus confounding differences due to the experiences of emigrants after they had departed from their former residences with differences in effect prior to emigration. Such experiences may significantly affect behaviour: emigrants are taxed energetically (Holekamp, 1984a), exposed to high rates of predation (Metzgar, 1967; Ambrose, 1972), and stressed, physiologically, often causing deterioration in condition (Andrzejewski and Wrocławek, 1961). Emigrants are also likely to have experienced a series of defeats at the hands of resident animals. Repeated experiences of losing encounters may dramatically affect aggression and dominance in subsequent encounters (Scott and Frederikson, 1951; Lagerspetz, 1961; Farr *et al.*, 1978; Rasa, 1981; Rushen, 1982). All these effects combine to produce animals that are likely to be less aggressive and perhaps smaller than their philopatric counterparts (e.g. Garrett and Franklin, 1988), particularly as the period between emigration and immigration lengthens, solely as a result of emigration itself.

In contrast to most studies of murids and cricetids, direct observations of behaviour prior to emigration have been accomplished with sciurids (*Spermophilus, Cynomys*, and *Marmota*), lagomorphs (*Ochotona* and *Oryctolagus*), and hyracoids (*Heterohyrax* and *Procavia*), though the studies of Busser *et al.* (1974) on *Mus musculus*, King (1983) on *Peromyscus maniculatus*, and Calhoun (1962) on *Rattus norvegicus* are notable exceptions. All these studies indicate that aggression by siblings or parents plays little part in most emigration. Although sciurid adults may be aggressive to their daughters prior to juvenile dispersal, more juvenile males than females disperse (Holekamp, 1984b). Available evidence further suggests that emigrating sciurids are not likely to be subordinate to their resident siblings (Dunford, 1977a; Festa-Bianchet and King, 1984; Holekamp, 1984a).

Two studies of *Ochotona princeps* provide examples of different conclusions caused by different methodologies. Natal emigration is female-biased among *O. princeps*. Whitworth and Southwick (1984) studied seven mixed-sex litters of *O. princeps* that were reared in laboratory enclosures with their dams. Dominance was measured as the number of encounters between all individuals in which other animals retreated from the focal animal minus the number of encounters in which the focal animal retreated. Measurements were repeated on a weekly basis and pooled for analysis. Thus, their measures of sibling dominance (1) confounded interactions in which offspring apparently caused their dams to retreat with interactions in which one offspring caused the retreat of another, (2) included differences in interaction rate as a measure of relative dominance, and (3) were based on repeated measures of the same animals in an unnatural setting. They concluded that juvenile males always dominated females, and that dominance drives the dispersal system.

Brandt (1984) studied 18 bisexual family groups of *O. princeps* in the field. He determined the number of sibling pairs in which females retreated more often from their male siblings than vice versa while both resided on their dam's territory. This number, along with the juvenile sex ratio, was used to predict the maximum number of females expected to emigrate if emigration were predicated solely upon sibling dominance. Though slightly more males ranked above their sisters, significantly more females emigrated than could have been predicted based upon sibling dominance alone.

Another source of variation in the data in Table 5.1 is more fundamental: dominance, aggression, and dispersal are not likely to be related in a simple, straightforward manner. The expected relationships may be inferred using game theory as developed in the theory of Evolutionarily Stable Strategies (ESS) (Maynard Smith, 1982). Without going into details of the ESS models involved, general predictions derived from ESS theory as it relates to contests include: (1) larger animals should dominate smaller animals, all else being equal (Parker, 1974; Treisman and Collins, 1980; Parker and Rubenstein, 1981; Tullock, 1983); (2) an individual's level of resource-related aggression will increase as its resources dwindle or become relatively more valuable (Parker, 1974; Enquist, 1984); (3) levels of aggression will be highest when contestants are equally matched and the contested resource is of equal value to both individuals (Enquist, 1984); (4) individuals with adequate resources will be unlikely to keep more from others or force others out of areas already held unless the costs of future competition are high or the risk involved in such displacement is small (Parker and Knowlton, 1980; Rubenstein, 1982); (5) residents of an area will be more likely to win contests on the area than will intruders (Parker, 1974; Enquist, 1984).

The relationships between aggression, dominance and dispersal will thus depend in part on the amount and distribution of critical resources, on competition levels, and consequently on the spatial structure of the population, i.e. whether individuals defend territories or have overlapping home ranges with individuals organized into dominance hierarchies (Petrusewicz, 1966; Lidicker, 1985c). If males and females exhibit different spatial structures, dispersal regimes may differ between the sexes. Because residents of an area are likely to be dominant to potential immigrants, (1) most immigration will be confined to poorly defended areas such as those where residents are absent, and (2) immigrants will be unlikely to induce the emigration of residents, at least in territorial systems. Highly aggressive individuals will be unlikely to induce the emigration of less aggressive individuals unless (1) aggressiveness and dominance are positively correlated or (2) less aggressive individuals have lower emigration thresholds (Łomnicki, 1978, 1982; Pulliam and Caraco, 1984). If dominants have the lower emigration thresholds, dominant animals may emigrate rather than subordinates.

There is conflicting evidence regarding the relationship between sexual differences in spatial structure and sexual differences in emigration, at least for those species where aggression or dominance has been implicated in emigration. Table 5.2 presents the social structures of the 16 species in Table 5.1 for which aggression and/or dominance may be a cause of at least some emigration. In 10 of the species, either males only are territorial during the dispersal period, or both sexes are territorial, but males defend the larger area. Nine of these 10 species exhibit the predominant emigration of juvenile males; data on the emigration pattern of one species (*P. maniculatus*) are equivocal. Yet, even though only males defend territories among *Marmota flaviventris, M. monax*, and *S. columbianus*, male natal dispersal is apparently not associated with adult aggression for these three species. *Lepus americanus* exhibits co-temporal territoriality among both sexes with no sexual differences in territory size; no sexual differences in emigration have been found. Recent evidence indicates neither sex of *Microtus ochrogaster* may be territorial. Sexual bias in juvenile dispersal may be small, though males apparently predominate. For the two sciurids in Table 5.2 in which females are territorial during peak emigration, male natal emigration predominates, but adult aggression is not associated with male emigration in either species. Females appear to be the only territorial sex in two of the *Microtus* species reviewed, and juvenile emigration may be biased toward females in each case in at least some studies. However, many microtines exhibit seasonal and spatial differences in social organization and spacing (Madison and McShea, 1987; Karlsson, 1988) that may cause similar temporal and spatial differences in dispersal patterns.

5.2.2 Experiments on aggression and dispersal

Attempts to alter emigration through experimental manipulations of aggression have met with varying success. One of the most elegant studies was Savidge's (1974) study of dispersal in *Peromyscus maniculatus*. Savidge housed family groups in cages separated from an empty cage by an electrified grid that served to deter emigration. Defining aggressiveness as the tendency for females to attack strange juveniles, Savidge divided adult

Table 5.2 Relationship between sexual differences in spacing and sexual differences in emigration for species suggested to emigrate due to aggression

Male defends the larger or only territory	*No sexual differences in territory sizes*	*Female defends the larger or only territory*	*Predominant dispersing sex (juvenile)*
Rattus norvegicus (Calhoun, 1962)			Male (Calhoun, 1962)
Clethrionomys rufocanus (Viitala, 1977)			Male (Viitala 1977)
Microtus agrestis (Myllymäki, 1977a)			Male (Myllymäki 1977b)
Peromyscus leucopus (Metzgar, 1971, 1979)			Male (Krohne and Miner, 1985)
Peromyscus maniculatus (Mihok, 1979)			Male (Taitt, 1981) Female (Fairbairn, 1977)
Spermophilus parryii (Carl, 1971)			Male (Carl, 1971)
Spermophilus columbianus (Murie and Harris, 1984)*			Male (Festa-Bianchet and King, 1984)
Marmota flaviventris (McLean, 1984)†			Male (Downhower and Armitage, 1981)
Oryctolagus cuniculus (Mykytowicz, 1968)			Male (Daly, 1981)
	Lepus americanus (Dolbeer and Clark, 1975)		Both (Boutin, 1980)
	Microtus ochrogaster (Danielson and Gaines, 1987)		Male (Myers and Krebs, 1971)
		Microtus pennsylvanicus (Madison, 1980)	Female (Myers and Krebs, 1971) Male (Boonstra *et al.*, 1987)
		Microtus townsendii (Krebs, 1979)	Female (Taitt and Krebs, 1981) Male (Boonstra *et al.*, 1987)
		Spermophilus beldingi (Holekamp, 1983)†	Male (Holekamp, 1983)
		Spermophilus elegans (Pfeifer, 1982)†	Male (Pfeifer, 1982)

* Emigration associated with maternal, not paternal aggression.
† Female, not male emigration associated with parental aggression.

females into aggressive and non-aggressive groups. He found juveniles in the presence of their aggressive mothers crossed the electrified grid at higher rates than did juveniles with non-aggressive mothers.

In another study, Healey (1967) bred aggressive and docile strains of *P. maniculatus*, then released juveniles onto field plots with either of the two strains of adults. Emigration of juveniles was higher from the plot with aggressive adults. Healey's study serves better as evidence that aggressive differences can affect immigration rather than emigration since the co-introduction of unrelated adults and juveniles to an area hardly constitutes a normal scenario for the initiation of dispersal.

Direct manipulations of aggression by administration of drugs have been attempted principally with *Microtus townsendii* (Table 5.3). In general, none of the treatments significantly altered emigration parameters despite the fact that most treatments had some effect on aggressive behaviour as measured in the laboratory or in the field. In all cases, emigration could only be estimated very inaccurately as losses from the monitored population.

All such chemical alterations of aggression suffer from a major drawback; namely, one does not know precisely what aspects of behaviour were altered by the treatment. Drugs could affect one aspect of aggressive behaviour, such as may be involved in, say, mating, and have no effect on aggressive behaviour that is related to resource defence or intolerance of neighbours. Furthermore, the drugs used in these studies alter more than just aggressive behaviour, so it is not clear to what factor any change in emigration may be due.

Aggression and dominance can act to force the emigration of a portion of the population under certain circumstances, as a few laboratory and enclosure experiments have demonstrated. As yet no data directly implicate dominance or aggression in the emigration of any but a small portion of emigrants in any natural populations of small mammals. Most emigration is by juvenile individuals; parental behaviour has been implicated in only a few instances, sibling behaviour in even fewer cases. There are no data to suggest that residents of territories commonly force the emigration of neighbouring animals, at least among small mammals. Differences in spacing behaviour may be related to differences in emigration patterns in some cases. Emigration is often associated with changes from dominance hierarchies to territoriality such as occur with the onset of breeding or rearing periods. Sexual differences in spacing behaviour are correlated with sexual differences in emigration in some cases, though experiments have yet to be performed that causally link the two phenomena. Attempts to alter dispersal parameters through the use of drugs have met with little success: behaviour is generally altered, but not emigration.

Clearly, more research is needed concerning the relationship between

Table 5.3 Effects of drug treatments on aggressive behaviour and emigration among *Microtus townsendii*

Treatment (sex treated)	*Effects on aggressive behaviour (location of study)*	*Effects on emigration (location of study)*	*Reference*
Testosterone (males)	None (lab) None (field)	None (field)	Krebs *et al.* (1977)
Testosterone (males)	Increased fighting (lab) None (field)	None (field)	Gipps (1982)
Scopolamine (males)	Decreased fighting (lab)	None (field)	Gipps *et al.* (1981)
Testosterone (females)	Increased wounding (field)	Increased female disappearance (field)	Taitt and Krebs (1982)
Mestranol (females)	None (field)	None (field)	Taitt and Krebs (1982)

spatial structure and emigration. The influence of variation in resource levels on territory sizes and priority of access to critical resources needs to be examined. Manipulations of behaviour through alterations in resource levels or selective breeding are a promising approach. One method uses supplemental feeders in either centralized or evenly spaced depots to alter resource distributions (Stueck and Barrett, 1978; Noyes *et al.*, 1982; Maly *et al.*, 1985).

Field and laboratory studies of the effects of aggression and dominance must be based on hypotheses concerning the identities and spatial relationships of the individuals involved. Such studies should involve manipulations of behaviour or be able to make quantitative predictions regarding the fraction of the population emigrating as a result of such behaviour. Finally, aggression and dominance must be defined and measured in a manner consistent with the situation proposed to engender emigration. Dewsbury (1987) provides a valuable critique of experimental analyses of aggressive behaviour in rodents that may serve as a useful guide for future studies.

5.2.3 Aggression and immigration: territories

Differences in dominance may play an important role in limiting immigration, particularly when dominance depends upon where the interaction takes place, i.e. a territory (e.g. Anderson, 1964; Fretwell and Lucas, 1969; Selander, 1970; Dunford, 1977a; Gauthreaux, 1978; Perrin, 1979; Halpin, 1981; Hestbeck, 1982, 1988). Sex-dependent territorial behaviour has been implicated as a control of female immigration in *Microtus townsendii* (Krebs, 1979) and *M. pennsylvanicus* (Madison, 1980), and male immigration in *Cynomys ludovicianus* (King, 1955). As noted earlier, the ability of animals to win contests over resources often depends upon each contestant's relative familiarity with the site: residents of an area are more likely to win contests on the area than are intruders.

Evidence of the relationship between territorial behaviour and immigration takes several forms: (1) observations of patterns of aggressive behaviour, (2) observations of the types of areas settled by immigrants and the timing of immigration relative to territorial activities, (3) removal of residents or introduction of immigrants followed by observations of movement patterns, and (4) manipulation of resource or competition levels noting changes in spacing behaviour and immigration.

Aside from potential mating partners, animals are usually more aggressive to unrelated individuals than to their close kin [e.g. *Microtus pennsylvanicus* (Boonstra, 1984); *Peromyscus maniculatus* (Sadleir, 1965; Halpin, 1981); *Ochotona princeps* (Smith and Ivins, 1983); *Sciurus carolinensis* (Thompson, 1978), and are more aggressive to strangers than to familiar neighbours [e.g. *Lepus americanus* (Graf, 1981)]. Though aggressiveness *per se* has little to do with territory acquisition (e.g. van Zegeren, 1980), aggressive behaviour is usually employed to discourage intrusion into territories (e.g. Wilson, 1975; Kaufmann, 1983). Aggression by residents may even be a significant source of mortality among immigrants (Garrett and Franklin, 1988). Observations of aggressive behaviour by territorial individuals directed toward immigrant animals followed by the subsequent departure of the latter constitute strong evidence in favour of the hypothesis that territoriality limits immigration [*Clethrionomys glareolus* (Rajska-Jurgiel, 1976); *Mus musculus* (Lidicker, 1976); *Tamias striatus* Brenner *et al.*, 1978)].

Immigrants to territorial populations usually settle in preferred areas only if they can take over a territory left vacant by the resident or if the average territory is much larger than the needs of the resident. Otherwise, dispersing animals must settle in less favourable habitat (Fretwell and Lucas, 1970). In *Spermophilus parryii*, for example, immigrant males take over territories of dead males until all vacancies are filled. Further immigration then ceases and the remaining males settle in suboptimal areas (Carl, 1971; McLean, 1983).

The same apparently holds for immigrant male *S. columbianus* (Murie and Harris, 1984). Male and female *Marmota monax* either immigrate to colonies vacated by the death of residents of the same sex or else become solitary, though some displacements of residents by immigrants does occur (Armitage, 1984). Similar patterns of settlement have been observed in studies of *Mus musculus* (Lidicker, 1976), *Ochotona princeps* (Smith, 1974, 1978; Smith and Ivins, 1983, 1984), *Tamiasciurus* spp. (Smith, 1968; Kemp and Keith, 1970), *Cricelutus griseus* (Dasser, 1981), and several species of small carnivores (Lockie, 1966).

The removal of a number of residents from a monitored area generally results in unusually high rates of immigration (e.g. Smyth, 1968; Myers and Krebs, 1971; Krebs *et al.*, 1976, 1978; Tamarin, 1977; Fairbairn, 1978a,b; Gaines *et al.*, 1979; Boutin *et al.*, 1985; Johnson and Gaines, 1985). Such observations alone do not support the notion that it is agonistic behaviour that has kept potential immigrants away. Instead, potential immigrants could simply have assessed resource levels versus the number of potential competitors and avoided settled areas as a means to avoid competition (i.e. the 'ideal free' conditions described by Fretwell and Lucas, 1970). One must also show that individuals exclude others from an area through the use of agonistic behaviour.

One such experiment was performed by Conner (1984) on *Ochotona princeps*. Pikas are vocal animals, and their call has been suggested to be involved with territory defence. After removing resident pikas from their territories, Conner noted a significant rise in the rate of intrusions by neighbours into the vacated areas. When he removed a resident and replaced it by a playback of its short call, rates of intrusions were no different from when the resident itself was present. In this case, territorial behaviour mediated by vocalizations limits immigration. Odour may play a similar role in the territorial dynamics of many other small mammals (e.g. Mackintosh, 1973).

Variation in resource levels can greatly affect immigration. For example, resident *Peromyscus maniculatus* may share resources with immigrants when resource levels are high, but may exclude them from the territory when resource levels decline (Metzgar, 1979, 1980).

Some understanding of the effects of resource levels on spacing behaviour and immigration has been gained through experimental manipulations of food sources and population densities (Table 5.4). As food supplies are increased, the amount of space used by residents commonly declines and immigration to the area increases, while the converse is observed when food supplies are removed. The conclusion that spacing behaviour limits immigration cannot be drawn from such information, however, as it is possible that changes in territory size are wholly the result rather than the cause of changes in immigration or competitor pressure. However, several studies provide evidence that suggests territory sizes of several small mammals are adjusted primarily according to absolute rather than relative resource levels. Mares *et al.* (1982) removed resident *Tamius striatus* from an area while holding food levels constant and observed no corresponding change in the home range sizes of the remaining residents. Similarly, Boutin (1980) removed a number of adult female *Lepus americanus* from their ranges in early spring and found neighbouring adult females did not expand their home ranges into the vacated area.

Intruder pressure certainly plays some role in controlling territory or home range sizes, though no studies of mammals have fully explored this possibility. It is probable that increased competition from intruders results in some decrease in territory size (e.g. Tullock, 1983), though only to a point determined by the resource needs of the resident. Experiments on shrews (*Sorex araneus*; Barnard and Brown, 1984) have shown that, all else being equal, animals experiencing low resource levels in an area expend more effort in fights with other animals and are more likely to be successful in defending the area. This observation explains the common observation that immigration becomes less successful as territory sizes decline (e.g. Mazurkiewicz, 1981). Island populations of small mammals usually exhibit restricted dispersal relative to mainland populations, perhaps due in part to the smaller territories imposed upon animals living at high densities (e.g. *Clethrionomys glareolus*; Gliwicz, 1980). Similarly,

Table 5.4 Effects of food supply manipulations on home range sizes and immigration rates

Change in food supply	*Effect on home range size*	*Effects on immigration rate*	*Species*	*Reference*
Increased	Decreased	Increased	*Apodemus sylvaticus*	Krebs *et al.* (1977)
Increased	Decreased	Increased	*Peromyscus maniculatus*	Taitt (1981)
Increased	Decreased	Increased	*Microtus townsendii*	Taitt and Krebs (1981)
Increased	Decreased	Increased	*Tamias striatus*	Mares *et al.* (1976, 1982)
Increased	Decreased	Increased	*Tamiasciurus douglasi*	Sullivan and Sullivan (1982)
Increased	Decreased	Increased	*Spermophilus beecheyi*	Dobson (1979)
Increased	Decreased	Increased	*Lepus americanus*	Boutin (1984)
Decreased	Increased	Decreased	*Tamiasciurus douglasi*	Sullivan and Sullivan (1982)

Ims (1987) found that supplemental food resulted in decreased home range size in female *Clethrionomys rufocanus* and increased dominance over unfamiliar females.

5.2.4 Aggression and immigration: dominance hierarchies

Relatively little is known about the effects of aggression on immigration into non-territorial systems, at least among small mammals. Data on larger mammals, mostly dealing with primates, suggest most aggression is directed at immigrants, particularly to those of the same sex (e.g. Dittus, 1977; Packer, 1979; Packer and Pusey, 1979; Russell, 1981; Wolf and Schulman, 1984). Such aggression commonly restricts immigration to times when dominant residents die or are weakened. Co-operation among emigrants, particularly among siblings, may enhance the chances of successful immigration to hierarchically organized groups [*Cercopithecus aethiops* (Cheney and Seyfarth, 1983); *Panthera leo* (Bertram, 1976; Bygott *et al.*, 1979)] though the possibility of such co-operation among small mammals has yet to be explored.

The effects of competition and resource variation on home range size and exclusiveness in non-territorial populations have been little studied. Tamarin (1977) observed larger home range sizes among densely settled island populations of *Microtus breweri* compared to neighbouring mainland populations of *M. pennsylvanicus*, though such interspecific comparisons should be viewed with caution. One study of a small mammal that provides a comparison between strongly and loosely territorial systems is Metzgar's (1971) study of *Peromyscus leucopus*. Both sexes appeared to be territorial, but intrasexual overlap of home ranges was greater for males than females. As population density increased, home range exclusiveness increased, but the effect was more pronounced for females than for males. The number of female immigrants bore a strong negative correlation with the number of female residents, while the relationship between male immigrants and residents was only slightly negative.

To summarize this discussion on aggression and dispersal, aggression associated with spacing is likely to play an important role in limiting immigration for many small mammals. Resident advantages in contests makes immigration into territorial systems difficult. Aggression is directed most often toward non-kin and unfamiliar animals. Sex-dependent territoriality may limit immigration of one sex, though few good data exist. Some evidence from removal studies indicates immigration may be largely constrained by territorial behaviour. Manipulations of resource and competition levels indicate that territory sizes and immigration rates are inversely related to food supplies (where territories are food-based); competitor pressure may also act to decrease territory sizes. Successful immigration into territorial systems is thus probably related in part to resource levels and the number of immigrants. Aggression in hierarchical systems may be directed most to potential immigrants, though data are few. Further studies should attempt to elucidate the relative importance of resources and immigration pressure on territory and home range sizes (e.g. Gaines and Johnson, 1982), on home range overlap and permeability, and on the degree to which access to resources is controlled.

5.3 LACK OF SOCIAL BONDING

The relationship of aggression to dispersal is generally seen as one of enforcement: individuals are forced to leave an area or are prevented from entering an area by the aggressive behaviour of others. In marked contrast to this idea is the suggestion that individuals might leave an area (or family group) because of their own failure to develop or capitalize on some aspect of social behaviour. Two versions of this hypothesis have received the greatest attention: (1) dominant animals are more likely to emigrate than are subordinate individuals as a consequence of their failure to develop social bonds with their subordinate kin (Bekoff, 1977); and (2) the sex that is least able to co-operate or to gain from co-operation will emigrate (Gadgil *et al.*, 1983). These hypotheses address emigration from the place of birth (natal dispersal) rather than emigration from a breeding site (breeding dispersal).

Both hypotheses have been virtually untested by researchers on microtines, in part because of the difficulty in observing interactions among kin prior to emigration. Evidence from small mammals that bears on the hypothesis comes mainly from work on sciurids, and each study has given unique results. Prior to natal dispersal, adult *Marmota flaviventris* of both sexes behave cohesively with their daughters, but not with their sons (Armitage and Johns, 1982). Natal dispersal in this species is heavily biased toward males, with as much as 95% of males leaving their natal areas within two to three years of birth versus 51% of females (Schwartz and Armitage, 1980). Yearling males disperse regardless of the amount they play, and males play more than females (Jamieson and Armitage, 1987). However, both sexes remain in their natal areas longer when rates of amicable behaviour with adults are relatively high (Downhower and Armitage, 1981). In contrast, although natal dispersal among *Spermophilus richardsonii* is also biased toward males (Michener, 1979; Schmutz *et al.*, 1979), juvenile males are more amicable with their family members than are juvenile females, and share more of their mother's home range than do females prior to emigration (Michener, 1981). Mothers, however, may be less amicable with their sons than with their daughters (Michener and Michener, 1973). Finally, Festa-Bianchet and King (1984) found that sibling *S. columbianus* exhibited no sexual differences in the number or form of play interactions, though play declined in frequency as the time of dispersal approached. More males than females undergo natal dispersal in this species (Murie and Harris, 1982).

A promising technique known as 'mirror-image stimulation' has been extensively employed in studies of *M. flaviventris* to determine the reactions of individuals to their own images (Svendsen and Armitage, 1973; Armitage, 1974; Svendsen, 1974). Animals characterized as 'social' make frequent contact with their image (nuzzling and pawing), and feed and explore the test arena apparently at ease. 'Asocial' animals do not contact the image, do not feed, and spend most of their time in the portion of the arena furthest from the mirror. Armitage (1984) classified 29 adult females as asocial or social, then compared the two groups on the basis of the number of young they produced per year and the number of daughters settling near them. No differences were found between social and asocial females in terms of the number of female young produced, but social females had 41% of their daughters residing nearby, while asocial females had only 22% of their daughters in the area. Unfortunately, Armitage was not able to separate mortality from emigration, but these results are certainly suggestive.

The hypothesis that the sex less able to benefit from co-operation should emigrate has received little attention. As Gadgil *et al.* (1983) recognized, selection for co-operation favours low dispersal rates, and low dispersal rates allow the evolution of co-operation. However, for selection for co-operation to affect dispersal rates, philopatry must already have arisen in a portion of the population. The evolution of co-operation requires either the close proximity of kin or stable, long-term relationships with other individuals who need not be related (Trivers, 1971; Parker, 1978; Baker and Marler, 1980; Bekoff, 1981). Thus, sexual differences in dispersal are likely to precede, rather than follow from, sexual differences in co-operation. Many species exhibit sexual differences in co-operation, yet it is doubtful that such differences account for the sexually biased dispersal patterns found in many species. For example, male lions perform most natal dispersal (Bertram, 1976), yet the benefits males realize from co-operating with other males in taking over prides may be enormously greater than female benefits from communally rearing their young (Bygott *et al.*, 1979). The role of co-operation in limiting the evolution of dispersal will be discussed later.

In summary, there is thus little evidence supporting the notion that the inability to develop social bonds serves as a cause of emigration among small mammals. Convincing evidence requires more than negative correlations between dispersal and bonding behaviour. It is likely that sexual differences in spacing behaviour have arisen as a consequence of selection for sexual differences in resource use (Cockburn and Lidicker, 1983; Lidicker, 1985b) or mating strategies (Ostfeld, 1985), and that the process of severing social ties with relatives serves as part of the ontogeny of spacing. Sexual differences in co-operation are probably the result rather than the cause of sexual differences in emigration.

5.4 MATE ACCESS AND MATE CHOICE

The strongest pressures shaping the evolution of dispersal patterns are likely to be associated with reproduction, specifically with obtaining more mating opportunities (Greenwood, 1980; Dobson, 1982), avoiding competing with kin for mates (Hamilton, 1972), and avoiding close inbreeding (Bengtsson, 1978; Shields, 1982, 1983). These effects may operate by lowering the emigration threshold for one sex. This may be the primary cause of dispersal in those species in which emigrants are larger on average than their resident counterparts [e.g. *Microtus pennsylvanicus* (Dueser *et al.*, 1981; Keith and Tamarin, 1981); *Spermophilus beldingi* (Holekamp, 1984a); *S. tereticaudus* (Dunford, 1977a); *Lepus americanus* (Boutin *et al.*, 1985)]. However, aggression related to reproduction may serve as a proximal cause of emigration: aggression by mothers against their maturing sons, for example. Reproductive pressures also affect immigration to some extent. Competition for mates may be one of the causes of male intolerance of immigrant males; mating considerations may also influence the selection of areas as sites of immigration.

Male dispersal as a means to increase access to females is seen by some as the main ultimate cause of sexual differences in the natal and breeding dispersal of mammals (Greenwood, 1980; Dobson, 1982). Under this scenario, males will leave areas where local sex ratios are unfavourable or where competition is especially intense in favour of more favourable sex ratios or less effective mate competition (King, 1955, 1983; Harcourt, 1978; Dobson, 1979, 1982; Greenwood, 1980, 1983; Michener, 1983; Gaines and Johnson, 1984; Moore and Ali, 1984). Because female reproductive success is more closely tied to resources than is male reproduction, at least among most mammals (Greenwood, 1980; Ostfeld *et al.*, 1985; but see Brandt, 1989), female emigration for reproductive reasons will be unlikely in most cases. In polygynous mating systems, males, but not females, commonly have restricted access to mates, since all males in such systems cannot breed. Hence, males unlikely to be successful competitors for mates in their local areas should be more likely to emigrate than their more capable counterparts, and emigration should become less likely as local sex ratios become more biased toward females. Furthermore, males should immigrate to areas where they are more likely to mate, i.e. to areas with few larger, older males and/or to areas where more females and fewer males are present.

The possibility that dispersal is tied to the avoidance of competition between kin for mates (local mate competition) is closely related to the previous hypothesis: the sex experiencing the greater variation in reproductive success should leave its natal area in order to avoid competing with its close relatives for limited reproductive opportunities. For mammals, this means males rather than females should emigrate from their birth sites. The predictions from this hypothesis are that male natal emigration will be less likely as the sex ratio within litters becomes heavily biased toward females, and immigrant males will avoid settling where they will be in competition for mates with male relatives.

In normally outbreeding populations, occasional close inbreeding has deleterious effects on reproductive success caused by the expression of lethal recessive traits and loss of vigour associated with heterosis (Bengtsson, 1978; Shields, 1982). In such populations, sex-dependent dispersal may serve as one means to prevent inbreeding with parents or siblings, but as Moore and Ali (1984) recognized, it cannot prevent matings between cousins, aunts and nephews, or uncles and nieces. To prevent accidental inbreeding, recognition of close kin must also be involved.

Inbreeding avoidance has been implicated as the primary ultimate cause of sexual differences in natal dispersal. But which sex should emigrate in order to avoid inbreeding? It is generally assumed that the sex suffering the higher cost of inbreeding will be more likely to emigrate. However, suggestions that male mammals instead of females disperse to avoid inbreeding (e.g. Schwartz and Armitage, 1980; Holekamp, 1984a; Krohne *et al.*, 1984) are probably erroneous. In monogamous systems, inbreeding costs (e.g. Hill, 1974; Haigh, 1983) are likely to be identical for both sexes. Thus, either sex would be expected to emigrate in order to avoid inbreeding. In polygynous mating systems, however, inbreeding should be more expensive for females than for males (Bengtsson, 1978; Smith, 1979), hence, females would be expected to emigrate (Waser *et al.*, 1986). However, female-biased dispersal patterns are rare among mammals, especially among the more polygynous species (Greenwood, 1980).

The relationship between sexual differences in inbreeding costs and sexual differences in dispersal could depend on the nature of the mating system, i.e. on whether mates or resources are defended. In all mating systems, the sex experiencing the greater cost of inbreeding should be less likely to accept mating with a close relative. Among polygynous species, females generally should refuse to mate with their close relatives. However, females should emigrate only if there are no unrelated males available in their natal areas. In territorial systems where males defend resources required for reproduction, females remaining on their natal areas would generally have little opportunity to mate with unrelated males, hence, more females than males would be expected to emigrate. In contrast, where males defend females and not resources, as the degree of polygyny increases, the more effective will be female refusal to mate (Brandt, 1984), and more males should emigrate from their places of birth as a result. Most mammals defend mates rather than resources (Greenwood, 1980); thus, among most mammals, the degree of polygyny should be associated with the degree of male bias in emigration. We should also expect emigration to decline as the potential for inbreeding declines, and immigrants to settle in areas where outbreeding opportunities exist. These predictions are supported by evidence that oestrus female *Peromyscus leucopus* in mate preference tests discriminate against closely related males (Keane, 1990a). Males in this species disperse over greater distances than do females (Keane, 1990b).

Studies of these hypotheses generally have taken three lines of attack. Many studies have involved altering or observing natural variation in sex ratios and observing the resultant effects on migration parameters. Others have examined dispersal in relation to breeding behaviour, mate choice, and the effects of relatives on gonad development. Still others have examined the relationship between dispersal and sexual differences in variability of reproductive success.

Variation in sex ratios among family groups may affect dispersal in some cases. For example, Michener and Michener (1973) observed that emigration by *Spermophilus richardsonii* females was not affected by the presence or absence of their mother, but male emigration was less likely if their mother was no longer present. Such evidence is consistent with the inbreeding avoidance hypothesis, especially as parental aggression does not seem to be associated with juvenile emigration in this species (Table 5.1).

In another study, Cockburn *et al.* (1985) found that male emigration in the dasyurid marsupial genus *Antechinus* was independent of the proportion of males to females in litters, but was less likely when mothers were removed at the time of weaning. The authors concluded on this basis that male emigration was due entirely to inbreeding considerations enforced by maternal behaviour. This cannot fully account for the emigration patterns of *Antechinus*, however, because these co-workers also reported that when mothers were removed, only one offspring (either a son or daughter) inherited the maternal range while all others emigrated independently of variation in litter sex ratios. Since several males normally share the same home range in this species (Cockburn *et al.*, 1985), it is doubtful that male emigration in such circumstances is the result of competition for exclusive habitat space. Instead, the avoidance of inbreeding and local mate competition probably both act to cause male-biased emigration.

Boonstra and Rodd (1983) noted that male *Microtus pennsylvanicus* are more likely to emigrate from areas devoid of females. In an experiment using field populations of *M. townsendii*, Boonstra (1978) observed no effects on the sex ratio of disappearing juveniles caused by the removal of resident males from an area, though more males were recruited to such areas than normal. Among *Spermophilus tereticaudus*, a greater proportion of the members of litters that are all of one sex emigrate (Dunford, 1977a,b). Redfield *et al.* (1978a,b) manipulated sex ratios among *M. townsendii* and *M. oregoni* by reciprocally transplanting female residents from one grid and male residents from another, giving two grids of equal density but with either male-biased or female-biased sex ratios. They were not able to separate immigration from birth in the area ('recruitment') or death from emigration ('disappearance'). The alteration of sex ratios had no effect on the rates of disappearance of *M. townsendii* compared to control areas, but male recruitment was highest to areas where sex ratios were biased toward females. Results for *M. oregoni* were inconsistent: the sex ratio of recruits was unaffected by resident sex ratios, while both sexes had lower disappearance rates on the area with more females than males.

Hoogland (1982) has presented evidence commonly interpreted as indicating male *Cynomys ludovicianus* emigrate to avoid inbreeding *per se*: young males emigrate prior to maturity and most older males emigrate before their daughters mature. However, Hoogland (and Garrett and Franklin, 1988) also noted that young females do not breed if their fathers remain in the coterie, while oestrus females avoid their sons, their sires, and their brothers, if any are in the area. Hoogland's evidence is therefore more likely to be consistent with the possibility that males emigrate because their female relatives refuse to mate with them. *Microtus ochrogaster* females also are known to refuse to mate with their fathers or male siblings (McGuire and Getz, 1981), though recent evidence suggests males too will not mate with their litter mates (Gavish *et al.*, 1984). Ims (1990) has shown that female *Clethrionomys rufocanus* in litters with a predominance of males had the greatest tendency to emigrate, whereas male dispersal was unaffected by litter sex ratio. Males of this species thus appear unresponsive to the potential for inbreeding, while females seem to avoid the potential by emigration. Terman's (1980) studies indicate that continued contact with siblings inhibits the growth of reproductive organs in both sexes of *Peromyscus maniculatus*, hence inbreeding avoidance may be important to males as well.

Dobson (1982) reviewed sex-dependent dispersal patterns among numerous mammalian species in an attempt to correlate degree of bias in dispersal with degree of polygyny. He assumed that if male emigration is tied to the advantages of acquiring more mates, more males should disperse when males mate with several females than when males mate monogamously. His data indicate that emigration is indeed more biased for polygynous than for monogamous species, but his conclusion that mate access drives sexual differences in dispersal may be incomplete. As noted earlier, both the inbreeding hypothesis and the local mate competition hypothesis predict an increase in male natal dispersal as the degree of polygyny increases.

In summary, the widespread occurrence of male-biased dispersal among mammals independent of sexual differences in dominance suggests that female mate choice, the avoidance of local mate competition, and increased access to females may be the most important ultimate causes of emigration among most species in this taxon. There is no reason for, and no evidence to support, the notion that only one cause drives dispersal in any one species (Chapter 2). Several researchers have suggested that males disperse to avoid mating with their mothers or sisters; this is probably incorrect, since the costs of inbreeding are greater for females than for males in most cases. Maternal aggression toward sons may be related to the enforcement of outbreeding and the avoidance of local mate competition.

Future work should include the use of radiotracking methods to allow experimenters the opportunity to monitor changes in sexual surroundings made by emigrating animals. Comparisons of natal and breeding dispersal should be enlightening, as the importance of local mate competition and inbreeding avoidance should differ for animals of different ages.

5.5 SOCIAL CONSEQUENCES OF DISPERSAL

Not only are dispersal patterns affected by their social context, but the course of social evolution itself may be altered by the dispersal regime. Dispersal is thus tied to the social regime in a cycle of causation, whereby social behaviour affects dispersal, which in turn affects social behaviour. The direct effects of dispersal patterns on age structure, sex ratios, population growth rates, and consequently on patterns of social behaviour, have been adequately reviewed elsewhere (Petrusewicz, 1966; Lidicker, 1975, 1985c; Altmann and Atmann, 1979; Dunbar, 1979; Gliwicz, 1980) and so will not be covered in depth here. Of particular importance are spontaneous abortions following immigration of strange males (i.e. the Bruce effect; e.g. Labov, 1980; Schadler, 1981; Clulow *et al.*, 1982; Heske and Nelson, 1984), adaptive changes in primary sex ratios caused by local resource competition between mothers and their philopatric offspring (Clark, 1978), and secondary sex ratio changes as a consequence of sexual differences in dispersal and thus mortality (e.g. Myllymäki, 1977b). Each of these results of dispersal patterns may have especially large effects on mating systems, particularly as sex ratios become heavily biased.

Infanticide has been noted particularly as a response of immigrant males to unrelated juveniles [e.g. *Spermophilus richardsonii* (Michener, 1982); *S. parryii* (McLean, 1983); *S. beldingi* (Sherman, 1981a); *Mus musculus* (Labov, 1980; Huck *et al.*, 1982)]. Infanticide can cause rapid return of oestrus in the females losing young, and allows immigrant males to avoid providing benefits to unrelated young (Labov *et al.*, 1985). Preventing infanticide can be an important cause of territorial behaviour among reproductive adults of both sexes [e.g. *Dicrostonyx groenlandicus* (Mallory and Brooks, 1978); *S. parryii* (McLean, 1983); *S. beldingi* (Sherman, 1981a)].

Philopatry results in the close spatial association of kin, and kin are usually much less aggressive to each other than to unrelated individuals [e.g. *S. parryii* (McLean, 1982); *M. musculus* (Kareem and Barnard, 1982)]. The close association of kin is also a prerequisite for the evolution of altruism via kin selection (Hamilton, 1972; Baker and Marler, 1980; Bekoff, 1981; Waser and Jones, 1983). Altruism may also evolve via selection for reciprocal behaviour (Trivers, 1971). Such also requires the stable association of individuals, and hence low dispersal rates.

Sex differences in dispersal are likely to result in sex differences in sociality, except for cases in which siblings immigrate to the same areas (Hilborn, 1975; Sherman, 1977; Wrangham, 1980). Sexual differences in sociality are commonly observed among sciurids (Table 5.5), where natal dispersal is largely by males and male siblings are not known to disperse as a unit (Holekamp, 1984b). Both sexes of the bat *Phyllostomus hastatus* leave their natal areas, though only females form stable groups. Some altruistic behaviour is seen among females, presumably as a result of reciprocity rather than of kin selection (McCracken and Bradbury, 1981). In an interesting experiment on *Clethrionomys rufocanus*, Kawata (1987) determined that females were better able to acquire territories when the neighbouring female was a sister than when the neighbour was unrelated. In the African wild dog, *Lycaon pictus*, more females than males leave their natal areas (Frame and Frame, 1976), yet both male and female adults rear pups communally. Such co-operation among emigrant females is possible because female wild dogs disperse together from their birth places, hence the adult females in a pack are sisters (Malcolm and Marten, 1982).

Complex sociality, once evolved, may act to constrain emigration rates even further (Gadgil *et al.*, 1983), since the costs of leaving related and/or familiar individuals become greater with increasing benefits from sociality. Several examples of such a constraint on the emigration of sciurids have been published. Hoogland (1981) found that the reproductive success of female *Cynomys ludovicianus* declined as the number of females in the coterie increased, yet females did not leave the coterie in part because

Table 5.5 Sex differences in altruistic behaviour associated with the predominant dispersal of males

Species	*Behaviour*	*Reference*
Spermophilus beldingi	Only females give alarm calls, co-defend areas, and co-operate in defending young.	Sherman (1977, 1981a)
Spermophilus tereticaudus	Only females give alarm calls, even when no young are present.	Dunford (1977b)
Spermophilus parryii	Only females share their nests with related females.	McLean (1982)
Spermophilus richardsonii	Females with kin as neighbours had higher reproductive success, fewer fights, more feeding time, less time watching for predators, and shared more of their home ranges.	Davis (1984)
Spermophilus tridecimlineatus	Females share more of their home range with kin than with non-kin; males share no more of their ranges with kin than with non-kin.	Vestal and McCarley (1984)
Marmota flaviventris	Female kin share more of their foraging and core ranges than do non-kin.	Frase and Armitage (1984) Armitage (1982)
Cynomys ludovicianus	Alarm calls are given mainly by females even when young are not present.	Hoogland (1981)

predator detection was enhanced in larger groups. Such may also explain the lack of dispersal among female *Marmota flaviventris*, who suffer similarly depressed reproduction in harems (Downhower and Armitage, 1971). *Spermophilus beldingi* females do not undergo breeding dispersal unless they both failed to wean a litter the previous season and had no female relatives living nearby (Sherman, 1981a). Presumably the advantages of having close relatives near at hand outweigh the chance of future reproductive failure.

Several theories suggest that the movements of certain groups of individuals can have important consequences for the populations donating and receiving these individuals. One such theory is the Chitty–Krebs theory (Chitty, 1967; Krebs, 1978), which suggests that population density cycles are driven by fluctuations in the level of genetically determined aggressiveness caused by the emigration of docile types. This emigration leaves donating populations composed largely of aggressive individuals who subsequently suffer depressed reproduction because of their aggressive behaviour and decreased resilience to environmental stress. Depressed reproduction and enhanced mortality lead to a population crash and the subsequent influx and success of docile phenotypes.

One of the difficulties with this theory is deciding upon a refuge for docile phenotypes. That is, when one aggressive individual competes with another aggressive individual for space, one must lose. Why then are not all refuges for docile phenotypes colonized by aggressive phenotypes? Another difficulty stems from the habit of viewing aggressive behaviour as a unitary drive, rather than as a means to an end. Hence there is no *a priori* basis for assuming that aggressive intolerance between neighbours over habitat space would have deleterious consequences for reproductive success. Aggressive solitaires such as shrews (*Sorex* sp.) manage to mate and produce offspring despite their 'aggressive handicap'. Other issues pertinent to the lack of heritability of aggressive behaviour have been reviewed by Boonstra and Boag (1987).

The Charnov–Finerty (1980) hypothesis and Stenseth's (1983) hypothesis also propose effects on population dynamics as a consequence of dispersal. Both suggest that immigration will result in decreased relatedness among neighbours, which is presumed to favour increased aggressiveness and, in turn, increased dispersal. Mating effectiveness presumably suffers also as a consequence of aggressive tendencies.

Criticisms of these hypotheses include the unitary drive fallacy mentioned above, and decreased relatedness between neigbours should have no effects on aggressive behaviour directed toward mates or kin. Neither hypothesis specifies exactly how aggression enhances dispersal. Do 'aggressive' individuals enhance the emigration rates of their own young, their neighbour's young, or their neigbours? Current evidence suggests kinship may have limited effects on behaviour. *Spermophilus beldingi* females treat both unrelated animals and kin more distant than half-sisters equally (Sherman, 1981b). Experimental removal of mothers had no significant effect on space use, dispersal tendency, or social behaviour of weanlings *Clethrionomys rufocanus* (Ims, 1990). Infanticide among *Cynomys ludovicianus* may be at least as frequent between mothers and grandoffspring as between individuals with lower genetic relationships, though it is rare between mothers and their own young (Hoogland, 1985). Increased immigration need not result in selection for the displacement of neighbours either; the 'dear enemy' phenomenon is quite common among territorial animals (Wilson, 1975), including territorial mammals (e.g. Conner, 1984). Because most emigration is performed by juveniles, intolerance of neighbours will have little effect on the total number of emigrating individuals in any population. Finally, any animal is likely to be intolerant of its neighbour's young to some degree regardless to kinship unless it has no offspring or excess space is available in its territory relative to the number of its own young. In either case, the effects on emigration will be both small and ephemeral.

Dispersal patterns may have consequences for social behaviour simply because dispersal usually involves a non-random subset of any population. Of particular relevance are infanticide and co-operation among kin, particularly sexual differences in co-operation related to sexual differences in emigration. Through its effects on dispersal, a given social regime may constrain the course of further social evolution.

5.6 CONCLUSIONS

Hypotheses regarding the role of social behaviour in the processes of emigration and immigration have proliferated far beyond the data available to test them. Social behaviour associated with resource access and spacing may be a common proximal cause of emigration and constraint on immigration, acting as an intermediary between resources such as food and mates and their utilization. Mating considerations may also be the single most important ultimate cause of sexual differences in dispersal. Despite their importance, there has been relatively little work on any of these ideas besides simple correlations between behaviour and dispersal patterns. Experimental studies dealing with all aspects of social behaviour and dispersal are desperately needed.

The data available indicate that, among small mammals at least, aggression associated with spacing may be of primary importance in limiting immigration, though it may play a lesser role in inducing emigration. The bulk of the evidence is against aggression as an ultimate cause of dispersal. Ontogenetic differences between individuals, such as a lack of cohesive behaviour with kin, could be a proximal mechanism for natal dispersal in some cases, but convincing data are few. Relatively little attention has been paid to the causes of breeding dispersal, perhaps because it is less frequent than natal dispersal. Its causes, however, should be no less interesting, and are likely to be related to the search for mates and female avoidance of close inbreeding. Much has been made of the male bias in natal and breeding dispersal among mammals, but little is known about the causes of female emigration. The female bias in the natal dispersal patterns of *Ochotona princeps* constitute a rare exception to the mammalian pattern (Brandt, 1985); intensive study of this species could be exceptionally rewarding.

Finally, the social ramifications of dispersal are many. A few of the most obvious include the evolution of sociality via kin selection and reciprocal altruism, infanticidal behaviour of immigrants, and the alteration of population structure with secondary effects on social interactions. Several theories dealing with population dynamics have their roots in social processes, though all such theories fail to recognize that social behaviour comprises the actions of individuals, and is not an attribute of the population itself. Hence one cannot speak of an 'aggressive' population without delineating the type of aggressive behaviour, the timing of aggression, and the identities of the individuals involved. Individuals seldom behave in a similar manner toward dissimilar individuals, and it is through such distinctions that variation in dispersal patterns are likely to arise.

ACKNOWLEDGEMENTS

It is a pleasure to acknowledge the services and reference collections of the Duke University Biology Library, where much of the research into the social behaviour literature was conducted. This paper benefitted greatly from discussions with and criticisms by the other authors in this book, especially those of Bill Lidicker.

REFERENCES

Altmann, S.A. and Altmann, J. (1979) Demographic constraints on behavior and social organization, in *Primate Ecology and Social Organization* (eds I.S. Bernstein and E.O. Smith), Garland STPM Press, New York, pp. 47–63.

Ambrose, H.W. (1972) Effect of habitat familiarity and toe clipping on rate of owl predation in *Microtus pennsylvanicus. Journal of Mammalogy*, 53, 909–12.

Anderson, P.K. (1961) Density, social structure, and nonsocial environment in house-mouse populations and the implications for regulation of numbers. *Transactions of the New York Academy of Science. Ser. II*, 23, 447–51.

Anderson, P.K. (1964) Lethal alleles in *Mus musculus*: local distribution and evidence for isolation of demes. *Science*, **145**, 177–8.

Andrzejewski, R. and Wrocławek, H.W. (1961) Mortality of small rodents in traps as an indication of the diminished resistance of the migratory part of population. *Bulletin Academie Polonaise Science*, **9**, 91–492.

Armitage, K.B. (1974) Male behaviour and territoriality in the yellow-bellied marmot. *Journal of Zoology*, **172**, 233–65.

Armitage, K.B. (1982) Social dynamics of juvenile marmots: role of kinship and individual variability. *Behavioral Ecology and Sociobiology*, **11**, 33–6.

Armitage, K.B. (1984) Recruitment in yellow-bellied marmot populations: kinship, philopatry, and individual variability, in *The Biology of Ground-dwelling Squirrels* (eds J.O. Murie and G.R. Michener), University Nebraska Press, Lincoln, pp. 377–403.

Armitage, K.B. and Johns, D.W. (1982) Kinship, reproductive strategies and social dynamics of yellow-bellied marmots. *Behavioral Ecology and Sociobiology*, **11**, 55–63.

Baird, D.D. and Birney, E.C. (1982a) Characteristics of dispersing meadow voles, *Microtus pennsylvanicus*. *American Midland Naturalist*, **107**, 262–83.

Baird, D.D. and Birney, E.C. (1982b) Pattern of colonization in *Microtus pennsylvanicus*. *Journal of Mammalogy*, **63**, 290–3.

Baker, M.C. and Marler, P. (1980) Behavioral adaptations that constrain the gene pool in vertebrates, in *Evolution of Social Behavior: hypotheses and empirical tests* (ed. H. Markl), Verlag Chemie GmbH, Weinheim, pp. 59–80.

Balph, D.F. (1984) Spatial and social behavior in a population of Uinta ground squirrels: interrelations with climate and annual cycle, in *The Biology of Ground-dwelling Squirrels* (eds J.O. Murie and G.R. Michener), University of Nebraska Press, Lincoln, pp. 336–52.

Barnard, C.J. and Brown, C.A.J. (1984) A payoff asymmetry in resident – resident disputes between shrews, *Animal Behaviour*, **32**, 302–4.

Beacham, T.D. and Krebs, C.J. (1980) Growth rates of aggressive and docile voles, *Microtus townsendii*. *American Midland Naturalist*, **104**, 387–9.

Bekoff, M. (1977) Mammalian dispersal and the ontogeny of individual behavioral phenotypes. *American Naturalist*, **111**, 715–32.

Bekoff, M. (1981) Mammalian sibling interactions: genes, facultative environments, and the coefficient of familiarity, in *Parental Care In Mammals* (eds D.J. Gubernick and P.H. Klopfer), Plenum Press, New York, pp. 307–46.

Bengtsson, B.O. (1978) Avoiding inbreeding: at what cost? *Journal of Theoretical Biology*, **73**, 439–44.

Benton, D. (1981) The measurement of aggression in the laboratory, in *The Biology of Aggression* (eds P.F. Brain and D. Benton), Sijthoff & Noordhoff, Alphen, pp. 487–502.

Benton, D. and Dalrymple-Alford, J. (1981) Social dominance in the mouse, in *The Biology of Aggression* (eds P.F. Brain and D. Benton), Sijthoff & Noordhoff, Alphen, pp. 571–5.

Bertram, B.C.R. (1976) Kin selection in lions and evolution, in *Growing Points in Ethology* (eds P.P.G. Bateson and R.A. Hinde), Cambridge University Press, Cambridge, pp. 281–301.

Boonstra, R. (1978) Effect of adult townsend voles (*Microtus townsendii*) on survival of young. *Ecology*, **59**, 242–8.

Boonstra, R. (1984) Aggressive behavior of adult meadow voles (*Microtus pennsylvanicus*) towards young. *Oecologia*, **62**, 126–31.

Boonstra, R. and Boag, P.T. (1987) A test of the Chitty hypothesis: inheritance of life-history traits in meadow voles, *Microtus pennsylvanicus*. *Evolution*, **41**, 928–47.

Boonstra, R. Krebs, C.J., Gaines, M.S. *et al.* (1987) Natal philopatry and breeding systems in voles (*Microtus* spp.). *Journal of Animal Ecology*, **56**, 655–73.

Boonstra, R. and Rodd, R. (1983) Regulation of breeding density in *Microtus pennsylvanicus*. *Journal of Animal Ecology*, **52**, 757–80.

Boutin, S. (1980) Effect of spring removal experiments on the spacing behavior of female snowshoe hares. *Canadian Journal of Zoology*, **58**, 2167–74.

Boutin, S. (1984) Effect of late winter food addition on numbers and movements of snowshoe hare. *Oecologia*, **62**, 393–400.

Boutin, S., Gilbert, B.S., Krebs, C.J. *et al.* (1985) The role of dispersal in the population dynamics of snowshoe hares. *Canadian Journal of Zoology*, **63**, 106–15.

Bradbury, J.W. and Vehrencamp, S.L. (1977) Social organization and foraging in Emballonurid bats. III. Mating systems. *Behavioral Ecology and Sociobiology*, **2**, 1–17.

Braithwaite, R.W. (1979) Social dominance and habitat utilization in *Antechinus stuartii* (Marsupalia). *Australian Journal of Zoology*, **27**, 517–28.

Brandt, C.A. (1984) The evolution of sexual differences in natal dispersal: tests of three hypotheses. Unpublished Ph.D. Dissertation, Duke University, Durham.

Brandt, C.A. (1985) The evolution of sexual differences in natal dispersal: a test of Greenwood's hypothesis, in *Migration: mechanisms and adaptive significance* (ed. M.A. Rankin), Contributions in Marine Science, Suppl., 27, pp. 386–96.

Brandt, C.A. (1989) Mate choice and reproductive success of pikas. *Animal Behaviour*, **37**, 118–32.

Brenner, F.J., Gaetano, C.P., Manser, S.W. and Belowich, D.L. (1978) Body weight and social interactions as factors in determining dominance in captive eastern chipmunks *Tamias striatus*. *Animal Behaviour*, **26**, 432–7.

Bronson, F.H. (1964) Agonistic behaviour in woodchucks. *Animal Behaviour*, **12**, 470–8.

Busser, J., Zweep, A. and van Oortmerssen, G.A. (1974) Variability in the aggressive behaviour of *Mus musculus domesticus*, its possible role in population structure, in *The Genetics of Behaviour* (ed. J.H.F. Van Abeelen), North-Holland, Amsterdam, pp. 185–99.

Bygott, J.D., Bertram, B.C.R. and Hanby, J.P. (1979) Male lions in large coalitions gain reproductive advantages. *Nature*, **282**, 839–41.

Calhoun, J.B. (1962) *The Ecology and Sociology of the Norway Rat*, US Department Health, Education, & Welfare, Washington, DC.

Carl, E.A. (1971) Population control in arctic ground squirrels. *Ecology*, **52**, 395–413.

Charnov, E.L. and Finerty, J.P. (1980) Vole population cycles: a case for kin-selection? *Oecologia*, **45**, 1–2.

Cheney, D.L. and Seyfarth, R.M. (1983) Nonrandom dispersal in free-ranging vervet monkeys: social and genetic consequences. *American Naturalist*, **122**, 392–412.

Chitty, D. (1967) The natural selection of self-regulatory behavior in natural populations. *Proceedings of the Ecological Society of Australia*, **2**, 51–78.

Chitty, D. (1987) Social and local environments of the vole *Microtus townsendii*. *Canadian Journal of Zoology*, **65**, 2555–66.

Christian, J.J. (1970) Social subordination, population density, and mammalian evolution. *Science*, **168**, 84–90.

Clark, A.B. (1978) Sex ratio and local resource competition in a prosimian primate. *Science*, **201**, 163–5.

Clulow, F.V., Franchetto, E.A. and Langford, P.E. (1982) Pregnancy failure in the red-backed vole, *Clethrionomys gapperi*. *Journal of Mammalogy*, **63**, 499–500.

Clutton-Brock, T.H. and Harvey, P.H. (1976) Evolutionary rules and primate societies, in *Growing Points in Ethology* (eds P.P.G. Bateson and R.A. Hinde), Cambridge University Press, Cambridge, pp. 195–237.

Cockburn, A. and Lidicker, W.Z. Jr (1983) Microhabitat heterogeneity and population ecology of an herbivorous rodent, *Microtus californicus*. *Oecologia*, **59**, 167–77.

Cockburn, A., Scott, M.P. and Scotts, D.J. (1985) Inbreeding avoidance and male-biased natal dispersal in *Antechinus* spp. (Marsupalia: Dasyuridae). *Animal Behaviour*, **33**, 908–15.

Conner, D.A. (1984) The role of an acoustic display in territorial maintenance in the pika. *Canadian Journal of Zoology*, **62**, 1906–9.

Daly, J.C. (1981) Effects of social organization and environmental diversity on determining the genetic structure of a population of the wild rabbit, *Oryctolagus cuniculus*. *Evolution*, **35**, 689–706.

Danielson, B.S. and Gaines, M.S. (1987) Spatial patterns in two sympatric species of microtines: *Microtus ochrogaster* and *Synaptomys cooperi*. *Journal of Mammalogy*, **68**, 313–22.

Dasser, V. (1981) Dispersal tendencies and social behaviour of young Chinese hamsters (*Cricelutus griseus*). *Behaviour*, **78**, 1–20.

Davis, L.S. (1984) Kin selection and adult female Richardson's ground squirrels: a test. *Canadian Journal of Zoology*, **62**, 2344–8.

Dewsbury, D.A. (1981) Social dominance, copulatory behavior, and differential reproduction in deermice (*Peromyscus maniculatus*). *Journal of Comparative and Physiological Psychology*, **95**, 880–95.

Dewsbury, D.A. (1987) Laboratory research on behavioral interactions as generators of population phenomena in rodents. *American Zoologist*, **27**, 941–51.

Dienske, H. (1979) The importance of social interactions and habitat in competition between *Microtus agrestis* and *M. arvalis*. *Behaviour*, **71**, 1–126.

Dittus, W.P.J. (1977) The social regulation of population density and age-sex distribution in the toque monkey. *Behaviour*, **63**, 281–322.

Dobson, F.S. (1979) An experimental study of dispersal in the California ground squirrel. *Ecology*, **60**, 1103–9.

Dobson, F.S. (1982) Competition for mates and predominant juvenile male dispersal in mammals. *Animal Behaviour*, **30**, 1183–92.

Dobson, F.S. (1983) Agonism and territoriality in the California ground squirrel. *Journal of Mammalogy*, **64**, 218–25.

Dolbeer, R.A. and Clark, W.R. (1975) Population ecology of snowshoe hares in the Central Rocky Mountains. *Journal of Wildlife Management*, **39**, 535–49.

Downhower, J.F. and Armitage, K.B. (1971) The yellow-bellied marmot and the evolution of polygamy. *American Naturalist*, **105**, 355–70.

Downhower, J.F. and Armitage, K.B. (1981) Dispersal of yearling yellow-bellied marmots (*Marmota flaviventris*). *Animal Behaviour*, **29**, 1064–9.

Dueser, R.D., Wilson, M.L. and Rose, R.K. (1981) Attributes of dispersing meadow voles (*Microtus pennsylvanicus*) in open-grid populations. *Acta Theriologica*, **26**, 139–62.

Dunbar, R.I.M. (1979) Population demography, social organization, and mating strategies, in *Primate Ecology and Human Origins* (eds I.S. Bernstein and E.O. Smith), Garland STPM Press, New York, pp. 65–88.

Dunford, C. (1977a) Behavioral limitation of round-tailed ground squirrel density. *Ecology*, **58**, 1254–68.

Dunford, C. (1977b) Kin selection for ground squirrel alarm calls. *American Naturalist*, **111**, 782–5.

Enquist, M. (1984) Game theory studies on aggressive behavior. Ph.D. Dissertation, University of Stockholm.

Fairbairn, D.J. (1977) The spring decline in deer mice: death or dispersal? *Canadian Journal of Zoology*, **55**, 84–92.

Fairbairn, D.J. (1978a) Dispersal of deer mice, *Peromyscus maniculatus*. Proximal causes and effects on fitness. *Oecologia*, **32**, 171–93.

Fairbairn, D.J. (1978b) Behaviour of dispersing deer mice (*Peromyscus maniculatus*). *Behavioral Ecology and Sociobiology*, **3**, 265–82.

Farr, L.A., Andrews, R.V. and Kline, M.R. (1978) Comparison of methods for estimating social rank of deer mice. *Behavioral Biology*, **23**, 399–404.

Festa-Bianchet, M. and King, W.J. (1984) Behavior and dispersal of yearling Columbian ground squirrels. *Canadian Journal of Zoology*, **62**, 161–7.

Frame, L.H. and Frame, G.W. (1976) Female African wild dogs emigrate. *Nature*, **263**, 227–9.

Francis, R.C. (1988) On the relationship between aggression and social dominance. *Ethology*, **78**, 223–37.

Frase, B.A. and Armitage, K.B. (1984) Foraging patterns of yellow-bellied marmots: role of kinship and individual variability. *Behavioral Ecology and Sociobiology*, **16**, 1–10.

Fretwell, S.D. and Lucas, H.L. (1970) On territorial behavior and other factors influencing habitat distribution in birds. 1. Theoretical development. *Acta Biotheoretica*, **19**, 16–36.

Gadgil, M., Joshi, N.V. and Gadgil, S. (1983) On the moulding of population viscosity by natural selection. *Journal of Theoretical Biology*, **104**, 21–42.

Gaines, M.S. and Johnson, M.L. (1982) Home range size and population dynamics in the prairie vole, *Microtus ochrogaster*. Oikos, **39**, 63–70.

Gaines, M.S. and Johnson, M.L. (1984) A multivariate study of the relationship between dispersal and demography in populations of *Microtus ochrogaster* in eastern Kansas. *American Midland Naturalist*, **111**, 223–33.

Gaines, M.S., Vivas, A.M. and Baker, C.L. (1979) An experimental analysis of dispersal in fluctuating vole populations: demographic parameters. *Ecology*, **60**, 814–28.

Garrett, M.G. and Franklin, W.L. (1988) Behavioral ecology of dispersal in the black-tailed prairie dog. *Journal of Mammalogy*, **69**, 236–50.

Gaulin, S.J.C. and FitzGerald, R.W. (1988) Home-range size as a predictor of mating systems in *Microtus*. *Journal of Mammalogy*, **69**, 311–9.

Gauthreaux, S.A. (1978) The ecological significance of behavioral dominance, in *Perspectives in Ethology* (eds P.P.G. Bateson and P.H. Klopfer), Plenum Press, New York, pp. 17–54.

Gavish, L., Hofman, J.E. and Getz, L.L. (1984) Sibling recognition in the prairie vole, *Microtus ochrogaster*. *Animal Behaviour*, **32**, 362–6.

Gipps, J.H.W. (1982) The effects of testosterone and scopolamine HBr on the aggressive behaviour of male voles, *Microtus townsendii*. *Canadian Journal of Zoology*, **60**, 946–50.

Gipps, J.H.W., Taitt, M.J., Krebs, C.J. and Dundjerski, Z. (1981) Male aggression and the population dynamics of the vole, *Microtus townsendii*. *Canadian Journal of Zoology*, **59**, 147–57.

Gleason, P.E., Michael, S.D. and Christian, J.J. (1980) Aggressive behavior during the reproductive cycle of female *Peromyscus leucopus*: effects of encounter site. *Behavioral and Neural Biology*, **29**, 506–11.

Gliwicz, J. (1980) Island populations of rodents: their organization and functioning. *Biological Review*, **55**, 109–38.

Graf, R.P. (1981) Some aspects of snowshoe hare behavioural ecology. Unpublished M.Sc. Thesis, University of British Columbia, Vancouver.

Greenwood, P.J. (1980) Mating systems, philopatry, and dispersal in birds and mammals. *Animal Behaviour*, **28**, 1140–62.

Greenwood, P.J. (1983) Mating systems and the evolutionary consequences of dispersal, in *The Ecology of Animal Movement* (eds I.R. Swingland and P.J. Greenwood), Clarendon Press, Oxford, pp. 116–31.

Gregory, M.J. and Cameron, G.N. (1988) Examination of socially induced dispersal in *Sigmodon hispidus*. *Journal of Mammalogy*, **69**, 251–60.

Gurnell, J. (1978) Seasonal changes in numbers and male behavioural interaction in a population of wood mice, *Apodemus sylvaticus*. *Journal of Animal Ecology*, **47**, 741–55.

Haigh, G.R. (1983) Effects of inbreeding and social factors on the reproduction of young female *Peromyscus maniculatus bairdii*. *Journal of Mammalogy*, **64**, 48–54.

Halpin, Z.T. (1981) Adult-young interactions in island and mainland populations of the deermouse *Peromyscus maniculatus*. *Oecologia*, **51**, 419–25.

Hamilton, W.D. (1972) Altruism and related phenomena, mainly in social insects. *Annual Review of Ecology and Systematics*, **3**, 193–232.

Harcourt, A.H. (1978) Strategies of emigration and transfer by primates, with particular reference to gorillas. *Zeitschrift für Tierpsychologie*, **48**, 401–20.

Healey, M.C. (1967) Aggression and self-regulation of population size in deermice. *Ecology*, **48**, 377–92.

Heske, E.J. and Nelson, R.J. (1984) Pregnancy interruption in *Microtus ochrogaster*: laboratory artifact or field phenomenon? *Biology of Reproduction*, **31**, 97–103.

Hestbeck, J.B. (1982) Population regulation of cyclic mammals: the social fence hypothesis. *Oikos*, **39**, 157–63.

Hestbeck, J.B. (1988) Population regulation of cyclic mammals: a model of the social fence hypothesis. *Oikos*, **52**, 156–68.

Hilborn, R. (1975) Similarities in dispersal tendency among siblings in four species of voles (*Microtus*). *Ecology*, **56**, 1221–5.

Hill, J.L. (1974) Peromyscus: effect of early pairing on reproduction. *Science*, **186**, 1042–4.

Hoeck, H.N. (1982) Population dynamics, dispersal and genetic isolation in two species of hyrax (*Heterohyrax brucei* and *Procavia johnstoni*) on habitat islands in the Serengeti. *Zeitchrift für Tierpsychologie*, **59**, 177–210.

Hoeck, H.N., Klein, H. and Hoeck, P. (1982) Flexible social organization in hyrax. *Zeitschrift für Tierpsychologie*, **59**, 265–98.

Holekamp, K.E. (1983) Proximal mechanisms of natal dispersal in Belding's ground squirrel (*Spermophilus beldingi beldingi*). Unpublished Ph.D. Dissertation, University of California, Berkeley.

Holekamp, K.E. (1984a) Natal dispersal in Belding's ground squirrels (*Spermophilus beldingi*). *Behavioral Ecology and Sociology*, **16**, 21–30.

Holekamp, K.E. (1984b) Dispersal in ground-dwelling sciurids, in *The Biology of Ground dwelling Squirrels* (eds J.O. Murie and G.R. Michener), University of Nebraska Press, Lincoln, pp. 297–320.

Hoogland, J.L. (1981) Nepotism and cooperative breeding in the black-tailed prairie dog (Schiuridae: *Cynomys ludovicianus), in Natural Selection and Social Behavior* (eds R.D. Alexander and D.W. Tinkle), Chiron Press, New York, pp. 283–310.

Hoogland, J.L. (1982) Prairie dogs avoid extreme inbreeding. *Science*, **215**, 1639–41.

Hoogland, J.L. (1985) Infanticide in prairie dogs: lactating females kill offspring of close kin. *Science*, **230**, 1037–40.

Huck, U.W., Soltis, R.L. and Coopersmith, C.B. (1982) Infanticide in male laboratory mice: effects of social status, prior sexual experience, and basis for discrimination between related and unrelated young. *Animal Behaviour*, **30**, 1158–65.

Ims, R.A. (1987) Responses in spatial organization and behaviour to manipulations of the food resource in the vole *Clethrionomys rufocanus. Journal of Animal Ecology*, **56**, 585–96.

Ims, R.A. (1990) Determinants of natal dispersal and space use in grey-sided voles, *Clethrionomys rufocanus*: a combined field and laboratory experiment. *Oikos*, **57**, 106–13.

Jamieson, S.H. and Armitage, K.B. (1987) Sex differences in the play behavior of yearling yellow-bellied marmots. *Ethology*, **74**, 237–53.

Johnson, M.L. and Gaines, M.S. (1985) Selective basis for emigration of the prairie vole, *Microtus ochrogaster*: open field experiment. *Journal of Animal Ecology*, **54**, 399–410.

Kareem, A.M. and Barnard, C.J. (1982) The importance of kinship and familiarity in social interactions between mice. *Animal Behaviour*, **30**, 594–601.

Karlsson, A.F. (1988) Social organization of a low density spring population of the bank vole, *Clethrionomys glareolus. Oikos*, **52**, 19–26.

Kaufmann, J.H. (1983) On the definitions and functions of dominance and territoriality. *Biological Review*, **50**, 1–20.

Kawata, M. (1987) The effect of kinship on spacing among female red-backed voles, *Clethrionomys rufocanus bedfordiae. Oecologia*, **72**, 115–22.

Keane, B. (1990a) The effect of relatedness on reproductive success and mate choice in the white-footed mouse, *Peromyscus leucopus. Animal Behaviour*, **39**, 264–73.

Keane, B. (1990b) Dispersal and inbreeding avoidance in the white-footed mouse, *Peromyscus leucopus. Animal Behaviour*, **40**, 143–52.

Keith, T.P. and Tamarin, R.H. (1981) Genetic and demographic differences between dispersers and residents in cycling and noncycling vole populations. *Journal of Mammalogy*, **62**, 713–25.

Kemp, G.A. and Keith, L.B. (1970) Dynamics and regulation of red squirrel (*Tamiasciurus hudsonicus*) populations. *Ecology*, **51**, 763–79.

King, J.A. (1955) Social behavior, social organization, and population dynamics in a black-tailed prairiedog town in the Black Hills of South Dakota. *Contributions of the Laboratory of Vertebrate Biology. University of Michigan*, **67**, 1–123.

King, J.A. (1983) Seasonal dispersal in a seminatural population of *Peromyscus maniculatus. Canadian Journal of Zoology*, **61**, 2740–50.

Krebs, C.J. (1978) A review of the Chitty hypothesis of population regulation. *Canadian Journal of Zoology*, **56**, 2463–80.

Krebs, C.J. (1979) Dispersal, spacing behaviour, and genetics in relation to population fluctuations in the vole *Microtus townsendii. Fortschritte der Zoologie*, **25**, 61–77.

Krebs, C.J., Halpin, Z.T. and Smith, J.N.M. (1977) Aggression, testosterone, and the spring decline in populations of the vole *Microtus townsendii. Canadian Journal of Zoology*, **55**, 430–7.

Krebs, C.J., Redfield, J.A. and Taitt, M.J. (1978) A pulsed-removal experiment on the vole *Microtus townsendii. Canadian Journal of Zoology*, **56**, 2253–62.

Krebs, C.J., Wingate, I., LeDuc, J. *et al.* (1976) *Microtus* population biology: dispersal in fluctuating populations of *Microtus townsendii. Canadian Journal of Zoology*, **54**, 79–95.

Krebs, J.R. and Perrin, C. (1978) Behaviour and population regulation in the great tit, in *Population Control by Social Behaviour* (eds F.J. Ebling and D.M. Stoddart) Cambridge University Press, London, pp. 23–47.

Krohne, D.T., Dubbs, B.A. and Baccus, R. (1984) An analysis of dispersal in an unmanipulated population of *Peromyscus leucopus. American Midland Naturalist.* **112**, 146–56,

Krohne, D.T. and Miner, M.S. (1985) Removal trapping studies of dispersal in *Peromyscus leucopus. Canadian Journal of Zoology*, **63**, 71–5.

Labov, J.B. (1980) Pregnancy blocking in rodents: adaptive advantages for females. *American Naturalist*, **118**, 361–71.

Labov, J.B., Huck, U.W., Elwood, R.W. and Brooks, R.J. (1985) Current problems in the study of infanticidal behavior of rodents. *Quarterly Review of Biology*, **60**, 1–20.

Lagerspetz, K. (1961) Genetic and social causes of aggressive behaviour in mice. *Scandinavian Journal of Psychology*, **2**, 167–79.

Lidicker, W.Z. Jr (1975) The role of dispersal in the demography of small mammals, in *Small Mammals: their productivity and population dynamics* (eds F.B. Golley, K. Petrusewicz and L. Rysczkowski), Cambridge University Press, Cambridge, pp. 103–28.

Lidicker, W.Z. Jr (1976) Social behaviour and density regulation in house mice living in large enclosures. *Journal of Animal Ecology*, **45**, 677–97.

Lidicker, W.Z. Jr (1985a) Dispersal, in *Biology of New World Microtus*, (ed. R.H. Tamarin), Special Publication No. 8, American Society of Mammalogists, pp. 420–54.

Lidicker, W.Z. Jr (1985b) Population structuring as a factor in understanding microtine cycles. *Acta Zoologica Fennica*, **173**, 23–7.

Lidicker, W.Z. Jr (1985c) An overview of dispersal in non-volant small mammals, in *Migration: mechanisms and adaptive significance* (ed. M.A. Rankin), Contributions in Marine Science, Suppl. **27**, 369–85.

Lockie, J.D. (1966) Territory in small carnivores. *Symposium of the Zoological Society of London*, **18**, 143–65.

Łomnicki, A. (1978) Individual differences between animals and the natural regulation of their numbers. *Journal of Animal Ecology*, **47**, 461–75.

Łomnicki, A. (1982) Individual heterogeneity and population regulation, in *Current Problems in Sociobiology* (ed. Kings College Sociobiology Group), Cambridge University Press, Cambridge, pp. 153–67.

Mackintosh, J.H. (1973) Factors affecting the recognition of territory boundaries by mice (*Mus musculus*). *Animal Behaviour*, **21**, 464–70.

Madison, D.M. (1980) Space use and social structure in meadow voles, *Microtus pennsylvanicus*. *Behavioral Ecology and Sociobiology*, **7**, 65–71.

Madison, D.M. and McShea, W.J. (1987) Seasonal changes in reproductive tolerance, spacing, and social organization in meadow voles: A microtine model. *American Zoologist*, **27**, 899–908.

Malcolm, J.R. and Marten, K. (1982) Natural selection and the communal rearing of pups in African wild dogs (*Lycaon pictus*). *Behavioral Ecology and Sociobiology*, **10**, 1–13.

Mallory, F.F. and Brooks, R.J. (1978) Infanticide and other reproductive strategies in the collared lemming, *Dicrostonyx groenlandicus*. *Nature*, **273**, 144–6.

Maly, M.S., Knuth, B.A and Barrett, G.W. (1985) Effects of resource partitioning on dispersal behavior of feral house mice. *Journal of Mammalogy*, **66**, 148–53.

Mares, M.A., Adams, R., Lacher, T.E. Jr and Willig, M.R. (1980) Home range dynamics in chipmunks: responses to experimental manipulation of population density and distribution. *Annals of the Carnegie Museum of Natural History*, **49**, 193–201.

Mares, M.A., Lacher, T.E. Jr, Willig, R.R. *et al.* (1982) An experimental analysis of social spacing in *Tamias striatus*. *Ecology*, **63**, 267–73.

Mares, M.A., Watson, M.D. and Lacher, T.E. Jr (1976) Home range perturbations in *Tamias striatus*. Food supply as a determinant of home range and density. *Oecologia*, **25**, 1–12.

Maynard Smith, J. (1982) *Evolution and the Theory of Games*, Cambridge University Press, Cambridge.

Mazurkiewicz, M. (1981) Spatial organization of a bank vole (*Clethrionomys glareolus*) population in years of small or large numbers. *Acta Theriologica*, **26**, 31–46.

McCracken, G.F. and Bradbury, J.W. (1981) Social organization and kinship in the polygynous bat *Phyllostomus hastatus*. *Behavioral Ecology and Sociobiology*, **8**, 11–34.

McGuire, M.R. and Getz, L.L. (1981) Incest taboo between sibling *Microtus ochrogaster*. *Journal of Mammalogy*, **62**, 213–5.

McLean, I.G. (1982) The association of female kin in the arctic ground squirrel *Spermophilus parryii*. *Behavioral Ecology and Sociobiology*, **10**, 91–9.

McLean, I.G. (1983) Paternal behaviour and killing of young in arctic ground squirrels. *Animal Behaviour*, **31**, 32–44.

McLean, I.G. (1984) Spacing behavior and aggression in female ground squirrels, in *The Biology of Ground-dwelling Squirrels* (eds J.O. Murie and G.R. Michener), University Nebraska Press, Lincoln, pp. 321–35.

Metzgar, L.H. (1967) An experimental comparison of screech owl predation on resident and transient white-footed mice (*Peromyscus leucopus*). *Journal of Mammalogy*, **48**, 387–91.

Metzgar, L.H. (1971) Behavioral population regulation in the woodmouse, *Peromyscus leucopus*. *American Midland Naturalist*, **86**, 434–48.

Metzgar, L.H. (1979) Dispersion patterns in a *Peromyscus* population. *Journal of Mammalogy*, **60**, 129–45.

Metzgar, L.H. (1980) Dispersion and numbers in *Peromyscus* populations. *American Midland Naturalist,* **103**, 26–31.

Michener, G.R. (1979) Yearly variations in the population dynamics of Richardson's ground squirrels. *Canadian Field-Naturalist*, **93**, 363–70.

Michener, G.R. (1981) Ontogeny of spatial relationships and social behaviour in juvenile Richardson's gound squirrels. *Canadian Journal of Zoology*, **59**, 1666–76.

Michener, G.R. (1982) Infanticide in ground squirrels. *Animal Behaviour*, **30**, 936–8.

Michener, G.R. (1983) Kin identification, matriarchies, and the evolution of sociality in ground-dwelling Sciurids. *Special Publications of the American Society of Mammalogists*, 7, 528–72.

Michener, G.R. and Michener, D.R. (1973) Spatial distribution of yearlings in a Richardson's ground squirrel population. *Ecology*, **54**, 1138–42.

Mihok, S. (1979) Behavioral structure and demography of subarctic *Clethrionomys gapperi* and *Peromyscus maniculatus*. *Canadian Journal of Zoology*, **57**, 1520–35.

Mihok, S. (1981) Chitty's hypothesis and behavior in subarctic red-backed voles, *Clethrionomys gapperi*. *Oikos*, **36**, 281–95.

Moore, J. and Ali, R. (1984) Are dispersal and inbreeding avoidance related? *Animal Behaviour*, **32**, 94–112.

Murie, J.O. and Harris, M.A. (1982) Annual variation in spring emergence and breeding in Columbian ground squirrels (*Spermophilus columbianus*). *Journal of Mammalogy*, **63**, 431–9.

Murie, J.O. and Harris, M.A. (1984) The history of individuals in a population of Columbian ground squirrels: source, settlement, and site attachment, in *The Biology of Ground-dwelling Squirrels* (eds J.O. Murie and G.R. Michener), University of Nebraska Press, Lincoln, pp. 353–73.

Myers, J.H. and Krebs, C.J. (1971) Genetic, behavioral, and reproductive attributes of dispersing field voles *Microtus pennsylvanicus* and *Microtus ochrogaster*. *Ecological Monographs*, **41**, 53–78.

Mykytowicz, R. (1968) Territorial marking by rabbits. *Scientific American*, **218**, 116–26.

Myllymäki, A. (1977a) Intraspecific competition and home range dynamics in the field vole *Microtus agrestis*. *Oikos*, **29**, 553–69.

Myllymäki, A. (1977b) Demographic mechanism in the fluctuating populations of the field vole *Microtus agrestis*. *Oikos*, **29**, 468–93.

Nadeau, J.H., Lombardi, R.T. and Tamarin, R.H. (1981) Population structure and dispersal of *Peromyscus leucopus* on Muskeget Island. *Canadian Journal of Zoology*, **59**, 793–9.

Noyes, R.F., Barrett, G.W. and Taylor, D.H. (1982) Social structure of feral house mouse populations: effects of resource partitioning. *Behavioral Ecology and Sociobiology*, **10**, 157–63.

Oakeshott, J.G. (1974) Social dominance, aggressiveness and mating success among male house mice (*Mus musculus*). *Oecologia*, **15**, 143–58.

Ostfeld, R.S. (1985) Limiting resources and territoriality in microtine rodents. *American Naturalist*, **126**, 1–15.

Ostfeld, R.S., Lidicker, W.Z. Jr and Heske, E.J. (1985) The relationship between habitat heterogeneity, space use, and demography in a population of California voles. *Oikos*, **45**, 433–42.

Packer, C. (1979) Inter-troop transfer and inbreeding avoidance in *Papio anubis*. *Animal Behaviour*, **27**, 1–36.

Packer, C. and Pusey, A.E. (1979) Female aggression and male membership in troops of Japanese macaques and olive baboons. *Folia Primatologica*, **31**, 212–8.

Parker, G.A. (1974) Assessment strategy and the evolution of fighting behaviour. *Journal of Theoretical Biology*, **47**, 223–43.

Parker, G.A. (1978) Selfish genes, evolutionary games, and the adaptiveness of behaviour. *Nature*, **274**, 849–55.

Parker, G.A. and Knowlton, N. (1980) The evolution of territory size – some ESS models. *Journal of Theoretical Biology*, **84**, 445–76.

Parker, G.A. and Rubenstein, D.A. (1981) Role assessment, reserve strategy, and acquisition of information in asymmetrical animal conflicts. *Animal Behaviour*, **29**, 221–40.

Perrin, M.R. (1979) The roles of reproduction, survival, and territoriality in the seasonal dynamics of *Clethrionomys gapperi* populations. *Acta Theriologica*, **24**, 475–500.

Perrin, M.R. (1981) Seasonal changes in agonistic behavior of *Clethrionomys gapperi* in southeastern Manitoba and its possible relation to population regulation. *American Midland Naturalist*, **106**, 102–10.

Petrusewicz, K. (1966) Dynamics, organization and ecological structure of population. *Ekologia Polska, Seria A*, **25**, 413–36.

Pfeifer, S. (1982) Disappearance and dispersal of *Spermophilus elegans* juveniles in relation to behavior. *Behavioral Ecology and Sociobiology*, **10**, 237–43.

Pulliam, H.R. and Caraco, T. (1984) Living in groups: is there an optimal group size? in *Behavioural Ecology. An Evolutionary Approach*, 2nd edn (eds J.R. Krebs and N.B. Davies), Sinauer, Sunderland, Massachusetts, pp. 122–47.

Rajska-Jurgiel, E. (1976) Interactions between individuals of a population of the bank vole, *Clethrionomys glareolus* (Schreber, 1780). *Ekologia Polska*, **24**, 3–35.

Rasa, O.A.E. (1981) Ethological aspects of aggression in sub-human animals, in *The Biology of Aggression* (eds P.F. Brain and D. Benton), Sijthoff & Noordhoff, Alphen, pp. 585–601.

Redfield, J.A., Taitt, M.J. and Krebs, C.J. (1978a) Experimental alteration of sex ratios in populations of *Microtus townsendii*, a field vole. *Canadian Journal of Zoology*, **56**, 17–27.

Refield, J.A., Taitt, M.J. and Krebs, C.J. (1978b) Experimental alterations of sex ratios in populations of *Microtus oregoni*, the creeping vole. *Journal of Animal Ecology*, **47**, 55–69.

Reich, L.M., Sood, K.M., Rothstein, B.E. and Tamarin, R.H. (1982) Aggressive behaviour of male *Microtus breweri* and its demographic implications. *Animal Behaviour*, **30**, 117–22.

Reinhardt, V. and Reinhardt, A. (1975) Dynamics of social hierarchy in a dairy herd. *Zeitschrift für Tierpsychologie*, **38**, 315–23.

Richards, S.M. (1974) The concept of dominance and methods of assessment. *Animal Behaviour*, **22**, 914–30.

Rose, R.K. and Gaines, M.S. (1976) Levels of aggression in fluctuating populations of the prairie vole, *Microtus ochrogaster*, in eastern Kansas. *Journal of Mammalogy*, **57**, 43–57.

Rowell, T.E. (1974) The concept of social dominance. *Behavioral Biology*, **11**, 131–54.

Rubenstein, D.A. (1981) Population density, resource patterning, and territoriality in the everglades pygmy sunfish. *Animal Behaviour*, **29**, 155–72.

Rubenstein, D.A. (1982) Risk, uncertainty, and evolutionary strategies, in *Current Problems in Sociobiology* (ed. Kings College Sociobiology Study Group), Cambridge University Press, Cambridge, pp. 91–111.

Rushen, J. (1982) The peck orders of chickens: how do they develop and why are they linear? *Animal Behaviour*, **30**, 1129–37.

Russell, J.K. (1981) Exclusion of adult male coatis from social groups: protection from predation. *Journal of Mammalogy*, **62**, 206–8.

Sadleir, R.M.F.S. (1965) The relationship between agonistic behaviour and population changes in the deermouse, *Peromyscus maniculatus* (Wagner). *Journal of Animal Ecology*, **34**, 331–52.

Savidge, I.R. (1974) Social factors in dispersal of deer mice (*Peromyscus maniculatus*) from their natal site. *American Midland Naturalist*, **91**, 395–405.

Schadler, M.H. (1981) Postimplantation abortion in pine vole (*Microtus pinetorum*) induced by strange males and pheromones of strange males. *Biology of Reproduction*, **25**, 295–7.

Schmutz, S.M., Boag, D.A. and Schmutz, J.K. (1979) Causes of the unequal sex ratio in populations of adult Richardson's ground squirrels. *Canadian Journal of Zoology*, **57**, 1849–55.

Schwartz, D.A. and Armitage, K.B. (1980) Genetic variation in social mammals: the marmot model. *Science*, **207**, 665–7.

Scott, J.P. and Frederikson, E. (1951) The causes of fighting in mice and rats. *Physiological Zoology*, **24**, 273–309.

Selander, R.K. (1970) Biochemical polymorphism in populations of the house mouse and old-field mouse. *Symposium of the Zoological Society of London*, **26**, 73–91.

Sherman, P.W. (1977) Nepotism and the evolution of alarm calls. *Science*, **197**, 1246–53.

Sherman, P.W. (1981a) Reproductive competition and infanticide in Belding's ground squirrels and other animals, in *Natural Selection and Social Behavior* (eds R.D. Alexander and D.W. Tinkle), Chiron Press, New York, pp. 311–31.

Sherman, P.W. (1981b) Kinship, demography, and Belding's ground squirrel nepotism. *Behavioral Ecology and Sociobiology*, **8**, 251–9.

Shields, W.M. (1982) *Philopatry, Inbreeding, and the Evolution of Sex*, SUNY Press, Albany.

Shields, W.M. (1983) Optimal inbreeding and the evolution of philopatry, in *The Ecology of Animal Movement* (eds I.R. Swingland and P.J. Greenwood), Clarendon Press, Oxford, pp. 132–59.

Slade, N.A. and Balph, D.F. (1974) Population ecology of Uinta ground squirrels. *Ecology*, **55**, 989–1003.

Smith, C.C. (1968) The adaptive nature of social organization in the genus of three squirrels *Tamiasciurus*. *Ecological Monographs*, **38**, 31–63.

Smith, A.T. (1974) The distribution and dispersal of pikas: influences of behavior and climate. *Ecology*, **55**, 1368–76.

Smith, A.T. (1978) Comparative demography of pikas (*Ochotona*): effect of spatial and temporal age-specific mortality. *Ecology*, **59**, 133–9.

Smith, A.T. and lvins, B.L. (1983) Colonization in a pika population: dispersal vs. philopatry. *Behavioral Ecology and Sociobiology*, **13**, 37–47.

Smith, A.T. and Ivins, B.L. (1984) Spatial relationships and social organization in adult pikas: a facultatively monogamous mammals. *Zeitschrift für Tierpsychologie*, **66**, 37–47.

Smith, R.H. (1979) On selection for inbreeding in polygynous animals. *Heredity*, **43**, 205–11.

Smyth, M. (1968) The effects of the removal of individuals from a population of bank voles *Clethrionomys glareolus*. *Journal of Animal Ecology*, **37**, 167–83.

Stenseth, N. Chr. (1983) Causes and consequences of dispersal in small mammals, in *The Ecology of Animal Movement* (eds I.R. Swingland and P.J. Greenwood), Clarendon Press, Oxford, pp. 63–101.

Stueck, K.L. and Barrett, G.W. (1978) Effects of resource partitioning on the population dynamics and energy utilization strategies of feral house mice

(*Mus musculus*) populations under experimental field conditions. *Ecology*, **59**, 539–51.

Sullivan, T.P. and Sullivan, D.S. (1982) Population dynamics and regulation of the Douglas squirrel (*Tamiasciurus douglasi*) with supplemental food. *Oecologia*, **53**, 264–70.

Svendsen, G.E. (1974) Behavioral and environmental factors in the spatial distribution and population dynamics of a yellow-bellied marmot population. *Ecology*, **55**, 660–771.

Svendsen, G.E. and Armitage, K.B. (1973) An application of mirror-image stimulation to field studies. *Ecology*, **54**, 623–7.

Taitt, M.J. (1981) The effect of extra food on small rodent populations: I deermice (*Peromyscus maniculatus*). *Journal of Animal Ecology*, **50**, 111–24.

Taitt, M.J. and Krebs, C.J. (1981) The effect of extra food on small rodent populations: II voles (*Microtus townsendii*). *Journal of Animal Ecology*, **50**, 125–37.

Taitt, M.J. and Krebs, C.J. (1982) Manipulation of female behaviour in field populations of *Microtus townsendii*. *Journal of Animal Ecology*, **51**, 681–90.

Tamarin, R.H. (1977) Dispersal in island and mainland voles. *Ecology*, **58**, 1044–54.

Terman, C.R. (1980) Social factors influencing delayed reproductive maturation in prairie deermice (*Peromyscus maniculatus bairdii*) in laboratory populations. *Journal of Mammalogy*, **61**, 219–23.

Thompson, D.C. (1978) The social system of the grey squirrel. *Behaviour*, **64**, 305–28.

Treisman, M. and Collins, J.S. (1980) Prior fitness and the evolutionarily stable strategy for intraspecific conflict. *Journal of Theoretical Biology*, **84**, 567–73.

Trivers, R.L. (1971) The evolution of reciprocal altruism. *Quarterly Review of Biology*, **46**, 35–57.

Tullock, G. (1983) Territorial boundaries: An economic view. *American Naturalist*, **121**, 440–2.

Turner, B.N. and Iverson, S.L. (1973) The annual cycle of aggression in male *Microtus pennsylvanicus*, and its relation to population parameters. *Ecology*, **54**, 967–81.

van Zegeren, K. (1980) Variation in aggressiveness and the regulation of numbers in house mouse populations. *Netherlands Journal of Zoology*, **30**, 635–770.

Vestal, B.M. and McCarley, H. (1984) Spatial and social relations of kin in thirteen-lined and other ground squirrels, in *The Biology of Ground-dwelling Squirrels* (eds J.O. Murie and G.R. Michener), University of Nebraska Press, Lincoln, pp. 404–23.

Viitala, J. (1977) Social organization in cyclic subarctic populations of the voles *Clethrionomys rufocanus* (Sund.) and *Microtus agrestis* (L.). *Annales Zoologica Fennici*, **14**, 53–93.

Waser, P.M., Austad, S.N. and Keane, B. (1986) When should animals tolerate inbreeding? *American Naturalist*, **128**, 529–37.

Waser, P.M. and Jones, W.T. (1983) Natal philopatry among solitary mammals. *Quarterly Review of Biology*, **58**, 355–90.

Whitworth, M.R. and Southwick, C.H. (1984) Sex differences in the ontogeny of social behavior in pikas: possible relationships to dispersal and territoriality. *Behavioral Ecology and Sociobiology*, **15**, 175–82.

Wiley, R.H. and Wiley, M.S. (1980) Territorial behavior of a blackbird: mechanisms of site-dependent dominance. *Behaviour*, **73**, 130–54.

Wilson, E.O. (1975) *Sociobiology: the new synthesis*, Belknap Press, Cambridge.

Wolf, K. and Schulman, S.R. (1984) Male response to 'stranger' females as a function of female reproductive value among chimpanzees. *American Naturalist*, **123**, 163–74.

Wolfe, J.L. (1966) Agonistic behavior and dominance relationships of the eastern chipmunk, *Tamias striatus*. *American Midland Naturalist*, **76**, 190–200.

Wrangham, R.W. (1980) An ecological model of female-bonded primate groups. *Behaviour*, **75**, 262–300.

Wynne-Edwards, V.C. (1962) *Animal Dispersion in Relation to Social Behaviour*, Hafner, New York.

Yasukawa, K. (1979) Territory establishment in red-winged blackbirds: importance of aggressive behavior and experience. *Condor*, **81**, 258–64.

Yeaton, R.I. (1972) Social behavior and social organization in Richardson's ground squirrel (*Spermophilus richardsonii*) in Saskatchewan. *Journal of Mammalogy*, **53**, 139–47.

PART THREE

Patterns: cyclic versus stable populations

The demographic causes and consequences of dispersal are major issues in population biology. In Part III, two world authorities on the subject review the evidence comparing small mammal populations that show two strongly contrasting patterns of density fluctuations. On the one hand, we have populations that exhibit multi-annual cycles in numbers, reviewed by Charles Krebs, and on the other hand, populations that do not show this pattern, reviewed by Joanna Gliwicz. Such a comparison is a deliberate strategy to search for insights into the role that dispersal may play in generating these diverse demographic patterns.

As it turns out, the distinction between multi-annual cycles and other patterns is not as clear as one might suppose *a priori*. In fact, both authors found ample opportunities to link the two together. Primarily using data from bank voles (*Clethrionomys glareolus*) and wood mice (*Apodemus sylvaticus*), Gliwicz (Chapter 6) analyses seasonal and within-season dispersal dynamics, and then connects them to longer-term density changes. Krebs (Chapter 7), in addition to reviewing the evidence relating dispersal behaviour to multi-annual cycles in microtine rodents, draws attention to some methodological problems which we also return to in Appendix 1. In addition, he focuses on 'the fence effect', a phenomenon of critical importance in linking dispersal to demography, and a topic to which Krebs has made many contributions.

Along with Part II, these two chapters provide the primary database for all that follows. The original symposium contained a contribution on lemmings by H. Hentonnen of the University of Helsinki. Unfortunately, it was not possible for his review to be included in this book, and so lemmings will for a while longer continue to retain that aura of mystery that has characterized them since their spectacular dispersal movements become widely known outside of Fennoscandia.

6

Patterns of dispersal in non-cyclic populations of small rodents

J. GLIWICZ

6.1 INTRODUCTION

Theoretical and empirical work on dispersal was stimulated by investigations of population cycles. Most concepts and data, therefore, deal with dispersal patterns in species and populations whose numbers change drastically in multiannual cycles. In non-cyclic or stable populations, numbers, breeding activity, and social behaviour of animals change much more significantly in the course of one year than between successive years. Also, the most clearly observed pattern of dispersal in such populations is a seasonal pattern. Studies on seasonal dispersal and on characteristics of dispersers in non-cyclic rodent populations may contribute to a better understanding of density regulation in these populations as well as to further development of views on dispersal. Yet, this chapter of the state of art in this field proved to be difficult, because only very few studies have dealt with dispersal in non-cyclic populations.

This chapter is based mainly on the author's own data on seasonal dispersal in non-cyclic populations of *Clethrionomys glareolus* (Gliwicz, 1988, 1989, 1990a,b) and *Apodemus flavicollis* (Gliwicz, 1988), and on data published by other authors on seasonal dispersal in non-cyclic and cyclic populations. Since microtine rodents have been more thoroughly studied than other small rodents, most of the data cited and conclusions drawn concern dispersal in voles. Dispersal studies are usually conducted by the 'removal grid' method (Chapter 7 and Appendix 1), and therefore give characteristics of immigrants onto depopulated areas, and not of dispersers from home areas. What the differences are between these two groups of animals is not clear.

This chapter first describes dispersing rodents which leave home populations in two distinct seasonal waves. Then the causes of their dispersal is speculated upon; and finally, a hypothesis is presented that integrates seasonal dispersal, spatial heterogeneity, and population dynamics.

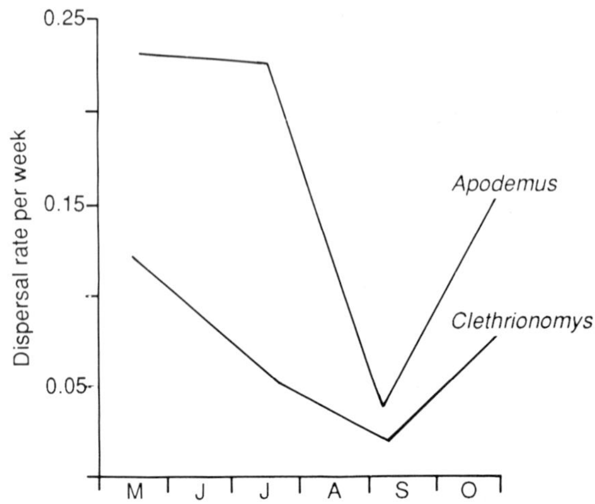

Figure 6.1 Dynamics of seasonal dispersal in non-cyclic populations of *Clethrionomys glareolus* and *Apodemus flavicollis* (after Gliwicz, 1988).

6.2 PATTERNS OF SEASONAL DISPERSAL

In populations of bank voles (*Clethrionomys glareolus*) and wood mice (*Apodemus flavicollis*) (Gliwicz, 1988) two different periods of increased dispersal occur: the first in the spring to early summer, and the second in the autumn (Figure 6.1). From data presented by other authors (Andrzejewski, 1963; Myers and Krebs, 1971; Fairbairn, 1978; Beacham, 1980; Wolff and Lidicker, 1980; Nadeau *et al.*, 1981; Krohne *et al.*, 1984) for non-cyclic and cyclic populations it appears that spring to early summer and autumn are the usual periods of increased dispersal in many rodent species, although both peaks were not observed in all populations (Table 6.1). This bimodal pattern of dispersal suggests that the two waves of dispersers may occur for different reasons, and that characteristics of individuals dispersing in different periods of the year may differ.

In the study of dispersal in bank vole and wood mouse populations it was found that dispersal rates, calculated separately for adults, subadults and juveniles (as the proportion between number of dispersers of a given age category and the density of individuals of the same category in the local population) were age-specific and changed in relation to each other in the course of the season (Figure 6.2).

In *C. glareolus*, during spring and early summer subadults dispersed at the highest rate and adults at the lowest rate. Later in the summer dispersal rates of all age groups decreased dramatically. In the autumn, after

Table 6.1 Seasons of increased dispersal in rodent populations

Species	*Spring–early summer*	*Autumn*	*Cyclic/non-cyclic*	*References*
Microtus ochrogaster		+	c	Myers and Krebs (1971)
Microtus pennsylvanicus		+	c	Myers and Krebs (1971)
Microtus townsendii	+	+	c	Beacham (1979)
Microtus xanthognathus		+	n	Wolff and Lidicker (1980)
Clethrionomys glareolus	+	+	n	Gliwicz (1988)
Apodemus flavicollis	+	+	n	Gliwicz (1988)
Peromyscus maniculatus	+	+	n	Fairbairn (1978)
Peromyscus leucopus	+	+	n	Krohne *et al.* (1984) Nadeau *et al.* (1981)

c, Cyclic population; n, non-cyclic population.

a seasonal cessation of breeding, dispersal of juveniles increased again, whereas older individuals remained settled (Figure 6.2a). Due to these differences in age-specific dispersal rates, spring–summer dispersers represented a different sample of the local population than did autumn dispersers.

In *A. flavicollis*, during spring and early summer relative dispersal rates in the three age categories were the same as in *C. glareolus*, but autumn dispersal in wood mice ceased to be age-specific. All age categories dispersed at similar rates (Figure 6.2b). Thus, in this species too the spring–summer dispersers differed from those migrating in the autumn.

Data of this kind, dealing with changes in dispersal tendencies of different age groups, or with age composition of dispersers leaving their home areas in different seasons, are not available for other rodent species. However, for several species, combined age characteristics of dispersers (without discrimination between spring and autumn dispersers) are given. In some cases only two age categories are distinguished: adult and young; in other cases a subadult category is also recognized. Generally, a higher than expected proportion of subadult or young individuals is often found among dispersers, e.g. in non-cyclic populations of *Apodemus sylvaticus* (Watt, 1970), *Clethrionomys gapperi* (Bondrup-Nielsen and Karlsson,

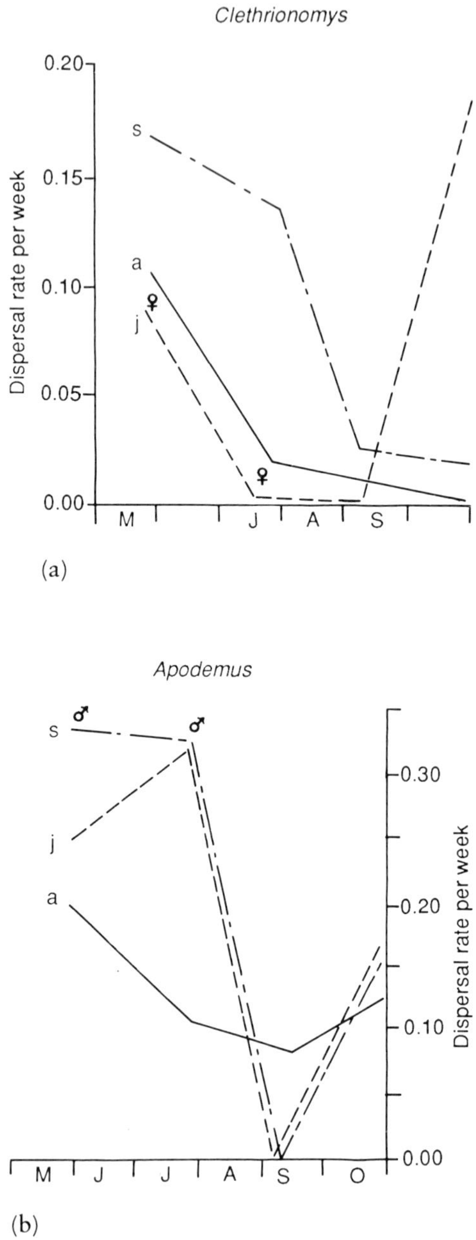

Figure 6.2 Dispersal dynamics of different age categories in (a) *Clethrionomys glareolus* and (b) *Apodemus flavicollis*. a, Adults; s, subadults; j, juveniles. The sex symbol indicates a surplus of one sex among dispersers (after Gliwicz, 1988).

Table 6.2 Territorial sex and predominant dispersing sex in different rodent species

Species	*Territoriality**	*Dispersal†*	*References*
Microtus pennsylvanicus	Female	Female	*Madison (1980) †Myers and Krebs (1971)
Clethrionomys glareolus	Female	Female	*Bujalska (1970) †Gliwicz (1988)
Peromyscus maniculatus	Male	Male	*Howard (1949) †Fairbairn (1978)
Apodemus flavicollis	Male?	Male	*Gliwicz (unpublished) †Gliwicz (1988)

1985), and *C. glareolus* (Smyth, 1966); in cyclic populations of *Synaptomys cooperi* (Gaines *et al.*, 1979), *Microtus pennsylvanicus* and *M. ochrogaster* (Myers and Krebs, 1971), and in fluctuating populations of *M. townsendii* (Beacham, 1979, 1980).

Although in non-cyclic species the sex ratio in residents and dispersers is claimed to be similar (Gaines and McClenaghan, 1980), in the two studied populations (*C. glareolus* and *A. flavicollis*) a significant surplus of one sex was found among subadult dispersers in spring and summer (Figure 6.2). In the bank vole subadult females dispersed at a higher rate, and in wood mice, subadult males. In cyclic populations one sex often predominates among dispersers (Myers and Krebs, 1971; Gaines *et al.*, 1979; Krohne *et al.*, 1984). Combined data on dispersal and spatial characteristics of four small rodent species suggest that the sex over-represented among dispersers is the territorial sex (Table 6.2; see also Chapter 5).

On the other hand, Greenwood (1980) suggests that in mammals male-biased dispersal should predominate, and the results of some vole studies support his view, even in the case of species with territorial females [see Cockburn (1988) for review]. In view of these contradictory findings I suggest that, since in rodents as in other mammals, female reproductive success depends strongly on access to food and male reproductive success on access to mates (Ostfeld, 1985), the sex ratio among dispersers may vary between populations (even of the same species) depending on the degree of intrasexual competition under local conditions (Gliwicz, in review).

Another property of dispersers that is often studied is their breeding condition. In the *Clethrionomys* and *Apodemus* study a higher proportion of immature individuals was found among subadult dispersers than among residents. The reason for such analysis is to establish, following the reasoning of Lidicker (1975), whether dispersers are individuals of high social rank and in good physiological condition (then they should attain puberty earlier than residents, be more active breeders, etc.) or whether they are subordinate weak individuals (then they should be mostly immature and non-breeding animals). However, results given by various authors are very confusing [see Gaines and McClenaghan (1980) for review]. To solve this problem it would be necessary to catch and examine dispersers-to-be in their home population. However, they are usually caught after immigration onto experimentally emptied areas, and after shorter or longer settlement there. Moreover, it is well known that non-breeding individuals may rapidly become reproductive after changing their social environment. Thus, our results are strongly affected by the method and schedule of trapping and cannot give a conclusive answer to the question about the breeding condition of dispersers.

In summary, seasonal dispersal seems to take place at two different times of the annual cycle: early in the breeding season and in the autumn. Some data indicate that the two groups of dispersers are different. In cyclic and non-cyclic populations the highest dispersal rates are found among young and subadult rodents. Intrasexual competition is hypothesized to influence the sex ratio of dispersers.

6.3 WHY DO RODENTS DISPERSE EARLY IN THE BREEDING SEASON?

The increased dispersal early in the breeding season may be a specific phenomenon in voles and mice, associated with specific life histories of these species. In the temperate zone these are the only vertebrates which commonly mature and reproduce in the year of their birth. Thus maturing young enter into competition with overwintered individuals established since spring in the local population.

For *C. glareolus* inhabiting patchy space, composed of optimal (rich alder forest) and suboptimal (poor small woodlots surrounded by vast pastures) habitats, it was found that young born early in the breeding season (Y1) in the optimal habitat, densely populated by overwintered individuals, dispersed soon after becoming independent into much less crowded (by overwintered adults) suboptimal habitat. Only 20% of Y1 young settled in the natal population. It was also found that the probability of maturing in the year of birth was higher in the suboptimal than optimal habitat, and between-habitat differences were greatest for young born early in the breeding season. Young born later in the reason (Y2) did not disperse, and their chances of maturing (much lower) were similar in optimal and suboptimal habitats. This leads to the conclusion that the driving force for pre-breeding dispersal of young born early in the breeding season may be the much higher probability of maturing early in less crowded habitats (Gliwicz, 1989). This also explains why dispersal stops later in the summer.

Moreover, it was found that the roughly estimated reproductive success of young mature males and females in the suboptimal habitat was not lower than the success of those less numerous young which settled successfully and became mature in the optimal habitat (Gliwicz, 1990a). Thus, since chances to survive to the next breeding season are low (winter mortality is 70–80%; Bujalska, 1975, Gliwicz, 1983, 1989), as are the chances to mature in the year of birth in densely populated optimal habitats, successful dispersal to suboptimal habitats is a viable alternative for young voles.

The results of the study cited above, indicating gain in fitness of successful dispersers born early in the breeding season, give support to the body of theory, named by Anderson (1989) Emigrant Fitness Hypothesis, according to which dispersal evolved through a gain in fitness to emigrants. However, we do not know how successful dispersers are in travelling to and establishing in vacant habitats. There are many indications of imbalance between emigration and immigration, suggesting that dispersal often leads to death. Thus, it remains to be shown that on average the benefits of spring–summer dispersal (successful breeding in the year of birth) exceed the costs of dispersal mortality. Early summer dispersers may have a good chance to find suitable vacant habitats because the overall density of rodents is still low at that time, and many habitats which have been unattractive in the early spring (and avoided by overwintered rodents) become more suitable in early summer due to the increase in plant biomass.

Yet, the proportion of emigrants that successfully settle after dispersal have never been measured adequately. Therefore, Anderson (1989), who is skeptical about the adaptive value of dispersal to emigrants, proposed the Resident Fitness Hypothesis, according to which dispersal of young is controlled by parental behaviour with parents maximizing their fitness by manipulating their offspring into dispersal.

6.4 WHY DO RODENTS DISPERSE IN THE AUTUMN?

In non-cyclic populations of mice and voles of the temperate zone, the annual peak in density usually occurs in the autumn, at the end of the breeding season. According to various authors (Table 6.1) this is also the period of increased dispersal in many rodent populations. Later dispersal decreases and continues at a very low rate throughout the winter (Kozakiewicz, 1976; Fairbairn, 1978; Beacham, 1980).

The different composition of autumn dispersers in the two studied rodent populations, *Clethrionomys* and *Apodemus* (Figure 6.2), suggests that there may be different reasons for dispersal at this time of the year. Equal dispersal rates of all age categories in *Apodemus* indicate that the motivation for dispersal was the same for all the animals. Most probably the observed increase in mobility of wood mice resulted from seasonal migration to another habitat, caused by seasonally changing habitat quality. (However, a return migration to breeding habitats by overwintered individuals was not recorded in this study). Such seasonal movements are known for other *Apodemus* species (Bergstedt, 1965; Zejda, 1967; Flowerdew, 1974; Viitala and Hoffmeyer, 1985), for *Mus musculus* (DeLong, 1967), for *Peromyscus maniculatus* (Fairbairn, 1978), and for lemmings (*Lemmus lemmus*; Kalela, 1961, 1971). It is not clear whether or not this kind of movement should be classified as dispersal, although it is not excluded from the concept of dispersal in the classification proposed by Lidicker and Stenseth (Chapter 2).

Another type of autumn dispersal was found in *Clethrionomys*. The dispersers were exclusively juvenile individuals, born at the end of the breeding season. Coincidence of annual peak in density with increased dispersal of young suggests that they are surplus individuals, forced to leave their home area before the onset of winter. Since carrying capacity of habitats decreases during winter, it may not be sufficient to accommodate all the individuals alive at the end of the breeding season. It is unlikely that young animals dispersing in the autumn gain anything from leaving their home area. They would probably survive better in the familiar site, and breed next year in the local population. Their dispersal must, therefore, be encouraged by older individuals, settling for winter and ensuring their own survival in the local population by getting rid of surplus young animals.

This type of dispersal may be expected in those years with higher autumn densities in all species characterized by high site tenacity during the whole year. The autumn dispersal may be considered an additional mechanism (besides the mechanisms effectively regulating the number of mature individuals in the breeding season) for density regulation in non-cyclic populations.

6.5 SPRING–SUMMER DISPERSAL, SPATIAL HETEROGENEITY, AND POPULATION CYCLES

Dispersal as an important factor, perhaps even a driving force behind population cycles, was indicated long ago (Chitty-Krebs hypothesis, Krebs, 1978; Lidicker, 1975; Tamarin, 1978; Charnov and Finerty, 1980; Hestbeck, 1982; Stenseth, 1983). It never has been clearly demonstrated, however, that dispersal behaviour of individuals is different in cyclic and in non-cyclic populations.

Recently, I proposed (Gliwicz, 1990b) a hypothesis according to which prebreeding dispersal of young born early in the season (Y1) from their natal (optimal) habitats into vacant (suboptimal) habitats may generate population cycles. According to this hypothesis, the dispersal behaviour of young (described here as spring–summer dispersal) does not vary between cyclic and non-cyclic populations. It results in multiannual cycles only under a certain set of external conditions (see below), otherwise it results in annual dynamics.

In cyclic populations the following events may be expected: after the population crash, the numbers of overwintered individuals inhabiting survival pockets (optimal habitats) steadily increase from one spring to the next. The more of them, the more Y1 young are born and emigrate into vacant habitats. The more dispersers, the more vacant habitats are colonized and the faster they are filled up. This goes on until no more vacant habitats are left for the current-year cohort of Y1 young. Frustrated dispersers wander in the population space densely populated by residents, aggravating social relations, increasing stress level, and/or carrying diseases between various habitat patches. One or all of these factors then cause the population crash in all habitats used by the population.

This hypothesis is based on two assumptions: (1) The dispersal behaviour of Y1 young is common among small rodents and obligatory (independent of the density of overwintered rodents). They may either always be encouraged by their genes or by older residents to leave the natal habitat; or they may each year experience shortage of space due to diffuse distribution of spring home ranges of overwintered rodents. (2) Most rodent populations live in a patchy space. Spatial heterogeneity of habitats as an important extrinsic factor shaping rodent dynamics has attracted much attention (e.g. Anderson, 1970; Hansson, 1977; Hestbeck, 1982; Stenseth, 1983; Lidicker, 1985, 1988; Bondrup-Nielsen and Ims, 1988). For my hypothesis of population cycles, habitat dichotomy is important. The cycle is produced by the feedback operating between two types of habitats: donor (optimal, where intrinsic factors cause small changes in density but rapid production of Y1 young) and recipient (suboptimal, where the surplus is fed unidirectionally and multiplies there for several years). In the last year of the cycle, the feedback from the second habitat into the first one regulates density in both parts of the system.

In conclusion, if (1) the proportion or location of donor and recipient habitats does not assure successful immigration of Y1 dispersers; or (2) vacant habitats are not suitable for settlement and reproduction of dispersers, but nevertheless act as dispersal sinks; or (3) during the nonbreeding season some factors such as shortage of suitable overwintering sites (Fuller, 1984; Mihok, 1988) or high predator pressure (Hansson, 1979; Erlinge, 1987) result in equally low spring densities, regardless of the autumn densities, then the population numbers will change in an annual pattern.

Evidently the set of extrinsic factors required for population cycles does not occur too often. It seems to occur regularly in northern Europe (Hansson and Henttonen, 1985), and less regularly in North America (Taitt and Krebs, 1985).

6.6 CONCLUSIONS

1. Seasonal dispersal patterns in non-cyclic populations of temperate zone rodents consist of two peaks: one early in the breeding season, and the other in the autumn.
2. Early in the breeding season young animals of the territorial sex, physiologically ready to reproduce, must compete with older residents for breeding territories. Some of them leave their home population to reproduce elsewhere. Later in the season, when breeding activity is suppressed, this type of dispersal ceases.
3. Another effect of seasonal changes in breeding activity is seasonal change in population growth rate; hence the positive correlation between density growth rate and breeding season dispersal rate. This kind of dispersal may be considered as seasonal presaturation dispersal.
4. Breeding season dispersal in small rodents may be a specific phenomenon, as voles and mice are the only vertebrates in the temperate zone whose young commonly mature and reproduce in the year of their birth.
5. Autumn dispersal has various patterns in different species: in some it takes the form of seasonal migration to better winter habitats, in others it is forced emigration of surplus individuals. The latter may be considered seasonal saturation dispersal.
6. Multiannual cycles result from the consequences of young born early in the breeding season dispersing (or attempting to disperse) in a patchy habitat (donor and recipient parts) in which at least some successful breeding occurs in recipient patches.

ACKNOWLEDGEMENTS

Nils Chr. Stenseth and Bill Lidicker suggested the theme of this paper. I am grateful to them for stimulating, valuable comments on the manuscript. I would also like to thank Charles Brandt and an anonymous reviewer for criticism on an earlier version of the text.

REFERENCES

Anderson, P.K. (1970) Ecological structure and gene flow in small mammals. *Symposia of the Zoological Society of London*, **26**, 299–325.

Anderson, P.K. (1989) *Dispersal in Rodents: a resident fitness hypothesis*, Special Publication No. 9, The American Society of Mammalogists, 139 pp.

Andrzejewski, R. (1963) Process of incoming, settlement and disappearance of individuals and variation in numbers of small rodents. *Acta Theriologica*, 7, 169–213.

Beacham, T.D. (1979) Size and growth characteristics of dispersing voles, *Microtus townsendii*. *Oecologia*, **12**, 1–10.

Beacham, T.D. (1980) Dispersal during fluctuations of the vole, *Microtus townsendii*. *Journal of Animal Ecology*, **49**, 867–77.

Bergstedt, B. (1965) Distribution, reproduction, growth and dynamics of the rodent species *Clethrionomys glareolus* (Schreber), *Apodemus flavicollis* (Melchior), and *Apodemus sylvaticus* (Linne) in southern Sweden. *Oikos*, **16**, 132–60.

Bondrup-Nielsen, S. and Ims, R.A. (1988) Predicting stable and cyclic populations of *Clethrionomys*. *Oikos*, **52**, 178–85.

Bondrup-Nielsen, S. and Karlsson, F. (1985) Movements and spatial patterns in populations of *Clethrionomys* species: a review. *Annales Zoologica Fennici*, **22**, 385–92.

Bujalska, G. (1970) Reproduction stabilizing elements in an island population of *Clethrionomys glareolus* (Schreber, 1780). *Acta Theriologica*, **15**, 381–412.

Bujalska, G. (1975) Reproduction and mortality of the bank vole and changes in size of an island population. *Acta Theriologica*, **20**, 41–56.

Charnov, E.L. and Finerty, J.P. (1980) Vole population cycles: a case for kin-selection? *Oecologia*, **45**, 1–2.

Cockburn, A. (1988) *Social Behaviour in Fluctuating Populations*, Croom Helm, London, 239 pp.

DeLong, K.T. (1967) Population ecology of feral house mice. *Ecology*, **48**, 611–34.

Erlinge, S. (1987) Predation and noncyclicity in a microtine population in southern Sweden. *Oikos*, **50**, 347–52.

Fairbairn, D.J. (1978) Dispersal of deer mice *Peromyscus maniculatus*. Proximal causes and effects on fitness. *Oecologia*, **32**, 171–94.

Flowerdew, J.R. (1974) Field and laboratory experiments on the social behaviour and population dynamics of the wood mouse (*Apodemus sylvaticus*). *Journal of Animal Ecology*, **43**, 499–511.

Fuller, W.E. (1984) Demography of a subarctic population of *Clethrionomys gapperi*: can winter mortality be predicted? Special Publication, Carnegie Museum of Natural History, **10**, 51–7.

Gaines, M.S., Baker, C.L. and Vivas, A.M. (1979) Demographic attributes of dispersing southern bog lemmings (*Synaptomys cooperi*) in eastern Kansas. *Oecologia*, **40**, 91–101.

Gaines, M.S. and McClenaghan, L.R. (1980) Dispersal in small mammals. *Annual Review of Ecology and Systematics*, **11**, 163–96.

Gliwicz, J. (1983) Survival and life span, in *Ecology of the Bank Vole* (ed. K. Petrusewicz), *Acta Theriologica*, **28**, Suppl. 1, pp. 161–72.

Gliwicz, J. (1988) Seasonal dispersal in non-cyclic populations of *Clethrionomys glareolus* and *Apodemus flavicollis*. *Acta Theriological*, **33**, 263–72.

Gliwicz, J. (1989) Individuals and populations of the bank vole in optimal, suboptimal and insular habitats. *Journal of Animal Ecology*, **58**, 237–47.

Gliwicz, J. (1990a) Habitat-dependent reproductive success in bank voles, in *Social Systems and Population Cycles in Voles* (eds R.H. Tamarin *et al.*), Birkhauser Verlag, Basel, pp. 169–80.

Gliwicz, J. (1990b) The first born, their dispersal, and vole cycles. *Oecologia*, **83**, 519–22.

Greenwood, P.J. (1980) Mating systems, philopatry and dispersal in birds and mammals. *Animal Behaviour*, **28**, 1140–62.

Hansson, L. (1977) Spatial dynamics of field voles *Microtus agrestis* in heterogeneous landscape. *Oikos*, **29**, 539–44.

Hansson, L. (1979) Condition and diet relation to habitat in bank voles *Clethrionomys glareolus*: population or community approach? *Oikos*, **33**, 55–63.

Hansson, L. and Henttonen, H. (1985) Regional differences in cyclicity and reproduction in *Clethrionomys glareolus* species: are they related? *Annales Zoologica Fennici*, **22**, 277–88.

Hestbeck, J. (1982) Population regulation in cyclic mammals; the social fence hypothesis. *Oikos*, **39**, 157–63.

Howard, W.E. (1949) Dispersal, amount of inbreeding, and longevity in a local population of prairie deer mice on the George Reserve, southern Michigan. *University of Michigan Contributions, Laboratory of Biology*, **43**, 1–52.

Kalela, O. (1961) Seasonal change of habitat in the Norwegian lemming, *Lemmus lemmus* (L.). *Annales Academic Scientinm Fennicae (A IV)*, **55**, 1–72.

Kalela, O. (1971) Seasonal differences in habitats of the Norwegian lemming, *Lemmus lemmus* (L.), in 1959 and 1960 at Kilpisjarvi, Finnish Lapland. *Annales Academic Scientinm Fennicae (A IV)*, **178**, 1–22.

Kozakiewicz, M. (1976) Migratory tendencies in a population of bank voles and description of migrants. *Acta Theriologica*, **21**, 321–38.

Krebs, C.J. (1978) A review of the Chitty hypothesis of population regulation. *Canadian Journal of Zoology*, **56**, 2463–80.

Krohne, D.T., Dubbs, B.A. and Baccus, R. (1984) An analysis of dispersal in an unmanipulated population of *Peromyscus leucopus. American Midland Naturalist*, **112**, 146–56.

Lidicker, W.Z. Jr (1975) The role of dispersal in the demography of small mammals, in *Small Mammals: their productivity and population dynamics* (eds F.B. Golley, K. Petrusewicz and L. Ryszkowski), Cambridge University Press, London, pp. 103–28.

Lidicker, W.Z. Jr (1985) Dispersal, in *Biology of New World Microtus* (ed. R.H. Tamarin), Special Publication No. 8, The American Society of Mammalogists, pp. 420–54.

Lidicker, W.Z. Jr (1988) Solving the enigma of microtine 'cycles'. *Journal of Mammalogy*, **69**, 225–35.

Madison, D.M. (1980) Space use and social structure in meadow voles *Microtus pennsylvanicus. Behavioral Ecology and Sociobiology*, 7, 65–71.

Mihok, S. (1988) The role of mortality in models of small rodent population dynamics. *Oikos*, **52**, 215–8.

Myers, J.H. and Krebs, C.J. (1971) Genetic, behavioral, and reproductive attributes of dispersing field voles, *Microtus pennsylvanicus* and *M. ochrogaster. Ecological Monographs*, **41**, 53–78.

Nadeau, J.H., Lombardi, R.T. and Tamarin, R.H. (1981) Population structure and dispersal of *Peromyscus leucopus* on Muskeget Island. *Canadian Journal of Zoology*, **59**, 793–9.

Ostfeld, R.S. (1985) Limiting resources and territoriality in microtine rodents. *American Naturalist*, **126**, 1–15.

Smyth, M. (1966) Winter breeding in woodland mice, *Apodemus sylvaticus* and voles *Clethrionomys glareolus* and *Microtus agrestis*, near Oxford. *Journal of Animal Ecology*, **35**, 471–85.

Stenseth, N.C. (1983) Causes and consequences of dispersal in small mammals, in *The Ecology of Animal Movement* (eds I.R. Swingland and P.J. Greenwood), Oxford University Press, Oxford, pp. 63–101.

Taitt, M.J. and Krebs, C.J. (1985) Population dynamics and cycles, in *Biology of New World Microtus* (ed. R.H. Tamarin), Special Publication No. 8, The American Society of Mammalogists, pp. 567–620.

Tamarin, R.H. (1978) Dispersal, population regulation and K-selection in field mice. *American Naturalist*, **112**, 545–55.

Viitala, J. and Hoffmeyer, I. (1985) Social organization in *Clethrionomys* compared with *Microtus* and *Apodemus*: social odours, chemistry and biological effects. *Annales Zoologica Fennici*, **22**, 359–371.

Watt, C.H.S. (1970) Long distance movement of bank voles and wood mice. *Journal of Zoology*, **161**, 247–56.

Wolff, J.O. and Lidicker, W.Z. Jr (1980) Population ecology of the taiga vole, *Microtus xanthognathus*, in interior Alaska. *Canadian Journal of Zoology*, **58**, 1800–12.

Zejda, J. (1967) Habitat selection in *Apodemus agrarius* (Pall. 1778) (Mammalia: Muridae) on the border of the area of its distribution. *Zoologicke Listy*, **16**, 221–38.

7

The role of dispersal in cyclic rodent populations

C.J. KREBS

7.1 INTRODUCTION

The hypothesis that dispersal causes the cyclic fluctuations of small mammals is perhaps the oldest hypothesis in small mammal ecology, dating from the last century. If lemmings really do march to the sea to drown, we have a perfect form of altruistic population control through dispersal-caused mortality. Unfortunately, neither lemmings nor other voles do such dramatic things, and thus very early in research on cyclic small mammals the dispersal hypothesis was rejected (Elton, 1942).

Populations of many species of voles and lemmings are known to fluctuate dramatically in size with a period of 3–4 years between peak populations. These fluctuations are loosely termed 'cycles' even though they are not constant in period or amplitude. Cycles occur in many different species of voles and lemmings in the north temperate and arctic regions of North America and Eurasia (Elton, 1942; Finerty, 1980). Their amplitude (maximum/minimum) is difficult to estimate; it is probably at least 15:1 and more often in the range of 50:1 or 100:1.

Not all populations of voles and lemmings show cycles. Within the same species some populations may cycles while others are stable or show only annual fluctuations (Hansson and Henttonen, 1985; Henttonen *et al.*, 1985). One of the best examples is from *Clethrionomys glareolus* populations which fluctuate cyclically in northern Sweden but are stable in southern Sweden. *Clethrionomys rutilus* is another species that seems to be cyclic in Europe but not in North America. Some *Microtus* populations also show cycles at some times and annual fluctuations in other periods (Taitt and Krebs, 1985).

Dispersal clearly plays many roles in the biology of small mammals (Lidicker, 1975), and this chapter is confined to the simple question: What is the role of dispersal in causing numerical changes in small mammal populations? We should be clear about several difficulties before we try to answer this question.

7.2 IMMIGRATION AND EMIGRATION ARE ALWAYS AREA-DEPENDENT

Some specific area which contains a population as the experimental unit must be defined before the role of dispersal can be discussed. Figure 7.1 illustrates the problem. Unless the study grid coincides exactly with the boundaries of a deme, it will not be possible to separate dispersal movements within a local population (A, B in Figure 7.1) from those between local populations (C in Figure 7.1). In operational terms movements B and C in Figure 7.1 will both be labelled as immigration (dispersal) while A will not. This is another illustration of the edge-effect problem in trapping. It can be countered only by trapping an entire patch or trapping a very large study area to minimize the edge.

Unfortunately the habitat of small mammals is rarely as tidy as the patches in Figure 7.1 suggest, and there is at present no way to relate the theoretical world of dispersal among patches (Chapter 1) to the observed world of movements on and off live-trapping areas. At present only dispersal of the easily visible small mammals, like ground squirrels and marmots, can be analysed unambiguously and related directly to theory. The use of radiotelemetry in voles (Appendix 2) may help us to overcome some of these problems.

7.3 DISPERSAL CAN AFFECT A POPULATION DIRECTLY OR INDIRECTLY

Direct effects occur when individuals immigrate or emigrate. Indirect effects occur when immigrants interact with residents and thereby change the local birth or death rates. For example, immigrants may introduce new genetic combinations into a local population. Or they may cause infanticide among the resident juveniles. These indirect interactions may be more critical than the direct effects and are the reason why simple density-dependent-density-independent dichotomies are not sufficient for understanding the role of dispersal in causing population changes.

The importance of dispersal in causing changes in numbers can be demonstrated in three ways.

1. Dispersal is density-dependent, i.e. one might show that the emigration rate or immigration rate is positively or negatively correlated with population density. This is the classical approach in population regulation studies.
2. Preventing dispersal causes changes in population density. The 'fence effect' reported by Krebs *et al.* (1969) is the paradigm of this approach.
3. Manipulating immigration or emigration rates causes changes in

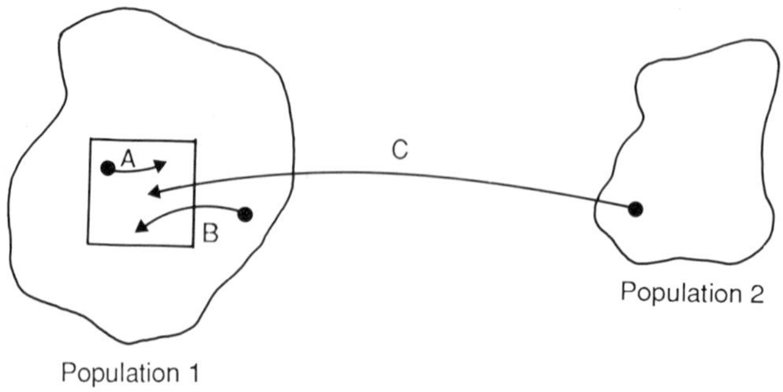

Figure 7.1 Hypothetical illustration of one problem in measuring dispersal. Two patches of suitable habitat are shown. The square represents the study area. •, place of birth; →, place of reproduction. Dispersal events A and B are within a population and should be recognized as distinct from dispersal C which is between populations. Most theoretical discussions are about C whereas most field methods (including removal grids) measure B type of movements.

population density. This is a more sophisticated variant of (2) that requires experimental control.

Behind all the experimental approaches to studying dispersal is the central question of how to measure dispersal. There are two techniques in use. Removal areas were first used to measure dispersal by Myers and Krebs (1971), although the idea was clearly much in use in pest control operations (e.g. Chitty and Southern, 1954). Dispersers are operationally defined as individuals that enter the removal area. The second technique is to label offspring in the nestling stage and locate their place of breeding (as done in birds). This technique utilizes radiotelemetry, radioactive markers, or other marking techniques (Beacham, 1980; Tamarin *et al.*, 1983; Appendices 2 and 3). These two approaches produce very different data sets. The first produces two categories of animals (dispersers and residents), while the second produces a continuum of dispersers from individuals that move a short distance to those that move long distances before becoming reproductively active. One can subdivide this continuum into two classes (Figure 7.1): short-distance movers that move within a patch (residents?) and long-distance dispersers that move between patches (dispersers?), but it is important to note that these two classes may not be equivalent to the two classes defined operationally by the first approach using removal areas. This is because the first technique is a 'vertical', or time-specific cross-section, while the second technique is a 'horizontal' or longitudinal cross-section. If we understood the social system completely for a particular species, we could specify how these two techniques would compare. Unfortunately, both techniques have not been used simultaneously on a small mammal population.

The removal area technique has been criticized by several authors (Dobson, 1981; Cockburn, 1985). Dobson (1981) concluded that removal areas did not measure the normal dispersal pattern of California ground squirrels (*Spermophilus beecheyi*), but his results are confounded by an inadequate distance between control and removal grids and by a failure to appreciate that successful dispersal and colonization are expected to vary between an occupied and an unoccupied area. If dispersal produces vagrants in search of a new home range (Lidicker, 1975), success of colonization may well differ between an occupied and a vacant habitat.

A classification system for the proximate factors motivating mammals to disperse is presented by Lidicker and Stenseth (Chapter 2). The diversity of dispersal categories and the fact that they are not mutually exclusive makes it difficult to isolate hypotheses to test. From a demographic perspective, Gaines and McClenaghan (1980) reviewed dispersal in small mammals and recognized four hypotheses about the proximate role of dispersal in small mammals.

1. Social subordination hypothesis.
2. Genetic–behavioural polymorphism hypothesis.
3. Presaturation–saturation dispersal hypothesis.
4. Social cohesion hypothesis.

These four hypotheses are most clearly separable by their predictions about the relationship between dispersal rate and population growth rate (Figure 7.2; see also Chapter 9). Stenseth (1983) has enlarged on this analysis and suggested a dispersal and kin selection hypothesis for explaining population cycles. The theoretical discussion about dispersal and population cycles (Stenseth, 1983; Warkowska-Dratnal and Stenseth, 1985) makes three critical assumptions that are reviewed here:

1. The fence effect demonstrates that dispersal is necessary for population regulation.
2. Dispersal from cyclic populations is largely presaturation dispersal.
3. Dispersing individuals are not a genetic or behavioural random sample from the population.

7.3.1 Assumption 1

The fence effect is one of the most striking but least understood effects in small mammal experimental work. Vole populations, enclosed in the field in a mouse-proof fence, increase in density to abnormally high levels, overgraze the habitat and decline from starvation. It was first described by

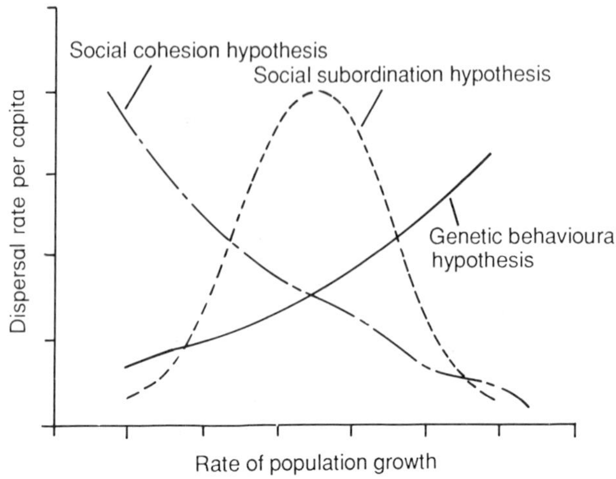

Figure 7.2 Relationship between dispersal rate and population growth rate as predicted by three hypotheses outlined by Gaines and McClenaghan (1980). Stenseth (1983) labels the curve for the genetic–behavioural hypothesis adaptive dispersal, and the curve for the social cohesion hypothesis non-adaptive dispersal. The curve for the genetic–behavioural hypothesis is essentially that for presaturation dispersal and that for the social cohesion hypothesis is essentially that for saturation dispersal.

Krebs *et al.* (1969) and observed again by Boonstra and Krebs (1977). Several laboratory studies of *Microtus* had recognized in the 1950s that enclosed vole populations reached abnormally high densities (references in Lidicker, 1985). But since laboratory populations were provided with extra food and protected from predation, dispersal was not isolated as a key process in producing the fence effect until 1969 (Krebs *et al.*, 1969).

Putting up a mouse-proof fence around a population eliminates both emigration and immigration, and the fence effect can be restated more simply as: an isolated population with no possible immigration and no possible emigration cannot regulate its growth below the limit set by starvation. One immediate question is which of these two processes, immigration or emigration, is the critical one. Abramsky and Tracy (1979) suggested that immigration was necessary for populations of *Microtus ochrogaster* to cycle and that emigration was not essential. One way to separate the effects of immigration and emigration is to build enclosures with one-way exit doors. Gaines *et al.* (1979) did this for *M. ochrogaster* and found that the fence effect was eliminated when one-way exit doors were allowed. Tamarin *et al.* (1984) concluded that emigration is the essential demographic parameter affected by fencing.

Does the fence effect occur in all populations of cyclic rodents? Only six microtine species have been studied in large enclosures for the fence effect: *M. ochrogaster*, *M. pennsylvanicus*, *M. townsendii*, *M. californicus*, *M. montanus* and *M. pinetorum*. The most comprehensive analysis of the fence effect comes from a 5-year study on *M. ochrogaster* and *M. pennsylvanicus* in central Illinois by Verner and Getz (1985). In this study they monitored the movement of voles through a 1 ha permeable enclosure with 50 entrance and exit gates, and employed drift fences to intercept emigrants. They observed similar density cycles in *M. ochrogaster* and *M. pennsylvanicus* on control and enclosed areas, and thus did not observe a fence effect. But of all the losses of voles in the enclosures, they estimated only 17% were due to dispersal in *M. ochrogaster* and 7% in *M. pennsylvanicus*. These loss rates were so low, Verner and Getz (1985) concluded that dispersal was not a significant factor in the demographic fluctuations in either species. There is unfortunately no way to judge this conclusion from their analysis; it implies that if all the entrance and exit gates were sealed on their enclosure, no fence effect would occur. This experiment remains to be done for their population. Krebs *et al.* (1969) did this experiment on the same two species in a site 185 km distant, and observed a strong fence effect. There would seem to be only three possibilities: (1) that populations of these voles are highly variable in dispersal tendencies, so that dispersal is necessary for normal population regulation in southern Indiana (Krebs *et al.*, 1969) but not necessary for regulation in central Illinois (Verner and Getz, 1985); or (2) that small loss rates due to dispersal are still necessary for normal population regulation, contrary to what Verner and Getz (1985) concluded; or (3) that Verner and Getz (1985) underestimated emigration rates. It would be useful to find out if the fence effect can occur in these Illinois populations.

Ford and Pitelka (1984) did not observe the fence effect in confined populations of *M. californicus*. However they used very small enclosures (0.07–0.13 ha) and their results might be interpreted to suggest that there is a lower limit to enclosure size below which one does not observe the fence effect, or that severe dry-season resource limitation can overwhelm the fence effect in the short term (<12 months). There was also a problem with leakage from their enclosures.

If the fence effect occurs and if dispersal is the critical parameter affected by fencing, one should be able to eliminate the fence effect by providing a dispersal sink (Lidicker, 1975) inside the fence. Gaines *et al.* (1979) showed that enclosures with exit doors allowed normal population fluctuations in *M. ochrogaster*. Beacham (1980, 1981) was the first to show that the fence effect could be eliminated in *M. townsendii* by providing a dispersal sink inside the fence (Figure 7.3). (Beacham did not correct his control population densities for the edge effect. But since his control

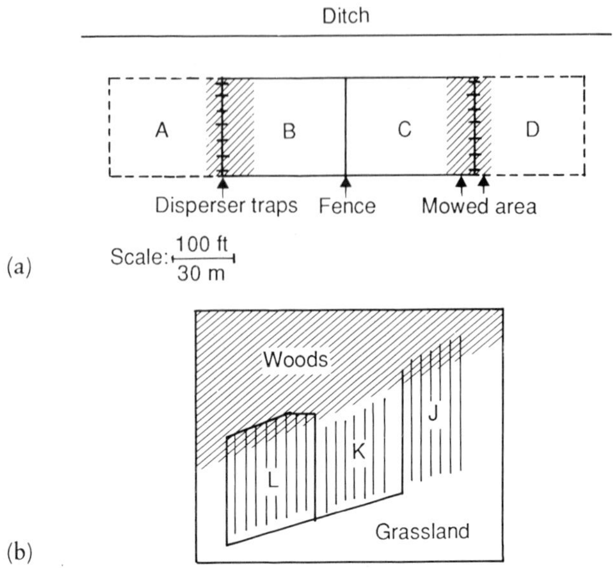

Figure 7.3 Two designs for fenced grids that provide a dispersal sink for voles. (a) Beacham (1980) design. Each grid is a 7 × 7 checker board of traps, 7.6 m apart. The shaded area was mowed. (b) Tamarin *et al.* (1984) design. Dots are trapping stations, 7.6 m apart. Each grid is 0.8 ha.

grids were positioned to have only two edges, his control grids could sample at most 1.15 times the area of his enclosed grids. This correction does not affect his conclusions.) Tamarin *et al.* (1984) used a clever design (Figure 7.3) of partially and fully fenced grids that included woodland areas as dispersal sinks. They concluded that the fence effect could be eliminated in *M. pennsylvanicus* with these types of dispersal sinks.

The fence effect is important because it shows that dispersal is necessary for normal population regulation. There is no reason to suspect that other non-cyclic species or populations of small mammals will not also show the fence effect. Showing a fence effect would appear to be one clear operational way of demonstrating that dispersal is involved in population regulation in any species. McMahon and Tash (1988) show, for example, that desert pupfish populations also show a 'fence effect' in isolated desert ponds.

We have far too few examples of the fence effect in small mammals, and there is no reason yet to assume that all populations must show the fence effect (Abramsky and Tracy, 1979). There is a tendency to assume complete generality with only a few examples of experimental manipulations and we should be cautioned against haste. More demonstrations of the fence effect are needed, particularly in the cyclic lemmings, in order to determine if the fence effect is common and general or uncommon and rare.

If the fence effect is widespread and significant, a whole series of further experiments are suggested. The fence effect may be caused by reduced immigration, reduced emigration, or both. Gaines *et al.* (1979) suggested the use of three enclosures with one-way exit doors, one-way entrance doors, and two-way doors could help to distinguish whether immigration or emigration was involved. The fence effect ought to occur on small islands as well, but the problem has been to find islands that are not subject to immigration and emigration (cf. Pokki, 1981). The same problem occurs in agricultural 'islands'. For example, Nabagło (1981) studied 'confined' populations of *Microtus arvalis* in agricultural crops but there is no evidence that his two populations were in fact closed to immigration or emigration.

7.3.2 Assumption 2

If dispersal is necessary for population regulation, it is important to specify exactly how and when it occurs. Lidicker (1975) pointed out the demographic consequences of saturation dispersal and how they differ from those of presaturation dispersal. There is a general consensus that dispersal in cyclic microtines is mostly presaturation dispersal (Gaines and McClenaghan, 1980, p. 172; Stenseth, 1983). The evidence for this belief comes largely from removal experiments. Very little of the loss that occurs during the decline phase seems to be due to dispersal (Myers and Krebs, 1971). Similar results have been obtained in analyses of the spring decline (Krebs and Boonstra, 1978). In moderate spring declines most of the loss is due to emigration, while in severe spring declines no emigration is found (Figure 7.4).

There is some direct evidence on dispersal during cyclic declines in enclosures. Beacham (1980) showed that some of the loss in a fenced *M. townsendii* population was due to dispersal in the spring (Figure 7.5) but none of the losses in the following year of severe decline were caused by dispersal. Tamarin *et al.* (1984) found about one-third of the losses in *M. pennsylvanicus* could be due to dispersal in a fenced population of *M. pennsylvanicus*.

There are, however, some dissenting data suggesting that in some severe declines dispersal losses are large. Desy and Thompson (1983) showed that emigration was an important component of loss in a *M. pennsylvanicus* decline. Hilborn and Krebs (1976) inferred that emigration caused one severe decline in *M. townsendii* because most of the radioactively tagged individuals disappeared.

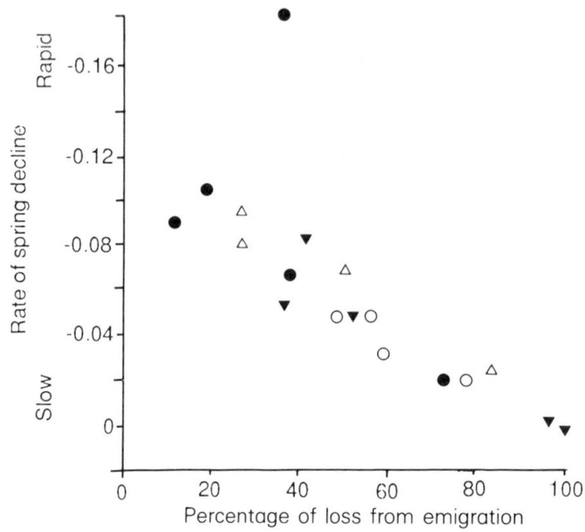

Figure 7.4 Relationship between the percentage of losses due to successful emigration and the rate of the spring decline in *Microtus townsendii*. ●, Grid I; O, grid C; ▼, grid E; △, other grids (from Krebs and Boonstra, 1978).

One way to test for the effect of dispersal during cyclic declines would be to fence a vole population when it is in the peak phase of the cycle. If no emigration or no immigration is essential for the decline to occur, we would predict the same decline inside and outside the fence. This experiment has not yet been carried out. Ford and Pitelka (1984) reported that enclosures did not affect the decline phase of *M. californicus* populations, and that no fence effect was present in their study. There may be a lower limit to enclosure size in studies of this kind at which results become artefacts of enclosure size because normal social processes cannot operate. It is not known if the enclosures used by Ford and Pitelka (1984) were too small to produce a fence effect, or whether this population does not show any effects from fencing.

Gaines and McClenaghan (1980) concluded that most dispersal in microtine rodents was density-independent. They recognized the problem that many workers do correlations with the number of dispersers rather than with the proportion of dispersers versus density (McClenaghan and Gaines, 1976). There is another hidden problem in the published data on the relationship between dispersal rate and population density – the time element. Over what time unit should the data be obtained? In most of our studies we have grouped data over seasonal 'periods', either two per year (e.g. Krebs *et al.*, 1976; Gaines *et al.*, 1979) or up to four per year (e.g.

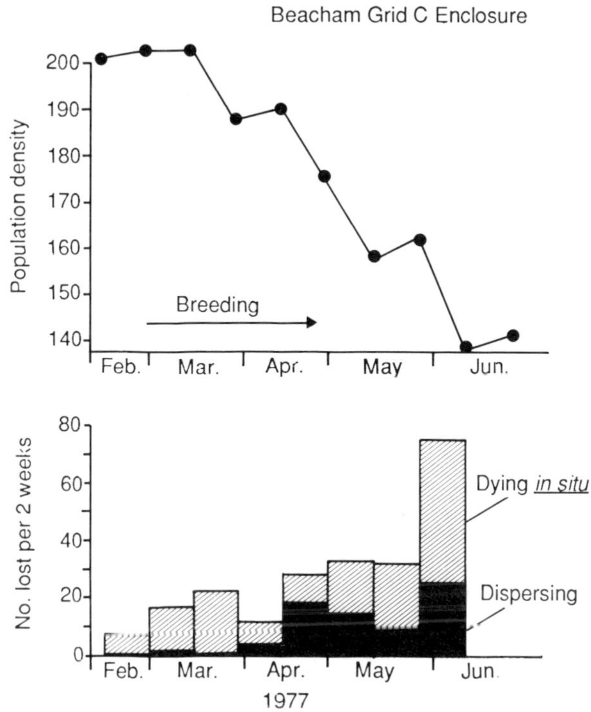

Figure 7.5 Partitioning of losses during a spring decline in a peak population of *Microtus townsendii* in an enclosure (Figure 7.3a). Data of Beacham (1980).

Myers and Krebs, 1971). The shorter the time period of grouping, the greater the problems of pseudoreplication (Hurlbert, 1984).

To investigate the sensitivity of the conclusions, the data given in Krebs *et al.* (1976) for *M. townsendii* are reanalysed, adding two subsequent years of additional data. Two pairs of control and removal grids (C and D, E and F grids) were live-trapped at biweekly intervals from May 1971 to October 1975. The areas and the trapping techniques are described in Krebs *et al.* (1976).

(a) Is dispersal rate density-independent in M. townsendii?

The recovery ratio is used as an index of the per capita dispersal rate (Krebs *et al.*, 1976). The regression is run between the recovery ratio and population density over five time groupings – 1, 2, 4, and 8 trapping periods and finally two seasonal periods (winter, summer) defined by breeding intensity. These data cover periods of increasing density as well as periods of declining density. Table 7.1 gives the correlation coefficients.

Table 7.1 Correlation coefficients for the relationship between the recovery ratio (a measure of dispersal rate) and population density of breeding *Microtus townsendii* for data from 1971 to 1975. Sample size in parentheses

Time grouping (trapping periods)	*Grids E* and F*	*Grids C and D*
1	0.49 (108)	0.50 (102)
2	0.55 (54)	0.53 (51)
4	0.68 (27)	0.50 (25)
8	0.78 (13)	0.45 (12)
2 Seasons/year	0.95 (9)	0.72 (10)

* Grids E and F are paired control-removal grids at Ladner Airbase, and grids C and D are paired control-removal grids at Serpentine Fen. These areas are described in detail in Krebs *et al.* (1976).

It is encouraging that they are all positive and, by the usual statistical criteria, significant. Moreover, the two replicates show the same relationship with breeding density (Figure 7.6). It is important to use breeding density rather than total density in these comparisons (i.e. the density of breeding adult animals, excluding immatures). This was not done in the past and Verner and Getz (1985) repeat this mistake in their analysis. It is concluded, contrary to Gaines and McClenaghan (1980) and contary to Stenseth (1983, p. 84), that dispersal rate is density-dependent in *M. townsendii.*

It is important to point out here that the *M. townsendii* populations analysed in Table 7.1 are not cyclic but fluctuate irregularly.This result suggests a generalization:

1. cyclic populations – presaturation dispersal predominates, density-independent;
2. non-cyclic populations – saturation dispersal predominates, density-dependent.

This dichotomy matches the extreme cases predicted by Stenseth and Lidicker (Chapter 9, Figures 9.5b and 9.6a).

(b) Is dispersal rate a function of population growth rate?

Figure 7.7 gives the regression for *M. townsendii* between dispersal rate and population growth rate. The correlation is not significant, although

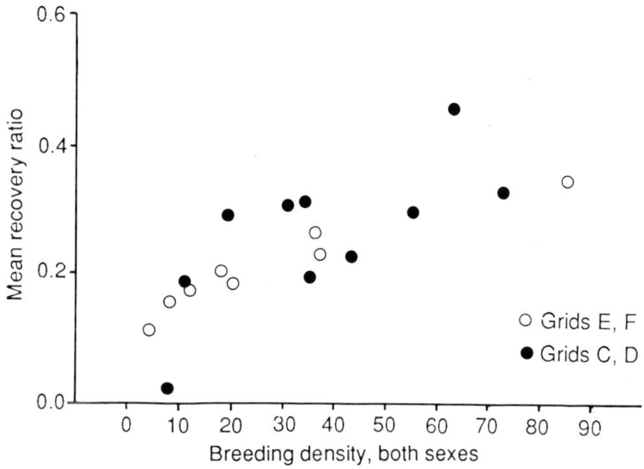

Figure 7.6 Relationship between the mean recovery ratio and breeding population density in *Microtus townsendii* near Vancouver, Canada. Data grouped over summer and winter periods, 1971–75. $r = 0.78$, $n = 19$.

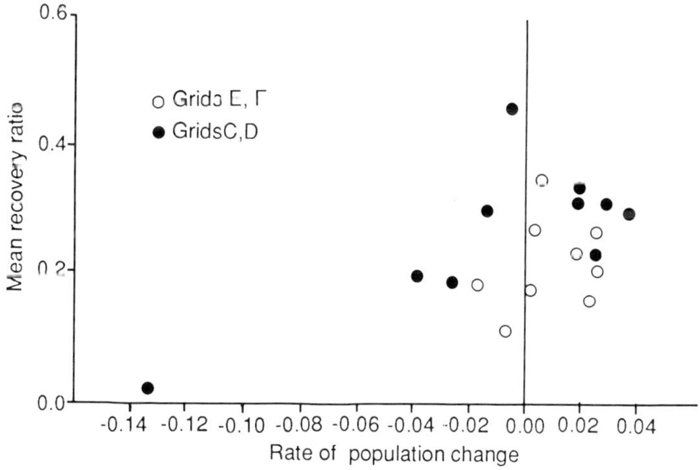

Figure 7.7 Relationship between the mean recovery ratio and rate of population growth per week in *Microtus townsendii* near Vancouver, Canada. Data grouped over summer and winter periods, 1971–75. $r = 0.25$, $n = 19$.

there is a suggestion of a positive trend. Stenseth (1983), following Krebs *et al.* (1976), reported a positive relationship for 1971–73 data; this now seems statistically non-significant.

In conclusion, *M. townsendii* does not fit the generalizations of Gaines and McClenaghan (1980), that microtine rodents typically show more presaturation dispersal of the type predicted by the genetic-behavioural hypothesis (cf. Figure 7.2). *Microtus townsendii* populations are typically characterized by saturation dispersal (Beacham, 1980). However, since *M. townsendii* populations do not show 3–4 year cycles, it is suggested that the Gaines and McClenaghan (1980) conclusion may still apply to cyclic microtine populations.

7.3.3 Assumption 3

There is a general consensus that dispersing voles will not be a random sample of the population with respect to age, sex, genotype, physiology, or behaviour (Gaines and McClenaghan, 1980). Almost all theories about dispersal make these general predictions but none are specific so that experimental tests are impossible. The field has thus become a neutral, observationally oriented description of the attributes of dispersers. There are two overwhelming problems: first, to define an individual as a disperser; and second, to measure some relevant attributes of the individuals dispersing as well as some 'control' group of individuals. It is suggested that these descriptive studies are no longer useful and it would be better to shift the emphasis to the experimental analysis of behaviour as carried out by behavioural ecologists in the field. Verner and Getz (1985) also reach this conclusion. There are now enough good examples to show how this can be done (e.g. Madison, 1980; Sherman, 1981). The important point is to concentrate on the behavioural traits that affect the fitness of an individual. These traits may well be different for males and females, and we need to know if the most successful traits change with a population's density.

There is one overarching problem that also needs to be addressed: the heritability of dispersal tendency. All of the evolutionary models of dispersal assume that there is some genetic variance for dispersal tendency in field populations (Stenseth, 1983). Whether such genetic variance still exists in local populations is critical for matching these models to field situations. Gaines and McClenaghan (1980) have reviewed the progress that has been made in attacking this question. There is a great need to bring together the more sophisticated techniques of biometrical genetics and the field techniques of kinship recognition to answer these questions.

The suggestion that dispersal is a critical factor affecting the population dynamics of microtine rodents has received wide support in a variety of studies. However, the detailed mechanisms by which dispersal acts are far from clear. Dispersal is not density-dependent in the cyclic species studied, and there is thus a great contrast between red grouse (*Lagopus lagopus scoticus*), in which the cyclic decline is associated with more dispersal (Moss and Watson, 1985), and cyclic voles. Manipulating immigration and emigration would seem to be the most powerful approaches we now have for studying the demographic aspects of dispersal. However, if dispersal is as critical as is thought, it must operate through changes in the social system of these populations, perhaps in the way suggested by Stenseth (1983). The key to future understanding of population cycles is likely to be obtained by behavioural – ecological studies on social organization to determine how immigrants and emigrants are fitted into the social fabric of vole and lemming societies. For the immediate future five significant questions on dispersal need to be answered.

1. Do cyclic lemming populations show the fence effect?
2. Is there a high amount of genetic variance for dispersal tendency in both cyclic and non-cyclic populations?
3. How does dispersal affect social structure in cyclic and non-cyclic populations?
4. How frequently is dispersal successful at different population densities?
5. Where are the demographic refugia for cyclic populations, and is dispersal from these refugia continuous or episodic?

If we can answer one or two of these questions during the next 10 years, we will have progressed significantly along the road to understanding the complex linkages between dispersal and demography.

ACKNOWLEDGEMENTS

I thank my students and colleagues over the years who have assisted me in this small rodent research, particularly Judy Myers, Mike Gaines, Bob Tamarin, Barry Keller, Rudy Boonstra, Terry Beacham, Mary Taitt, Tom Sullivan, and Dennis Chitty. I thank an anonymous referee for suggesting Figure 1.

REFERENCES

Abramsky, Z. and Tracy, C.R. (1979) Population biology of a 'noncycling' population of prairie voles and a hypothesis on the role of migration in regulating microtine cycles. *Ecology*, **60**, 349–61.

Beacham, T.D. (1980) Dispersal during population fluctuations of the vole, *Microtus townsendii. Journal of Animal Ecology*, **49**, 867–77.

Beacham, T.D. (1981) Some demographic aspects of dispersers in fluctuating populations of the vole *Microtus townsendii. Oikos*, **36**, 273–80.

Boonstra, R. and Krebs, C.J. (1977) A fencing experiment on a high-density population of *Microtus townsendii. Canadian Journal of Zoology*, **55**, 1166–75.

Chitty, D. and Southern, H.N. (eds) (1954) *Control of Rats and Mice*, Clarendon Press, Oxford.

Cockburn, A. (1985) Does dispersal increase as populations expand? *Oikos*, **44**, 367–8.

Desy, E.A. and Thompson, C.F. (1983) Effects of supplemental food on a *Microtus pennsylvanicus* population in central Illinois. *Journal of Animal Ecology*, **52**, 127–40.

Dobson, F.S. (1981) An experimental examination of an artificial dispersal sink. *Journal of Mammalogy*, **62**, 74–81.

Elton, C. (1942) *Voles, Mice and Lemmings*, Clarendon Press, Oxford.

Finerty, J.P. (1980) *The Population Ecology of Cycles in Small Mammals*, Yale University Press, New Haven.

Ford, R.G. and Pitelka, F.A. (1984) Resource limitation in populations of the California vole. *Ecology*, **65**, 122–36.

Gaines, M.S. and McClenaghan, L.R. (1980) Dispersal in small mammals. *Annual Review of Ecology and Systematics*, **11**, 163–96.

Gaines, M.S., Vivas, A.M. and Baker, C.L. (1979) An experimental analysis of dispersal in fluctuating vole populations: demographic parameters. *Ecology*, **60**, 814–28.

Hansson, L. and Henttonen, H. (1985) Regional differences in cyclicity and reproduction in *Clethrionomys* species: are they related? *Annales Zoologica Fennici*, **22**, 277–88.

Henttonen, H., McGuire, A.D. and Hansson, L. (1985) Comparisons of amplitudes and frequencies (spectral analyses) of density variation in long-term data sets of *Clethrionomys* species. *Annales Zoologica Fennici*, **22**, 221–7.

Hilborn, R. and Krebs, C.J. (1976) Fates of disappearing individuals in fluctuating populations of *Microtus townsendii. Canadian Journal of Zoology*, **54**, 1507–18.

Hurlbert, S.H. (1984) Pseudoreplication and the design of ecological field experiments. *Ecological Monographs*, **54**, 187–211.

Krebs, C.J. and Boonstra, R. (1978) Demography of the spring decline in populations of the vole *Microtus townsendii. Journal of Animal Ecology*, **47**, 1007–15.

Krebs, C.J., Keller, B.L. and Tamarin, R.H. (1969) *Microtus* population biology: demographic changes in fluctuating populations of *M. ochrogaster* and *M. pennsylvanicus* in southern Indiana. *Ecology*, **50**, 587–607.

Krebs, C.J., Wingate, I., LeDuc, J. *et al.* (1976) *Microtus* population biology: dispersal in fluctuating populations of *M. townsendii. Canadian Journal of Zoology*, **54**, 79–95.

Lidicker, W.Z. (1975) The role of dispersal in the demography of small mammals, in *Small Mammals: their productivity and population dynamics* (eds F.B. Golley, K. Petrusewicz and L. Ryszkowski), Cambridge University Press, London, pp. 103–28.

Lidicker, W.Z. Jr (1985) Dispersal, in *Biology of New World Microtus* (ed. R.H.

Tamarin), American Society of Mammalogists, Special Publication No. 8, pp. 420–54.

Madison, D.M. (1980) Space use and social structure in meadow voles, *Microtus pennsylvanicus. Behavioural Ecology Sociobiology*, 7, 65–71.

McClenaghan, L.R. and Gaines, M.S. (1976) Density-dependent dispersal in *Sigmodon*: a critique. *Journal of Mammalogy*, **57**, 758–9.

McMahon, T.E. and Tash, J.C. (1988) Experimental analysis of the role of emigration in population regulation of desert pupfish. *Ecology*, **63**, 1871–83.

Moss, R. and Watson, A. (1985) Adaptive value of spacing behaviour in population cycles of red grouse and other animals, in *Behavioural Ecology* (eds R.M. Sibley and R.H. Smith), Blackwell, Oxford, pp. 275–94.

Myers, J. and Krebs, C. (1971) Genetic, behavioural, and reproductive attributes of dispersing field voles *Microtus pennsylvanicus* and *Microtus ochrogaster. Ecological Monographs*, **41**, 53–78.

Nabagło, L. (1981) Demographic processes in a confined population of the common vole *Microtus arvalis. Acta Theriologica*, **26**, 163–83.

Pokki, J. (1981) Distribution, demography and dispersal of the field vole, *Microtus agrestis* (L.), in the Tvärminne archipelago, Finland. *Acta Zoologica Fennica*, **164**, 1–48.

Sherman, P.W. (1981) Kinship, demography, and Belding's ground squirrel nepotism. *Behavioural Ecology and Sociobiology*, **8**, 251–9.

Stenseth, N.C. (1983) Causes and consequences of dispersal in small mammals, in *The Ecology of Animal Movement* (eds I.R. Swingland and P.J. Greenwood), Clarendon Press, Oxford, pp. 63–101.

Taitt, M.J. and Krebs, C.J. (1985) Population dynamics and cycles, in *Biology of New World Microtus* (ed. R.H. Tamarin), American Society of Mammalogists, Special Publication No. 8, pp. 567–620.

Tamarin, R.H., Sheridan, M. and Levy, C.K. (1983) Determining matrilineal kinship in natural populations of rodents using radionuclides. *Canadian Journal of Zoology*, **61**, 271–4.

Tamarin, R.H., Reich, L.M. and Moyer, C.A. (1984) Meadow vole cycles within fences. *Canadian Journal of Zoology*, **62**, 1796–804.

Verner, L. and Getz, L.L. (1985) Significance of dispersal in fluctuating populations of *Microtus ochrogaster* and *M. pennsylvanicus. Journal of Mammalogy*, **66**, 338–47.

Warkowska-Dratnal, H. and Stenseth, N.C. (1985) Dispersal and the microtine cycle: comparison of two hypotheses. *Oecologia*, **65**, 468–77.

PART IV

Applications

While this book is primarily devoted to the theoretical and empirical substrates basic to the study of dispersal, we felt it would be wise to include at least one chapter that focused on the applied or practical aspects of the field. Dispersal is fundamentally important to such applied fields as epidemiology, wildlife management, pest control, and conservation biology. Lennart Hansson has had extensive experience in both basic and applied aspects of small mammal ecology. His contribution here (Chapter 8) reflects a pervasive and rich perspective. It takes two approaches: (1) he uses the concepts and ideas generated by theoretical and non-applied research to address applied problems, specifically rodent pest management and conservation; (2) he suggests how applied research can provide challenging opportunities for addressing issues in basic research on dispersal. In both these respects, Hansson's chapter makes important contributions. For example, he calls attention to such topics as how the seasonal dynamics of dispersal affects pest management and conservation, why an understanding of exploratory excursions may be a critically important but neglected subject, why it is important to develop techniques for objectively measuring the quality of habitat patches, and how a landscape perspective may be needed for both basic and applied questions. Overall one comes away with the impression that, at least in this field, there is no meaningful borderline between applied and basic science. In this perspective, we concur enthusiastically.

Hansson's chapter also emphasizes that small mammals seem to be exceptionally good model organisms for experimental work in conservation biology, as well as other related fields. This is a specific example of a general theme to which this book is dedicated. In making this point, Hansson lends support to the use of small mammals in the emerging discipline of landscape ecology. There seems no doubt that dispersal will be the premier process of interest as our understanding at this level of complexity evolves.

Part IV completes our substantive overview of theory and facts as it relates to dispersal in small mammals. Remaining chapters cover the comments, reassessments, prognoses, and techniques for where we have been and where we hope to go.

– 8

Small mammal dispersal in pest management and conservation

L. HANSSON

8.1 INTRODUCTION

Applied ecology deals with damage to crop plants and their products, prevention of outbreaks of zoonotic diseases, exploitation of game species, protection of endangered species, etc. Most of these aspects are important with regard to small mammals although few such species are regarded as game animals. Dispersal has usually not been considered as an important factor in the management of small mammals, but I will here try to show that it is. Specific problems across the spectrum of population eradication to population conservation are addressed. Knowledge is lacking for important aspects of dispersals and I therefore endeavour to establish what aspect of dispersal need further study, both for specific types of problems and generally in applied ecology.

It is usually assumed that applied ecology benefits from general or theoretical ecology in a unidirectional way. However, I hope here also to be able to demonstrate that pure ecology may profit by its contact with the real world's practical problems!

8.2 POPULATION ERADICATION

Personnel in charge of rodent control often attempt to eradicate pest populations completely, especially commensal rats and mice in buildings (e.g. Brooks, 1973) that can form fairly distinct and semi-isolated populations. Eradication programmes have also been directed against free-living species causing damage in agriculture, horticulture or forestry (e.g. Myllymäki, 1979). Immigration rates vary considerably between these two situation; in urban areas it seems possible to obtain 'rat-free towns' by the use of permanent bait stations (Telle, 1969) while control is much more difficult in more open populations. Animals usually begin arriving at treated open sites within 2–3 weeks after rodenticide treatment, but up to six months may elapse before densities have again attained pre-treatment levels or levels found in untreated areas (Spencer, 1954; Gashwiler, 1969;

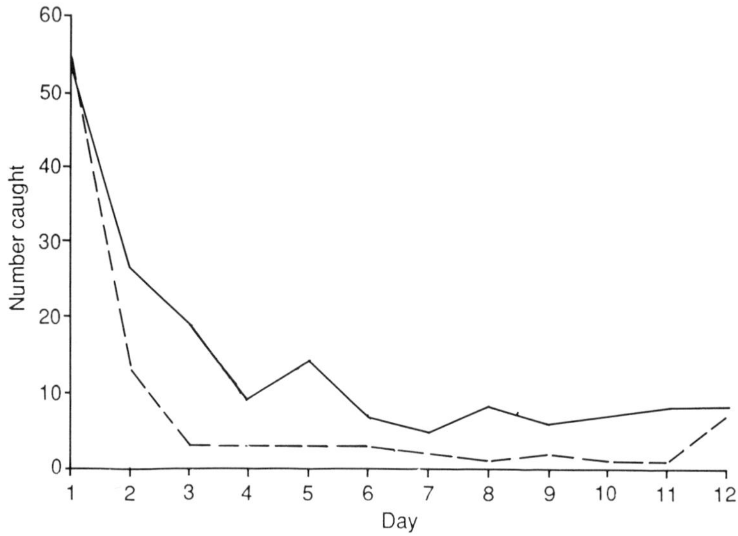

Figure 8.1 Removal from and reinvasion to a 5 ha reforestation area in southern Sweden continuously trapped with 512 snap traps. The removal plot was surrounded by areas of similar habitat. Solid line, *Microtus aggrestis*; dashed line, *Clethrionomys glareolus.*

Pank and Matschke, 1972; Radvanyi, 1974; Sullivan, 1979). The numbers of other small mammal species, some of them noxious, may increase in treated areas (Sullivan, 1979). Rebaiting is possible in some localities (Yeatman, 1969; Gashwiler, 1969) but is largely ineffective in areas with long-lasting snow cover and a negative economic return can often be expected. Permanent bait stations have been developed also for field (e.g. reforestation) conditions; Radvanyi (1974) reported that this strategy halted the invasion of animals into the treated areas. However, the efficiency of this method will probably vary, depending on the population density of the target species in the surrounding landscape.

Apparently, no thorough applied studies have been made on the factors influencing immigration into areas depopulated by trapping or rodenticide treatment. However, at least in part, this same problem applies to the standard estimation of rodent densities by removal trapping [see e.g. Grodziński *et al.* (1966) and Ryszkowski (1969) for studies during the International Biological Programme, and Stenseth and Hansson (1979) for general removal estimates]. In these cases the probability of capture was fairly constant during the first 3–4 days of removal capture, followed by a declining probability (Figure 8.1). This second phase is mainly caused by ingress of animals whose main activity areas lie outside the study (removal) plot, but who made temporary sallies into it (Holisova, 1968; Ryszkowski, 1971). However, animals with long residency in a particular area seemed less prone to move into removal areas (Andrzejewski and Wrocławek, 1962). The extent to which excursions are undertaken varies considerably between population categories and seasons (Wolton and Flowerdew, 1985). Settling in removal areas was much more rapid in autumn, with generally high density levels, than in spring with low numbers (Andrzejewski and Wrocławek, 1962). However, these observations come from representative study plots surrounded by areas of similar habitat. Eradication plots usually differ to some extent from their surroundings, although such differences are considerably less under field conditions than in urban areas. Thus, the amount of deviation in the demographic structure of the invaders should be directly related to the amount of difference between the treated and the surrounding habitats. This will be exemplified in sections 8.3 and 8.5.

Total eradication is rarely possible except in certain closed environments. All attempts to temporarily remove small mammal populations should be based on good knowledge of the probability of reinvasion. Damage in agriculture and forestry is usually most severe at the late increase or peak densities of cyclic populations. If presaturation and saturation dispersal are both important (Lidicker, 1975), eradication should have little chance of success in open and continuous treatment plots.

8.3 DAMAGE RELATED TO DISPERSAL FROM PERMANENT HABITATS

Microtine rodents avoid moving over exposed areas, probably due to the risk from predation. Thus, orchards (including seed orchards for the production of forestry seeds) are preferably kept free from sheltering vegetation by repeated soil treatment (e.g. Bärring, 1963; Davis, 1977). However, good protection only seems possible in areas where the soil is exposed almost all the year round, e.g. in Denmark (Bang, 1975) and Virginia (Byers *et al.*, 1976). In areas with much snow, considerable damage (e.g. debarking of seedlings and grafts by *Microtus agrestis*) occurs under the snow even in well-treated orchards (Larsson, 1975; Myllymäki, 1979). Obviously, animals can disperse in large numbers under a snow cover, as observed by Hansson (1981) while trapping under snow. They settle temporarily at very localized food sources; for example, voles will utilize single seedlings or trees with suitable bark. According to the amount of bark consumed, the small rodents appear to remain at such sites for weeks or even months. Movements under snow seem easiest at snow thicknesses between 10 and 30 cm (Hansson, 1986).

North and west European water voles (*Arvicola terrestris*) usually occur in reed-belts of swamps, lakes and rivers during summer but move to easily excavated mineral soil in autumn, remaining there for the winter where they live an almost completely fossorial life. During this time they often cause considerable damage by consuming bulbs and tree roots, e.g. in orchards (Ophof, 1970). It should be emphasized that this damage is related to their regular migration between two seasonal habitats.

The European common vole (*Microtus arvalis*) causes great damage to cereals and alfalfa, but its appearance within annually ploughed fields is usually a result of invasions originating in areas with perennial grass such as drainage ditches, verges, etc. (Frank, 1956; van Wijngaarden, 1957). Such conditions have to be taken into account in models forecasting outbreaks of this species (Spitz, 1985). The populations reach much higher densities, although usually temporarily, in the reception habitats than in the original donor habitats [see Hansson (1977) for terminology]. Invasions are facilitated by proximity of the two types of habitats, i.e. a limited distance of transition habitats. Settling is probably released both by shelter and food stimuli and by access to easily excavated soil for this partly fossorial species. Recent large-scale agriculture in central Europe, with few parcels of unused grassland, is thought to have lessened the risks for densities of the pest species reaching economically injurious levels (Myllymäki, 1979). An analogous situation was observed in *Peromyscus maniculatus* invading North American clearcuts from adjoining conifer forests (Martell, 1983). The immigrating animals were young and predominantly male; they reproduced intensively and within 3 years they had reached densities considerably higher than those in the original habitats. *Apodemus sylvaticus*, an ecological equivalent in Europe, is also especially abundant on southern Scandinavian clearcuts (Hansson, 1978) and may show similar behaviour.

Rats (e.g. *Rattus exulans*) invade sugar-cane plantations in Hawaii from nearby gulches (Hood *et al.*, 1970) and show a consecutive exploration–dispersal–settlement behaviour in this temporary but excellent habitat type. This behaviour is probably an adaptation to maximize short-term reproductive output (Chapter 4). House mice (*Mus musculus*) living in a mixture of agricultural crop lands show temporary residency and rapidly invade new suitable crops at phenological habitat changes (Zejda, 1975; Stickel, 1979). Similar patterns have been observed in the tropical rodent pest *Arvicanthis niloticus* (Poulet, 1985).

From a practical standpoint, it would be helpful if we could estimate the risks for invasion based on the proximity of various habitat types and determine the minimum requirements (threshold values) in food, shelter, etc., necessary to induce and sustain settling and breeding for pest species.

8.4 PEST OUTBREAKS

Small mammal populations fluctuate strongly in general, and cyclic populations occur in certain species and in certain regions (Hansson and Henttonen, 1985), e.g. in microtines generally, and at least in snowshoe hare (*Lepus americanus*) among the lagomorphs. Other species may show occasional population outbreaks, e.g. house mice.

Showshoe hares are cyclic in the northern parts of their range in North America (e.g. Keith, 1983) where habitats are relatively homogeneous; in contrast, fairly stable populations appear in southern areas (Dolbeer and Clark, 1975). In these southern areas, production (donor) habitats of the hare occur as islands in a matrix of unsuitable open environments. Generalist predators are thought to take a large proportion of the juvenile hares dispersing through this matrix (Dolbeer and Clark, 1975; Wolff, 1980), thereby preventing the build-up of very high numbers and cyclic fluctuations. In northern areas the animals disperse within fairly secure habitats and are possibly exposed to fewer generalist predators; consequently their numbers rise accordingly. Similar mechanisms were proposed for the European rabbit (*Oryctolagus cuniculus*) in Australia and New Zealand (Myers and Poole, 1963) and for the American black-tailed jackrabbit (*Lepus californicus*; Wagner, 1981).

Within the more homogeneous northern environment there are also differences in suitability between habitats causing the movement of snowshoe hares into thick, brushy areas during the early decline phase when predation is generally intense (Wolff, 1981; Keith *et al.*, 1984). Thus, the availability of protective shelter and the levels of predator pressure are probably the most important factors determining the fate of dispersing snowshoe hares.

House mice are famous for their outbreak numbers in field populations under certain environmental conditions (Crowcroft, 1966). Such outbreaks are common in semi-arid areas in Australia where a comprehensive study on the background of these outbreaks has been made (Newsome, 1969a,b, 1970). House mouse populations remain fairly stable in reed-beds along rivers in South Australia but many animals, especially young individuals and males, disperse during summer to neighbouring wheat fields where food supplies are larger. These dispersing animals are able to settle and reproduce only after heavy rains, which soften the clayey soil and make it possible for mice to burrow in soil cracks. The burrows protect the mice from excessive heat during hot periods which follow. Experimental studies (Newsome, 1970) verified the importance of soil conditions. Later studies (Redhead *et al.*, 1985) have indicated that physical conditions in certain permanent habitats (e.g. rubble piles and stubble fields) usually are improved by autumn and winter rains during the year prior to an outbreak on agricultural land. However, the length of time since this precipitation effect may vary between 3 months and one year for environments with different landscape heterogeneity. In certain more hostile areas, induced donor habitats must be created before an outbreak can occur, and thus temporary weather conditions may adversely affect populations for a series of years (Singleton, 1989).

These two studies demonstrate the importance of mortality amongst dispersing animals for the stability of the populations. However, the mortality agents differed between and within species and the relative importance of such factors deserves more consideration.

8.5 SETTLING IN HUMAN HABITATIONS

Rats (e.g. *Rattus norvegicus*) and mice (*M. musculus*) are well known for the damage they cause in and around houses. However, many individuals of both species live away from buildings during the summer, at least in temperate and subtropical areas, and may recolonize houses, stables, etc., in autumn (Leslie *et al.*, 1952; Rowe *et al.*, 1963; Jensen, 1966; Reimer and Petras, 1968; Twigg, 1975; Rowe and Swinney, 1977). Farmland, river valleys and verges form the main summer habitats. In earlier times, corn ricks functioned as winter refuges, and it has even been claimed that ricks support a great part of the overwintering field populations of *M. musculus* (Rowe *et al.*, 1963). The extent to which mice migrate back to houses in autumn and settle is unknown; it has been documented, however, that the majority of the immigrating *M. musculus* are males (Rowe *et al.*, 1963).

In the past, the Norway rat and house mouse were common pests in human habitations, at least for the winter, but nowadays they are often eradicated soon after arrival in many areas. Anderson (1970) stated that mice in buildings and other man-made structures formed permanent populations, sending emigrants to surrounding temporary habitats while showing only low levels of reimmigration. However, in regions with high sanitary standards there are hardly any house mice left inside the houses all the year and they are obviously replaced by 'colonists' from other places, e.g. fields. From the concentrated immigration in autumn, observed for example by Jensen (1966), there seems to be a large turnover after weaning of the last litter of the breeding season. Thus the pattern of spatial dynamics of mouse and rat populations seem to change along with changes in the designs of man-made structures and improvement in sanitary standards.

A large array of other species also invade human habitations but the lengths of their stays are probably shorter. Only immigration, not settlement, has been studied in these other species. Besides *M. musculus*, *Apodemus flavicollis* (Jensen, 1966) and *A. sylvaticus* (Berry and Tricker, 1969; Hansson, 1981) have been reported invading houses in western Europe; in eastern Europe, *Apodemus agrarius* (Walkowa *et al.*, 1982), *A. flavicollis* and *M. arvalis* (Vlasak and Porkert, 1973) are often unwanted house guests, while *Clethrionomys glareolus* (Hansson and Zejda, 1977), *Clethrionomys rutilus* (Lundquist *et al.*, 1973) and *Microtus oeconomus* (Tast, 1968) have been observed inside houses in northern Europe during the winter. In general, granivorous species – nutritionally similar to rats and house mice – enter houses in areas with milder climates, while folivorous species are more common invaders in northern areas where there are few real granivores. These folivores occur mainly in barns while *Apodemus* spp. often appear in human dwellings. Small numbers of other species, even insectivores such as *Sorex araneus* (Jensen, 1966; Vlasak and Porkert, 1973; Hansson, 1981) can appear indoors.

Immigration to human habitations in most of the rodent species occurs mainly in autumn (Figure 8.2) at the end of the reproductive season (Jensen, 1966; Vlasak and Porkert, 1973; Hansson, 1981). Soricids start immigrating earlier in summer, possibly because they terminate breeding earlier. When the field populations decline in late autumn–winter immigration also declines sharply except for shrews (Vlasak and Porkert, 1973). Immature individuals predominate among the immigrating animals.

To prevent or forecast immigration into households, it is necessary to know the seasonal patterns of dispersal and to determine the relationship between dispersal and density. Factors releasing a settling response should be examined for species which are not general commensals, i.e. species in genera other than *Rattus* and *Mus*.

8.6 DISEASE TRANSMISSION

Many small mammal species harbour diseases transferable to man (zoonoses) and his livestock. Such diseases are usually thought to appear endemically in certain 'foci of infection in which the infectious agent, its host and its vectors exist as a part of a biotic community' (Cox, 1979). However, it is admitted that such foci may be either of a restricted or a diffuse type (Rosicky, 1967). Certainly, some zoonotic diseases, e.g. tularemia, show a wide distribution and their outbreaks are related to the dynamics of hare and small rodent populations (Pearson, 1975; Hörnfeldt, 1978). Transmission to humans is generally supposed to occur either when people enter localized foci or when rats or house mice immigrate to human habitations (Kucheruk, 1965). Small mammal movements, e.g. dispersal (often called 'emigration' in this context), are considered to be generally important for the spread of such diseases (Rosickly, 1965; WHO, 1974).

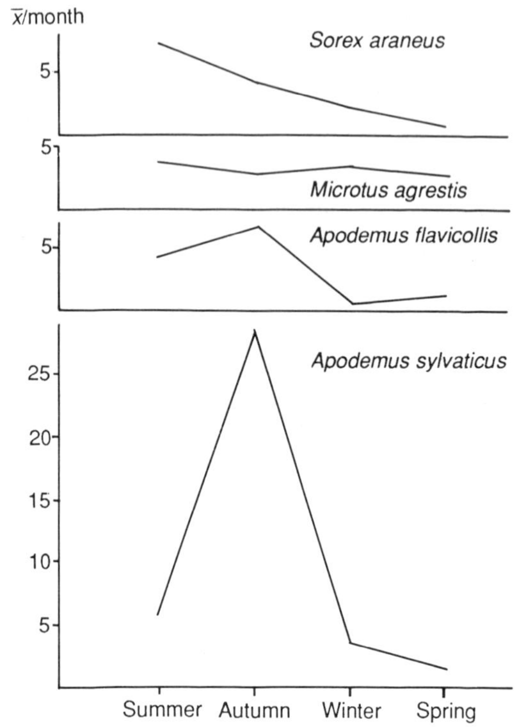

Figure 8.2 Seasonal invasion by certain species of small mammals into human habitations in southern Sweden. The year has been divided into summer (main breeding season, June–August), autumn (September–November), winter (December–February) and spring (March–May). Recalculated from a 9-year study reported in Hansson (1981).

Naumov (1965) argued that constant foci of infection cannot be maintained in large, continuously distributed populations since they should be easily swept away by epizootics. Instead, he proposed that constant foci occur in aggregated populations with permanent nesting sites, which may locally become extinct due to disease outbreaks but are latter repopulated by immigrants which have travelled long distances from other animal settlements. The latter situation should apply to, for example, gerbils harbouring bubonic plague.

In view of the great number of species actually entering human dwellings, these concepts might be incomplete. It is well known (Niklasson and LeDuc, 1984), for example, that *C. glareolus* is the principal vector of the renal disease *Nephropathia epidemica* (one type of haemorrhagic fever) and that these voles invade huts, barns and villages during peak vole years in northern Fennoscandia (Hansson and Zejda, 1977). During such periods there are also many cases of disease among humans (Nyström, 1977; Niklasson and LeDuc, 1987). Thus vole dispersal may well be one causative factor.

Epidemiological studies of small mammal-transmitted diseases should include information on the relative health of dispersing and resident animals and estimates of the rate of establishment of a disease at the settlement of dispersing animals.

8.7 MODELS IN CONSERVATION THEORY

Few species of small mammals are objects of conservation efforts as most of them are numerous and the individuals easily replaced due to their rapid population turnover. Some small mammal species thought of as endangered in Europe (Smit and van Wijngaarden, 1981) are hibernators, for example species in the families Gliridae and Zapodidae (e.g. *Muscardinus avellanarius* and *Sicista betulina*). Hibernators are characterized by especially low reproductive rates (Lord, 1960). However, the study of dispersal in common small mammals has found an important place in conservation theory, providing both a conceptual basis and test subjects.

Island biogeography (MacArthur and Wilson, 1967) has, in a general way, demonstrated the vulnerability of isolated populations. However, continuous immigration from larger source populations may enhance the demographic and genetic integrity of such small populations. This 'rescue effect' (Brown and Kodric-Brown, 1977) has been implied to have important effects on the numerical stability of populations of small mammals in a woodland mosaic in a farmland landscape (Henderson *et al.*, 1985). These findings were then extrapolated so as to apply to management of endangered species.

Wegner and Merriam (1979) and Merriam (1984) have discussed the still more profound role of 'connectivity' for patchily distributed populations. Connectivity is a species-specific measure of the interconnections among functionally related ecological elements of a landscape so that individuals of a certain species can move among them. It is defined both by environmental features and by the behaviour of a specific species. Small populations often go extinct and with isolation such extinctions can be permanent. However, connectivity increases numerical stability in such populations.

Merriam (1984) and Henderson *et al.* (1985) showed that the dispersal of white-footed mice (*Peromyscus leucopus*) and of chipmunks (*Tamias striatus*) in fence rows prevented extinctions in isolated woodlot populations of these species in a farmland matrix. Here the fence rows made up the crucial part of the connectivity. A digital model simulation (Merriam,

1984) generalized these results. Connectivity has important effects on the abundance, strength of fluctuations and probability of extinction for both individual populations and for the species as a whole over an entire landscape (Fahrig and Merriam, 1985). Thus, a larger environmental mosaic has to be considered before such a system is understandable and before small semi-isolated populations can be effectively managed.

The model can be applied directly to one species of small rodent in the World Red Data Book (IUCN, 1972), viz. the salt-marsh harvest mouse (*Reithrodontomys raviventris*). It occurs only in salt-marshes of the San Francisco Bay region and its habitat is becoming increasingly disjointed by urban and industrial development (Shellhammer, 1972, 1989). Small populations become isolated and there are obvious risks for inbreeding and extinction. Higher connectivity may be needed for this species as for several other small mammals on the Red list (and for rare animals at large).

These effects of connectivity are supposed to be very general and applicable to most species subjected to conservation efforts. Since the theoretical framework was based on the behaviour of small secretive mammals strongly dependent on shelter, the conclusions may only be generalized for similar species. However, other species move through a variety of habitats, including very open areas, thereby increasing connectivity (Baudry, 1984). The various species of small rodents differ with respect to this type of dispersal behaviour, viz. from the very secretive *Microtus* voles to the bolder mice such as *Apodemus* or *Peromyscus* (Hansson, 1987). Snails and birds may make up the end-points of this gradient! Thus, this model could be expanded to cover other types of animals, the populations of which should show different degrees of connectivity in the same landscape.

8.8 DISCUSSION AND GENERALIZATIONS

There are certain topics related to dispersal which are generally important in a practical context but are not studied very intensely in basic ecology. One such topic is the relationship between exploratory movements and dispersal proper. Explorations (or even more generally, enlargements of ranges) seem to pass over into dispersal under certain environmental conditions. However, these conditions have never been subject to serious examination. Experimental field studies on marked animals at various population densities, population sex ratios, food supply levels, etc., in adjacent areas are badly needed. The factors releasing settlement behaviour in dispersing animals also need to be determined. Such information could probably be obtained by offering mates, food, nests, etc., to natural populations and to manipulated populations with decreased or increased densities. Additionally, we need to determine the relative importance of various mortality factors among dispersing (and exploring) and resident animals. Predators have been reported to cause high mortality to dispersing animals in experiments (Metzgar, 1976; Ambrose, 1972; Smirin, 1975) but few good field data are available.

Some information could be obtained by measuring predation rates on the same small mammal species in donor, transition and reception habitats (cf. Hansson, 1977). However, the demographic conditions in such habitats need to be characterized carefully; little can be gained by giving only general descriptions of the landscape heterogeneity.

This brings us to the importance of landscape heterogeneity for population stability and damage related to high densities. Three important topics can be distinguished, at least two of which are also essential for the basic understanding of dispersal and its effects on population dynamics. There has recently been a general tendency to positively associate the degree of population stability to landscape heterogeneity (e.g. Taitt and Krebs, 1981). This association has been theoretically justified by Stenseth (1980) regarding predator–prey relationships and certain other aspects. However, environmental heterogeneities do not constitute fixed patterns and generalizations seem premature. The suitability and productivity of various habitats can instead be discussed by analogy with the fundamental and realized niches. There may be a fundamental suitability which can be measured, at least conceptually, in enclosures with, for example, predatory and competing species excluded. The basic measure could be the number of young weaned per unit area and unit time. If there are seasonal or annual changes within the habitat, e.g. in seed setting or regular or irregular flooding, the measurements should cover different seasons or years. Adequately located dispersal sinks (possibly traps along the perimeter of the enclosures) could estimate dispersal pressures. The realized suitability is affected by dispersal from other habitats and by the distribution and dynamics of natural mortality agents. The latter may be due to immigration of competitors (e.g. *A. terrestris* into *M. agrestis* habitats; Myllymäki, 1977) or to predators tracking spots of high prey densities with some delay (possibly small mustelids in cyclic vole populations; Hansson, 1984). Thus, the importance of landscape heterogeneity cannot be estimated without a fair knowledge of external influences on the habitat-specific population dynamics of the species examined. In future, certain generalizations may be possible for characteristic landscapes, but just knowing the proportions of open and sheltered habitats, as often discussed, will help little in predicting the risks for population outbreaks and the possible ensuing economic damage.

Another important question from both the basic and applied viewpoints regards the existence of survival and colonization habitats. These concepts were first developed by Russian authors (e.g. Naumov, 1965) but further elaborated upon by Anderson (1970). The latter model strictly distinguishes between survival habitats, with a fairly constant number of interbreeding animals, and colonization habitats, with widely, often seasonally, fluctuating densities. The majority of the colonizing animals are removed by various mortality agents before breeding and the breeding animals in the survival habitats are assumed to be replaced by the last cohort of young born there in late summer – early autumn. Few, if any, animals move back from colonization to survival habitats. These patterns are based on observations of certain desert rodents (Naumov, 1965) and house mice living in farm environments with less good sanitary standards (Anderson, 1970); they cannot be extrapolated to include small mammals in general.

This might be made clear by considering three cases of population outbreaks or cyclic peaks in small rodents: house mice in Australia, snowshoe hares in North America, and microtine voles in northern Fennoscandia. The mice (Newsome, 1969a,b) were dispersing from fairly stable populations in reed-beds to wheat fields with abundant food; however, they only settled and increased to plague conditions during periods characterized by a specific combination of weather conditions. The relative proportions of the various habitats would not predict the probability for an outbreak but the concept of survival/colonization habitats was very applicable.

Predators were assumed to remove dispersing snowshoe hares in less suitable habitats and only fairly homogenous aspen habitats supported outbreak populations (Wolff, 1980). However, even in aspen habitats populations declined to very low levels during bottom years and much dispersal there seemed to be caused by a general food shortage (Boutin *et al*., 1985). Thus, contrary to the Australian house mouse story, snowshoe hare populations did not conform to either the Anderson (1970) model or to any assumption of population stability inherent to a particular habitat.

The same comment can be made for the Fennoscandian voles: even in the most preferred habitats they decline to extremely low population levels after peak years (e.g. Hansson and Henttonen, 1985). Dispersing voles move securely under thick snow cover in most habitats but refrain from, or are removed from, open habitats during summer. Thus, the geographic location determines the relative suitability of the various habitats for the vole species. Predation may be fundamentally important for both hares and voles, but behavioural adaptations toward this predictable mortality agent obviously influence habitat choice by voles in winter. As there are both stable and cyclic hare and vole populations and cyclic populations decline even in preferred habitats, dispersing animals are not secure even in the best habitats during the late peak–decline phase, i.e. the dynamics of the mortality agent(s) have to be considered.

These considerations lead to a third, mainly practical, problem with implications for general population dynamics. The importance of landscape homogeneity and heterogeneity, i.e. monocultures and mixed crop lands, has been discussed widely for many pest animals, also from the point of view of rodent control or management (e.g. Myllymäki, 1979 Stenseth, 1981). No consensus has emerged, but this can be easily understood based on the present survey; species which are favoured (i.e. with low mortality rate) by habitat characteristics of the crop land may be most easily managed in heterogeneous areas. In contrast, those species with high mortality rates and only seasonal populations in crop lands will in the long run cause least trouble in extensive monocultures with no refuge. Certain deductions (Stenseth and Hansson, 1981) imply that in the former case most available resources for pest control ought to be spent on reducing immigration to crop lands where effective methods exist. Only when dispersal cannot be affected should we attempt manipulating birth and death rates, and then, only after understanding the demographic strategy of the pest species. Here, as in the case of disease transmission and connectivity, tests of such practically oriented ideas may influence basic understanding, but first the basic assumptions should be carefully evaluated as to their theoretical soundness. Through both theoretical and applied studies we might develop the means for preventing pest outbreaks and preserving endangered species. Simultaneously, we will come to better understand the dynamic effects of dispersal when related to a variety of ecological strategies.

8.9 CONCLUSIONS

Dispersal is a factor important in almost all aspects of applied ecology but immigration is usually more pertinent than emigration. Too little emphasis has been placed on immigration to depopulated areas, on movements (often together with disease organisms) to new habitats and to areas with extinct populations. The intensity and effects of dispersal will vary in these ecological settings and appropriate theoretical models should be developed and tested for a variety of demographic strategies.

ACKNOWLEDGEMENTS

I appreciate comments by Charles A. Brandt, William Z. Lidicker, Jr and Nils Chr. Stenseth.

REFERENCES

Ambrose, H.W. (1972) Effect of habitat familiarity and toe-clipping on rate of owl predation in *Microtus pennsylvanicus*. *Journal of Mammalogy*, **53**, 909–12.

Anderson, P.K. (1970) Ecological structure and gene flow in small mammals. *Symposium of the Zoological Society of London*, **26**, 299–325.

Andrzejewski, R. and Wrocławek, H. (1962) Settling by small rodents in a terrain in which catching out had been performed. *Acta Theriologica*, **6**, 257–74.

Bang, P. (1975) Damage by small mammals in Denmark. *Ecological Bulletin*, **19**, 13–15.

Bärring, U. (1963) Vole damage to pine and spruce seedlings in 1961–1962. *Svenska Skogsvårdsföreningens Tidskrift*, **61**, 1–40.

Baudry, J. (1984) Effects of landscape structure on biological communities: The case of hedgerow network landscape. *Proceedings of the First International Seminar on Methodology in Landscape Ecology, Research and Planning*, **1**, 55–65.

Berry, R.J. and Tricker, B.J.K. (1969) Competition and extinction: the mice of Foula, with notes on those of Fair Isle and St. Kilda. *Journal of Zoology*, **158**, 247–65.

Boutin, S., Gilbert, B.S., Krebs, C.J. *et al.* (1985) The role of dispersal in the population dynamics of snowshoe hares. *Canadian Journal of Zoology*, **63**, 106–15.

Brooks, J.E. (1973) A review of commensal rodents and their control. *Critical Review of Environmental Control*, **3**, 405–53.

Brown, J.H. and Kodric-Brown, A. (1977) Turnover rate in insular biogeography: effects of immigration on extinction. *Ecology*, **58**, 445–9.

Byers, R.E., Young, R.S. and Neely, R.D. (1976) Review of cultural and other control methods for reducing pine vole populations in apple orchards. *Proceedings of Vertebrate Pest Conferences*, **7**, 242–53.

Cox, F.E.G. (1979) Ecological importance of small mammals as reservoirs of disease, in *Ecology of Small Mammals* (ed. D.M. Stoddart), Chapman & Hall, London, pp. 213–38.

Crowcroft, P. (1966) *Mice all Over*, Foulis, London.

Davis, D.E. (1977) Advances in rodent control. *Zeitschrift für Angewandte Zoologie*, **64**, 193–211.

Dolbeer, R.A. and Clark, W.R. (1975) Population ecology of snowshoe hares in the central Rocky Mountains. *Journal of Wildlife Management*, **39**, 535–49.

Fahrig, L. and Merriam, G. (1985) Habitat patch connectivity and population survival. *Ecology*, **66**, 1762–8.

Frank, F. (1956) Grundlagen, Möglichkeiten und Methoden der Sanierung von Feldmausplagegebieten. *Nachrichtenblatt des Deutchen Pflanzenschutzdienstes*, **10**, 147–57.

Gashwiler, J.S. (1969) Deer mouse repopulation of a poisoned Douglas-fir clearcut. *Journal of Forestry*, **67**, 494–7.

Grodziński, W., Pucek, Z. and Ryszkowski, L. (1966) Estimation of rodent numbers by means of prebaiting and intensive removal. *Acta Theriologica*, **11**, 297–314.

Hansson, L. (1977) Spatial dynamics of field voles *Microtus agrestis* in heterogeneous landscapes. *Oikos*, **29**, 539–44.

Hansson, L. (1978) Small mammal abundance in relation to environmental variables in three Swedish forest phases. *Studia Forestalia Suecica*, **147**, 1–40.

Hansson, L. (1981) Immigration to human habitations by south Swedish small mammals. *Zeitschrift für Angewandte Zoologie*, **68**, 339–55.

Hansson, L. (1984) Predation as the factor causing extended low densities in microtine cycles. *Oikos*, **43**, 255–6.

Hansson, L. (1986) Vole snow trails and tunnels as density indicators. *Acta Theriologica*, **31**, 401–8.

Hansson, L. (1987) Dispersal routes of small mammals at an abandoned field in central Sweden. *Holarctic Ecology*, **10**, 154–9.

Hansson, L. and Henttonen, H. (1985) Gradients in density variations of small rodents: the importance of latitude and snow cover. *Oecologia*, **67**, 394–402.

Hansson, L. and Zejda, J. (1977) Plant damage by bank voles (*Clethrionomys glareolus*/Schreber) and related species in Europe. *EPPO Bulletin*, 7, 223–42.

Henderson, M.T., Merriam, G. and Wegner, J. (1985) Patchy environments and species survival: chipmunks in an agricultural mosaic. *Biological Conservation*, **31**, 95–123.

Holisova, V. (1968) Results of experimental baiting of small mammals with a marking bait. *Zoologicke Listy*, **17**, 311–25.

Hood, G.A., Nass, R.D. and Lindsay, G.D. (1970) The rat in Hawaiian sugarcane. *Proceedings of Vertebrate Pest Conferences*, **4**, 34–7.

Hörnfeldt, B. (1978) Synchronous population fluctuations in voles, small game, owls and tularemia in northern Sweden. *Oecologia*, **32**, 141–52.

IUCN (1972) Salt-marsh harvest mouse, *Reithrodontomys raviventris*. *Red Data Book*, **1**, 10.67.14.1.

Jensen, B. (1966) Mice indoors. *Flora og Fauna*, **73**, 130–6.

Keith, L.B. (1983) Role of food in hare population cycles. *Oikos*, **40**, 385–95.

Keith, L.B., Cary, J.R., Rongstad, O.J. and Brittingham, M.C. (1984) Demography and ecology of a declining snowshoe hare population. *Wildlife Monographs*, **90**, 1–43.

Kucheruk, V.V. (1965) Synanthropic rodents and their significance in the transmission of infections, in *Theoretical Questions of Natural Foci of Diseases* (eds B. Rosicky and K. Heyberger), Czechoslovak Academy of Sciences, Prague, pp. 353–67.

Larsson, T.B. (1975) Damage caused by small rodents in Sweden. *Ecological Bulletin*, **19**, 147–55.

Leslie, P.H., Venables, U.M. and Venables, L.S.V. (1952) The fertility and population structure of the brown rat (*Rattus norvegicus*) in corn ricks and some other habitats. *Proceedings of the Zoological Society of London*, **122**, 187–238.

Lidicker, W.Z. Jr (1975) The role of dispersal in the demography of small mammals, in *Small Mammals: their productivity and population dynamics* (eds F.B. Golley, K. Petrusewicz and L. Ryszkowski), Cambridge University Press, Cambridge, pp. 103–28.

Lord, R.D. (1960) Litter size and latitude in North American small mammals. *American Midland Naturalist*, **64**, 488–99.

Lundquist, L., Nilsson, A. and Hansson, L. (1973) Winter breeding observed in northern red-backed mouse. *Fauna*, **26**, 216–7.

MacArthur, R.H. and Wilson, E.O. (1967) *The Theory of Island Biogeography*, Princeton University Press, Princeton.

Martell, A.M. (1983) Demography of southern red-backed voles (*Clethrionomys gapperi*) and deer mice (*Peromyscus maniculatus*) after logging in north-central Ontario. *Canadian Journal of Zoology*, **61**, 958–69.

Merriam, G. (1984) Connectivity: a fundamental ecological characteristic of landscape pattern. *Proceedings of the First International Seminar on Methodology in Landscape Ecology, Research and Planning*, **1**, 5–15.

Metzgar, L.H. (1967) An experimental comparison of screech owl predation on resident and transient white-footed mice (*Peromyscus leucopus*). *Journal of Mammalogy*, **48**, 387–91.

Myers, K. and Poole, W.E. (1963) A study of the biology of the wild rabbit, *Oryctolagus cuniculus* (L.) in confined populations. IV. Population dynamics. *CSIRO Wildlife Research*, **8**, 166–203.

Myllymäki, A. (1977) Interaction between the field vole *Microtus agrestis* and its microtine competitors in Central Scandinavian populations. *Oikos*, **29**, 570–80.

Myllymäki, A. (1979) Importance of small mammals as pests in agriculture and stored products, in *Ecology of Small Mammals* (ed. D.M. Stoddart), Chapman & Hall, London, pp. 239–79.

Naumov, N.P. (1965) Microstructure and stability of natural foci of diseases, in *Theoretical Questions of Natural Foci of Diseases* (eds B. Rosicky and K. Heyberger), Czechoslovak Academy of Sciences, Prague, pp. 41–54.

Newsome, A.E. (1969a) A population study of house-mice temporarily inhabiting a south Australian wheat field. *Journal of Animal Ecology*, **38**, 341–59.

Newsome, A.E. (1969b) A population study of house-mice permanently inhabiting a reed-bed in south Australia. *Journal of Animal Ecology*, **38**, 361–77.

Newsome, A.E. (1970) An experimental attempt to produce a mouse plague. *Journal of Animal Ecology*, **39**, 299–312.

Niklasson, B. and LeDuc, J. (1984) Isolation of the *Nephropathia epidemica* agent in Sweden. *Lancet*, 1012–3.

Niklasson, B. and LeDuc, J. (1987) Epidemiology of *Nephropathia epidemica* in Sweden. *Journal of Infectious Disease*, **155**, 269–76.

Nyström, K. (1977) Incidence and prevalence of endemic benign (epidemic) nephropathy in AC county, Sweden, in relation to population density and prevalence of small rodents. *Acta Medica Scandinavica*, Suppl. 609.

Ophof, A.J. (1970) A modified 'fish-trap' for the control of the water vole (*Arvicola terrestris* L.). *EPPO Publications, Serie A*, **58**, 159–60.

Pank, L.F. and Matschke, G.H. (1972) Decline and reinvasion of deer mouse populations after baiting Douglas-fir clearcuts with 6-aminonicotinamide. *Journal of Forestry*, **70**, 678–80.

Pearson, A.D. (1975) Epidemiology of rodent tularemia in Norway and Sweden. *Ecological Bulletin*, **19**, 99–111.

Poulet, A.R. (1985) The ecological basis of forecasting rodent outbreaks in a Sahelian agrosystem. *Acta Zoologica Fennica*, **173**, 107–11.

Radvanyi, A. (1974) Small mammal census and control on a hardwood plantation. *Proceedings of Vertebrate Pest Conferences*, **6**, 9–19.

Redhead, T.D., Enright, N. and Newsome, A.E. (1985) Causes and predictions of outbreaks of *Mus musculus* in irrigated and non-irrigated cereal farms. *Acta Zoologica Fennica*, **173**, 123–7.

Reimer, J.D. and Petras, M.L. (1968) Some aspects of commensal populations of *Mus musculus* in southwestern Ontario. *Canadian Field-Naturalist*, **82**, 32–42.

Rosicky, B. (1965) Types of animal movement and their influence on natural foci of diseases, in *Theoretical Questions of Natural Foci of Diseases* (eds B. Rosicky and K. Heyberger), Czechoslovak Academy of Sciences, Prague, pp. 151–64.

Rosicky, B. (1967) Natural foci of disease, in *Infectious Diseases* (ed. A. Cockburn), Thomas, Springfield, Illinois, pp. 108–26.

Rowe, F.P., and Swinney, T. (1977) Population dynamics of small rodents in farm buildings and on arable land. *EPPO Bulletin*, **8**, 431–7.

Rowe, F.P., Taylor, E.J. and Chudley, A.A.J. (1963) The numbers and movements of house-mice (*Mus musculus* L.) in the vicinity of four corn-ricks. *Journal of Animal Ecology*, **32**, 87–97.

Ryszkowski, L. (1969) Operation of the Standard-Minimum methods, in *Energy Flow Through Small Mammal Populations* (eds K. Petrusewicz and L. Ryszkowski), Polish Scientific Publishers, Warsaw, pp. 13–24.

Ryszkowski, L. (1971) Estimation of small rodent density with the aid of coloured bait. *Annales Zoologica Fennici*, **8**, 8–13.

Shellhammer, H.S. (1972) *Rare and Endangered Fish and Wildlife of the United States*, USDI Fish and Wildlife Services, Resource Publ. 34.

Shellhammer, H.S. (1989) Salt marsh harvest mice, urban development, and rising sea levels. *Conservation Biology*, **3**, 59–65.

Singleton, G.R. (1989) Population dynamics of an outbreak of house mice (*Mus domesticus*) in the mallee wheatlands of Australia – hypothesis of plague formation. *Journal of Zoology*, **219**, 495–515.

Smirin, Y.M. (1975) Familiarity with home range in rodents and the successfulness of the predation on them by *Asio otus* and *Tyto alba*. *Bulletin of the Moscow Naturalist Society*, **80**, 14–20.

Smit, C.J. and van Wijngaarden, A. (1981) *Threatened Mammals in Europe*, Akademische Verlagsgesellschaft, Wiesbaden.

Spencer, D.A. (1954) Rodents and direct seeding. *Journal of Forestry*, **52**, 824–6.

Spitz, F. (1985) Further development of the forecasting model for *Microtus arvalis*. *Acta Zoologica Fennica*, **173**, 89–92.

Stenseth, N.C. (1980) Spatial heterogeneity and population stability: some evolutionary consequences. *Oikos*, **35**, 165–84.

Stenseth, N.C. (1981) How to control pest species: application of models from the theory of island biogeography in formulating pest control strategies. *Journal of Applied Ecology*, **18**, 773–94.

Stenseth, N.C. and Hansson, L. (1979) Correcting for the edge effect in density estimation: explorations around a new method. *Oikos*, **32**, 337–48.

Stenseth, N.C. and Hansson, L. (1981) The importance of population dynamics in heterogeneous landscapes: management of vertebrate pests and some other animals. *Agro-Ecosystems*, **7**, 187–211.

Stickel, L.F. (1979) Population ecology of house mice in unstable habitats. *Journal of Animal Ecology*, **48**, 871–87.

Sullivan, Th.P. (1979) Repopulation of clear-cut habitat and conifer seed predation by deer mice. *Journal of Wildlife Management*, **43**, 861–71.

Taitt, M.J. and Krebs, C.J. (1981) The effect of extra food on small rodent populations: II. Voles (*Microtus townsendii*). *Journal of Animal Ecology*, **50**, 127–37.

Tast, J. (1968) The root vole, *Microtus oeconomus* (Pallas) in man-made habitats in Finland. *Annales Zoologica Fennici*, **5**, 230–40.

Telle, H.-J. (1969) 12 Jahre grossräumige Rattenbekämpfung in Niedersachsen – ein kritischer Rück- und Ausblick. *Schriftenreihe des Vereins für Wasser-, Boden- und Lufthygiene*, **32**, 131–59.

Twigg, G. (1975) *The Brown Rat*, David & Charles, London.

van Wijngaarden, A. (1957) The rise and disappearance of continental vole plague zones in the Netherlands. *Verslagen van Landbouwkundige Onderzoekingen*, **63**, 1–21.

Vlasak, P. and Porkert, J. (1973) Immigration von Kleinsaugern in ein Wohngebaude abhängig von meteorologischen Bedingungen. *Lynx*, **14**, 70–98.

Wagner, F.H. (1981) Role of lagomorphs in ecosystems. *Proceedings of the World Lagomorph Conferences*, **1**, 668–94.

Walkowa, W., Adamczyk, K. and Chelkowska, H. (1982) Numbers and structure of rodent communities in the forest environment of Silesia. *Polish Ecological Studies*, **8**, 305–20.

Wegner, J.F. and Merriam, G. (1979) Movements by birds and small mammals between a wood and adjoining farmland habitats. *Journal of Applied Ecology*, **16**, 349–57.

WHO (1974) *Ecology and Control of Rodents of Public Health Importance*, WHO, Geneva.

Wolff, J.O. (1980) The role of habitat patchiness in the population dynamics of snowshoe hares. *Ecological Monographs*, **50**, 111–30.

Wolff, J.O. (1981) Refugia, dispersal and geographic variation in snowshoe hare cycles. *Proceedings of the World Lagomorph Conferences*, **1**, 441–9.

Wolton, R.J. and Flowerdew, J.R. (1985) Spatial distribution and movements of wood mice, yellow-necked mice and bank voles. *Symposium of the Zoological Society of London*, **55**, 249–75.

Yeatman, H.C. (1969) Population studies of seed-eating mammals. *Tennessee Academy of Sciences*, **35**, 32–48.

Zejda, J. (1975) Habitat selection in two feral house mouse (*Mus musculus* L.) lowland populations. *Zoologicke Listy*, **24**, 99–111.

PART FIVE

Looking backward and forward

– 9

Presaturation and saturation dispersal 15 years later: some theoretical considerations

N.C. STENSETH AND W.Z. LIDICKER JR.

9.1 INTRODUCTION

It is a common English idiom to 'look forward and back'. In this chapter we reverse this procedure and first look backward to 1975 when the idea of presaturation dispersal was first introduced by Lidicker (1975). We will review what has happened to this suggestion subsequently, and finally look forward to where it might lead us next.

Presaturation dispersers are those that leave their current place of residency before the patch's carrying capacity is reached (Figure 9.1). However, because carrying capacity is often difficult to determine and varies greatly over time in most species, Stenseth (1983) proposed, following an earlier idea of Krebs *et al.* (1973), to define presaturation dispersers more narrowly as those individuals who responded to an increasing specific (per capita) growth rate in a population so as to bring about an increasing specific rate of dispersal. More recently, Gaines *et al.* (1991) have made the similar suggestion that presaturation dispersal be defined as that occurring when population growth rates are positive. However one chooses to operationally define presaturation dispersers, the critically important point inherent in the presaturation dispersal concept is that individuals become motivated, that is they are selected, to disperse before their home habitat becomes filled to the capacity of its basic resources. Much data have been interpreted to demonstrate the existence of presaturation dispersal (e.g. Lidicker, 1975, 1985a,b; Stenseth, 1983; Chapters 1, 2 and 7). The reality of presaturation dispersal continues, however, to be questioned by some (e.g. Gaines and McClenaghan, 1980; Chapters 10 and 11).

Lidicker (1975) presented verbal arguments to explain the evolution of presaturation dispersal. Without referring explicitly to this idea, Hamilton and May (1977) presented a model predicting that dispersal behaviour can evolve even in a stable environment. Hence, the idea of presaturation

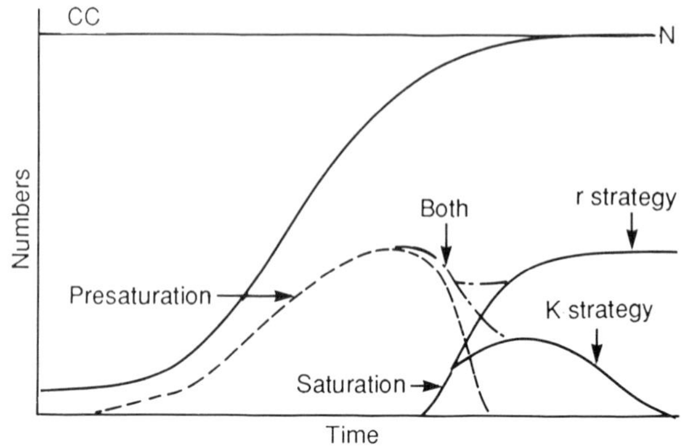

Figure 9.1 Presaturation and saturation dispersal as defined in Lidicker (1975). CC refers to estimated carrying capacity.

dispersal was in effect theoretically supported. Later, Comins *et al.* (1980) and Comins (1982) extended the analysis of Hamilton and May to biologically more general situations, though without altering the theoretical support for the idea of presaturation dispersal. These and other models have earlier been reviewed with special reference to small rodent cycles by Stenseth (1983).

Here we review available quantitative models relevant to evaluating the idea of presaturation dispersal. We pay special attention to possible age and sex differences with respect to this style of dispersal. Primarily we present the theoretical aspects of presaturation dispersal. While the subject has been discussed in several other chapters in this book, the present chapter differs from the others in its particular emphasis on the theoretical foundations of presaturation dispersal. After all, the status of the idea of presaturation dispersal was our immediate incentive to organize the symposium on which this book is based.

Having established the theoretical bases for the evolution of dispersal, we then discuss the population dynamics of dispersal. Again we pay particular attention to presaturation dispersal in this connection. This, too, has been alluded to in earlier chapters of the book, but here we discuss the ecological consequences *per se* of the relevant evolutionary models.

9.2 THE EVOLUTION OF DISPERSAL

9.2.1 Preamble

Fairy tales about lemmings rushing down to the sea and drowning themselves are, particularly in the older scientific literature (e.g. Allee *et al.*,

1949; Wynne-Edwards, 1962; but see Wynne-Edwards, 1986), discussed in group-selectionistic terms. Undoubtedly, dispersal sometimes benefits the population or species as a whole (e.g. Van Valen, 1971; Maynard Smith, 1976; Futuyma, 1986). However, dispersal and migration ought to be understood as individually beneficial if their evolution is to be understood as a result of natural selection on the individual level (Maynard Smith, 1976). From this viewpoint, it seems reasonable to assume that there may be very strong individual selective forces in favour of dispersal since so many lemmings drown during dispersal. Moreover, dispersal does in most cases tend to be a stabilizing element for populations (discussed below), and hence increases the stability of the community of which the species is a part. Therefore, even if there is higher level selection in favour of the evolution of dispersal, this would be in the same direction as individual selection and hence not be of critical importance to our arguments here.

Take the case of a patchy environment (Figure 1.1; Hansson, 1977; Anderson, 1980; Stenseth, 1983, 1988). Patches of suitable habitat may be bounded by areas that are less favourable or even unfavourable because of physical, biotic, or social reasons. Note that patch boundaries are not necessarily the result of abiotic factors as is sometimes implied or stated in such discussions. In such a heterogeneous or patchy environment, it is by no means obvious which is the better strategy for an individual to adopt: to remain in the patch of current residency or to undertake a potentially dangerous dispersal through some sort of transition habitat in search of a new home site and in most cases a mate as well (Figure 1.1).

The quality of the site in which the individual was born, grew up and/or currently resides is known to a certain extent to be a good quality site; in addition the individual also knows its way around in its current home range. By leaving the current home site, the risk of being taken by a predator increases greatly, and there is always the risk that another home site with a mate may not be found. There are also costs associated with the search for an unrelated mate; these are due to the increased exposure to predation in the transition habitat, difficulties in entering established demes (e.g. Myers, 1974; Getz, 1978), and difficulties in finding a mate in another patch. On the other hand, dispersal may offer a gain in inclusive fitness (*sensu* Hamilton, 1964) owing to increased access to critical resources and mates elsewhere, an opportunity to produce more fit offspring by avoiding inbreeding depression (Falconer, 1960), decreased exposure to predators and diseases in the current site, and reduced competition for those family members which do not disperse (Chapters 1 and 2).

As far as we know, Bengtsson (1978) was the first to formally treat the outcome of these conflicting interests (Smith, 1978; Maynard Smith, 1978). We utilize Bengtsson's framework (Chapter 1) as a starting point for developing our theoretical arguments.

All predictions mentioned in this chapter refer to specific (or per capita) rates, not total numbers dispersing. Unfortunately, there has been some confusion, or imprecise statements, in the literature on this point [see Gaines and McLenaghan (1980) for review]. Lidicker (1975), for example, referred to the number of dispersers under various conditions; what was important at that time, however, was to demonstrate that *some* dispersal occurred before saturation densities were reached. For this, it is sufficient to know that the number is different from zero; if it is, then the rate must also be different from zero. Obviously, it is now inadequate to know only that at a high density there are a lot of dispersers around. It is far more meaningful to know how the numbers of dispersers relate to the pool of potential dispersers, i.e. we need to know the percentage dispersing, or the per capita rate of dispersal.

9.2.2 Models for the evolution of dispersal

Several models have been presented in the literature in order to analyse possible mechanisms for the evolution of dispersal (some of these are, for instance, discussed in Stenseth, 1983). Here we present a synopsis.

(a) The Gadgil model

Assuming an asexually reproducing population, Gadgil (1971) developed a model for analysing the conditions under which dispersal behaviour is favoured by natural selection in a habitat complex. Generally speaking, dispersal was found to be favoured when there is a chance of colonizing another site more favourable than the one in which the individual presently finds itself (e.g. Chapter 1). A pronounced tendency to disperse will, according to Gadgil, be favoured under the following environmental conditions:

1. when habitats have low duration stability relative to the species' generation time (then conditions in the current place of residency frequently become unfavourable);
2. when highly suitable areas for establishing home sites are relatively close together (so as to make effective survival, p, during dispersal high); and,
3. when catastrophic mortality creates temporarily vacant areas (so as to make good areas available for dispersers to settle in).

Dispersal is furthermore favoured when the pattern of environmental fluctuations (affecting site suitability) varies from place to place; in such cases there will often exist some places in the habitat complex where conditions are better. However, if the changes in suitability are highly correlated from place to place (synchronous), then hard times in one site mean hard times in another site too, and site or patch tenacity is likely to be favoured over dispersal. Dispersal is therefore expected to be highest when spatial asynchrony in site quality is the highest. This model then supports the theoretical plausibility of the presaturation dispersal idea, especially for a particular suite of environmental circumstances.

Łomnicki (1978, 1980, 1987) refined Gadgil's model by incorporating hierarchical differences between individuals in a population. Łomnicki's refinement made it possible to differentiate predictions with respect to the individuals' hierarchical rankings. Specifically, low ranking individuals are those first to leave an area when competition becomes intense, and hence when conditions become unfavourable.

(b) The Hamilton–May model

It is no surprise that dispersal should be advantageous when local environmental conditions are fluctuating strongly (i.e. condition 1 of the Gadgil model). Hamilton and May (1977) argued, as did originally Gadgil (1971), that some dispersal may evolve even in temporally stable environments, even under high dispersal mortality. For an asexual population Hamilton and May derived, in the limiting case of a large number of habitat sites and a large number of offspring, that the optimal (or ESS; cf. Maynard Smith, 1982) per capita dispersal rate, d^*, may be expressed analytically as

$$d^* = 1/(2 - p_1), \tag{9.1}$$

where p_1 is the probability of a dispersing individual being able to find another site in which to reproduce (Figure 1.1). Specifically, this formula implies that the dispersal rate, d^*, will be highest when the habitat is homogeneous, dispersal confers no extra mortality, and a new site is always available. On the other hand, dispersal would, according to this model, be low when no dispersers survive long enough to locate another site (as might be the case at high overall densities).

The lowest dispersal rates would occur in an extremely patchy habitat with an effectively uncrossable transition habitat in between (either due to physical, biotic, or social conditions; Figure 1.1). Such extremes are rarely realized, but may for different reasons be approached at very low densities and at very high densities. At intermediate densities, dispersal is highest.

Even with the low dispersal scenario, however, a case can be made that some dispersal should occur. As advocated by Anderson (1980, 1989) and modeled by Łomnicki (personal communication), parents may force dispersal of offspring if this would improve conditions for the parents and/or siblings. The effectiveness of this would depend on the details of the species social system and the specifics of the environmental mosaic, including genetic environment, faced by the would-be dispersers. Again, we find that the Hamilton–May approach leads to the expectation that presaturation dispersal will evolve frequently.

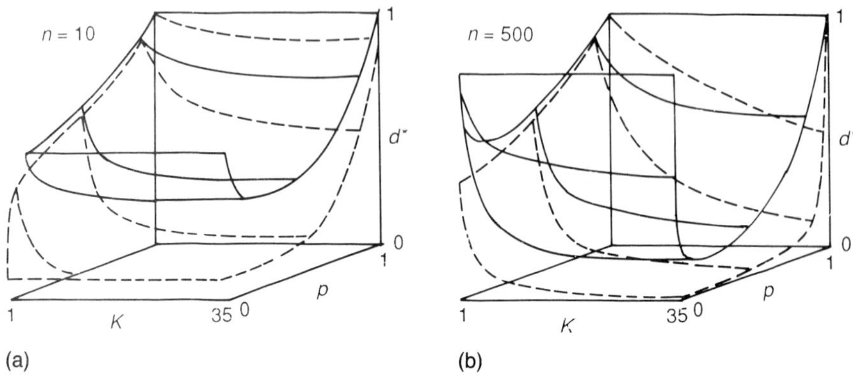

Figure 9.2 Three-dimensional plots for the evolutionarily stable strategy of dispersal probabilities (d^*), as functions of patch size (K) and the probability of surviving dispersal (p). Dashed lines represent the surfaces derived for situations without any extrinsically imposed extinction rate. The solid lines delimit the surfaces derived from unstable environments, i.e. with an imposed extinction rate of 0.1. (a) A population with an average number of offspring per site (n) equal to 10. (b) A population for which $n = 500$. Redrawn from Stenseth (1983, figure 5.4).

(c) Modifications of the Hamilton–May model

Comins *et al.* (1980) and Comins (1982) extended the model of Hamilton and May. The most general model presented by Comins *et al.* (1980) was a discrete generation island–type model in which reproduction, migration, and competition are stochastic processes. Reproduction is, as in the Gadgil model, asexual and individuals of each genotype have a certain probability of being born with a predisposition to disperse. This generalized model predicts that the following features will favour dispersal (Stenseth, 1983) (Figure 9.2):

1. increasing extrinsic disturbances (i.e. when habitats have short duration stability);
2. decreasing hostility of the transition habitat, $1 - p$; and
3. decreasing patch size.

Only feature (3) extends the conclusions derived from Gadgil's model (see above). As in the original Hamilton–May model, some dispersal is always found to be optimal. And again, presaturation dispersal is predicted to occur.

Motro (1982a) developed a slightly different model adopting a more standard population genetics approach analysing changes in frequencies of existing genes. Similar, albeit not identical, conclusions were found, but presaturation dispersal was again, in effect, predicted to occur.

(d) The Bengtsson model

These models all assume asexual reproduction. As mammals reproduce sexually,we need to extend the analysis to sexually reproducing organisms. According to Comins *et al.* (1980), the effect of introducing sexual reproduction is, however, only slight under most conditions. To a certain extent this was confirmed by Roff (1975), who developed and analysed a model for the evolution of dispersal in a sexually reproducing species. He found an optimal dispersal rate which was always slightly smaller than the one predicted by Hamilton and May (1977). This result is indeed consistent with the results of Comins *et al.* (1980), who found that the effective population size for sexual species needed to be increased by only half an individual to make them correspond to asexual conditions. Similar results were obtained by Motro (1982b, 1983), who also considered sexual populations.

Specifically these models suggest that the asexual models predict somewhat too high dispersal rates at the lowest density; at higher densities, the asexual and sexual models provide very similar results. Collectively these studies further strengthen the notion of presaturation dispersal.

Bengtsson's (1978) model differs somewhat from the sexual version of the model developed by Comins and co-workers. Although the general conclusions reached on the basis of the Comins–Hamilton–May models remain valid, Bengtsson found more pronounced effects of sexuality than did Comins *et al.* (1980). For sexually reproducing populations, a dispersing individual must (unless it is a pregnant female) find a mate in order to avoid effective sterility. This is particularly a problem at very low densities when no mate may be located, and at very high densities when entry to a social group may be strongly resisted. Hence, it is expected that the optimal rate of dispersal should be relatively low during periods of low population densities in the habitat complex. As densities increase, higher rates of dispersal will be favoured. Our previous conclusions regarding the theoretical plausibility of presaturation dispersal indeed remain valid.

9.2.3 Sex differences in natal dispersal

Which sex should disperse? This issue is discussed in some detail in Chapters 4, 5 and 11, and much speculation exists in the literature. As far as we know, no quantitative model has been developed specifically for answering this question. However, both the Bengtsson model (see above and Chapter 1) and the parental care model developed by Maynard Smith (1977, 1978, 1982) may be interpreted within this framework.

In Bengtsson's (1978) model and as implied by several earlier writers, the cost of inbreeding must exceed the risks of not finding another mate and another site in which to breed if dispersal is to be favoured by natural selection. This balance will depend on the social and territorial system of the species (or population) under consideration. Moreover, the costs of philopatry may include factors other than those of inbreeding. For example, there may be parental aggression to contend with or parentally induced suppression of reproductive maturation. All of these costs must be compared to potential benefits of the alternative behaviours. The problem is clearly complex, and formal analysis is badly needed. However, even without such an analysis, it is obvious that the cost of inbreeding will be experienced differently by the two sexes. For mammals, this will generally be lower for males than for females (e.g. Maynard Smith, 1978). This is because mammals are mostly polygynous with males contributing less to parental care. This implies that males experience greater variance in reproductive success, and also have less at risk in a given litter. Females, on the other hand, may produce only a few litters in a lifetime, and invest heavily in pregnancy, lactation, and parental care. For them, experiencing inbreeding depression would be strongly selected against. So, if inbreeding were the only issue in dispersal, we might expect females to predominate among dispersers. Generally in mammals this is not the case, although the extent of female dispersal is often underestimated. Obviously, other factors are involved. For females this is generally a sensitivity to the resource base of their habitat, and a greater potential benefit from nepotism. Both of these factors would tend to make females philopatric. Furthermore, if males tend to benefit more by searching for multiple mates and especially if this is concordant with reducing the costs of inbreeding, males will tend to be less philopatric. Given this, females can afford to reap the benefits of philopatry without much risk of inbreeding.

With the multiplicity of independent factors in the dispersal equations, it is not surprising that generalizations about sex bias in dispersal are few and even then contentious. Much more needs to be learned regarding the costs and benefits of various components of the problem so that formal modelling will be more rewarding. Also, a game theory approach could lead to substantial new insights.

9.2.4 Age dependencies in dispersal

To our knowledge, the model developed by Morris (1982) is the only one considering the evolution of age-specific dispersal rates in terrestrial vertebrates. Morris found that most dispersers should be either young animals which have not yet attained maximal reproductive value (and which have not yet established a breeding site), or much older individuals whose prospects for future reproductive output is low (but still greater than zero). This prediction is consistent with the result obtained by Engen and Stenseth (1989), analyzing another behavioural trait (predator risks in connection with optimal foraging). They found that, when everything else is the same, older animals should be 'willing' to take the greater risks. Everything else is, of course, not equal: younger animals have, after all, often greater difficulties in establishing territories, or finding suitable breeding sites (e.g. Chapter 6).

In summary we suggest on the basis of the analyses by Morris and by Engen and Stenseth that adult dispersal will increase in frequency with age, especially after peak reproductive value has been reached.

9.2.5 Combined evolution of dispersal and social systems

Bengtsson (1978) pointed out that the dispersal pattern adopted will, in addition to being influenced by the cost of inbreeding, also determine over time the magnitude of the inbreeding effects (Chapter 2). Specifically, if inbreeding is a common feature, the cost of inbreeding will over time be reduced due to very strong selection against genes causing inbreeding depression. Such conjoint evolution of dispersal and social behaviour was, in a general context, treated formally in some depth by Gadgil *et al.* (1983). They demonstrated a positive feedback between the evolution of greater co-operation and altruism, and the evolution of higher levels of population viscosity (i.e. a population in which an individual's neighbours are its close relatives, which in turn is brought about by low dispersal rates). Thus, when ecological conditions favour high levels of viscosity, the evolution of high levels of co-operation will be favoured. And, when ecological conditions favour high levels of co-operation, the evolution of high levels of viscosity will be favoured. Given this positive feedback, a population initially below a threshold of co-operation/viscosity will be expected to evolve in the direction of higher dispersal and lower co-operation, while a population above the threshold should evolve in the direction of higher co-operation and viscosity (Figure 9.3).

In state A of Figure 9.3, we would expect far more dispersal to characterize the population than in state S. In state A both presaturation as well as saturation dispersal is expected to take place in both sexes. In state S, on the other hand, presaturation dispersal is, for genetic reasons, likely to be particularly pronounced in only one of the sexes, whereas saturation dispersal may occur in both sexes. The genetic motivation for presaturation dispersal may be both more or less pronounced due to a positive feedback loop involving dispersal and site tenacity, and outbreeding versus inbreeding (Bengtsson, 1978).

Maynard Smith's (1982) model for parental care may be used for predicting the extent of adult dispersal (Figure 9.4). In mammals, it will

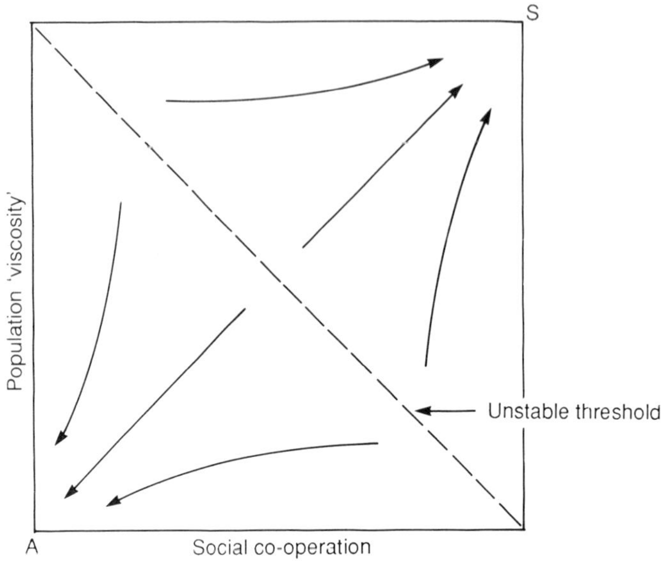

Figure 9.3 The evolution of dispersal or philopatric behaviour based on a positive feedback between social co-operation and the level of population 'viscosity'. Stable state A is characterized by high levels of dispersal and state S by low levels. After Gadgil *et al.* (1983).

have to be the female that remains with the litter for the nursing period, and generally provides most of the parental care. Then, males will disperse after having mated successfully with a female if

$$1 \approx P_1/P_2 > (1 + p_0)/(1 + p_1) \tag{9.2}$$

where P_1 and P_2 are the survival of litters with only females caring for the litter (P_1) and both parents providing care for the litter (P_2). The symbol p_1 is the probability of a dispersing male finding another mate somewhere else, and p_0 is the probability that he will find an additional female in the immediate vicinity of the first female (p_1 is equivalent to the p defined in connection with the discussion of the Bengtsson model). Conditions for optimal dispersal rate (equation 9.1) in this case will be satisfied if P_1/P_2 is close to one, and if it is easy to find mates elsewhere (i.e. p_1 is large), or if p_0 approaches zero (i.e. monogamy). This circumstance should lead to pronounced dispersal (presaturation) of males. Whether females disperse or not after having weaned their young, will be determined on the basis of the complex of considerations already discussed regarding sex and age influences on dispersal.

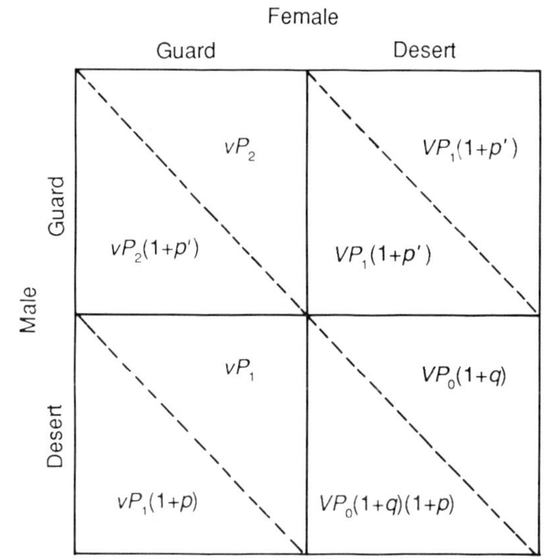

Figure 9.4 Payoff matrix of Maynard Smith (1982) for the asymmetric male versus female parental care game. Payoffs in the upper right corners of each box are the female's, and in the lower right corners are the male's. P_0, P_1, and P_2 are the probabilities of offspring survivorship with no, one, and two parents, respectively. v is the clutch size of a guarding female. V is that of a deserting female. p is the probability that a deserting male can obtain an additional mate, p' is the probability that a guarding male can obtain an additional mate, q is the probability, or fraction of an additional clutch (of size V), that a deserting female can lay during the time it would take a guarding female to guard. q' is the probability that a deserting female can lay an additional clutch of size V when males guard. There are four Evolutionary Stable Strategies (ESS): 'Both Guard', 'Both Desert', 'Male Guards, Female Deserts', and 'Female Guards, Male Deserts'. For mammals, the q and v functions are not relevant nor are the ESS 'Both Desert' and 'Male Guards, Female Deserts'.

9.2.6 Conclusion

Summarizing the results emerging from these models within the context of 'cyclic small rodent biology', Stenseth (1983) deduced the relation between the optimal per capita dispersal rate, d^*, and both numbers (N) and specific growth rate $(dN/dt)/N$ in the population. The patterns predicted are depicted in slightly modified form in Figure 9.5. In fact, these two functions constitute an operational definition of presaturation dispersal as long as the carrying capacity and equilibrium densities are the same. This means that if presaturation dispersal is the dominant type of dispersal, a positive correlation is expected between dispersal rates, d^*, and specific

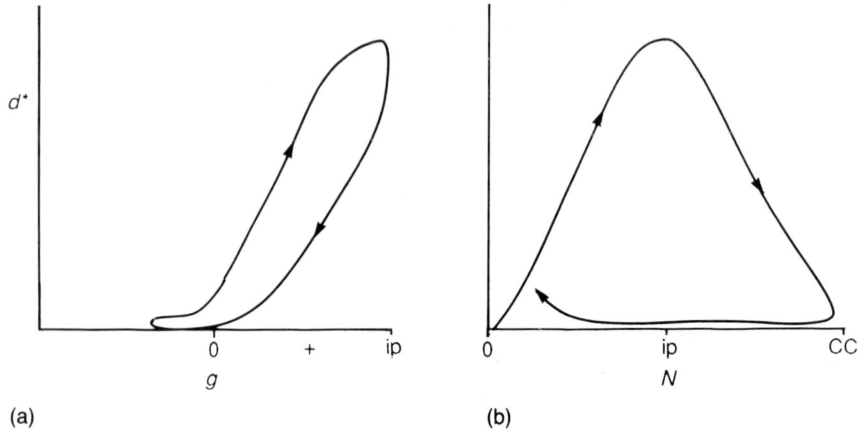

Figure 9.5 Predicted changes in the optimal dispersal rate d^* (a) as a function of the specific growth rate, $(dN/dt)/N = g$ and (b) density N. ip, inflection point; cc, carrying capacity. The optimal dispersal rate is predicted to be slightly higher in the early increase phase as compared with the late increase phase; thus, the depicted clockwise rotation (a). Redrawn from Stenseth (1983, figure 5.7).

growth rates, $(dN/dt)/N$, while no such correlation is expected between dispersal rates, d^*, and numbers, N. For many cyclic small mammals this pattern is in fact found. Krebs *et al.* (1973) were the first to report these patterns, and Stenseth (1983) summarized the available relevant data (see also Krebs, 1985). Indeed, it was observations of this kind that led to the idea of presaturation dispersal (*sensu* Lidicker, 1975) in the first place.

Three optimal dispersal scenarios with respect to the relative importance of presaturation and saturation are possible (Figures 9.5 and 9.6).

1. Presaturation dispersal is much more prevalent than saturation dispersal (Figure 9.5b).
2. Saturation dispersal is much more pronounced than presaturation dispersal (Figure 9.6a).
3. Presaturation and saturation dispersal are approximately equally pronounced (Figure 9.6b).

Note that for saturation dispersal, d^* cycles counterclockwise but goes clockwise when presaturation dispersal is present in significant amounts.

9.2.7 Addition of other factors

Until now we have made only rather superficial references to reproduction and demography in general. This, we believe, reflects the biological litera-

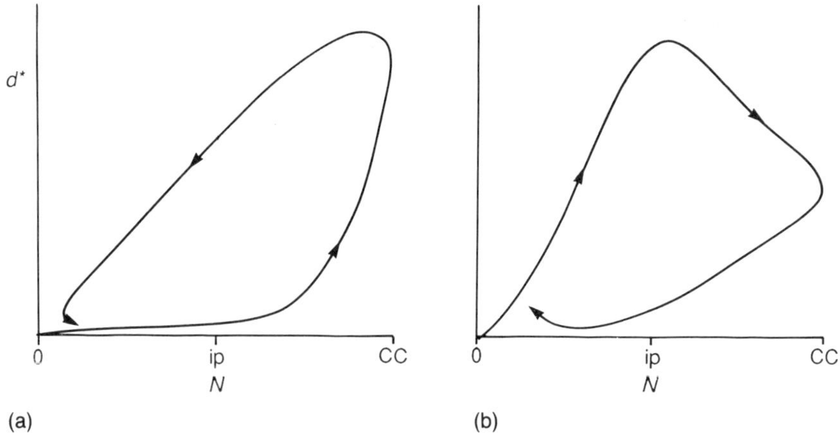

Figure 9.6 Optimal dispersal rates (d^*) as a function of population density (N) in situations where (a) saturation dispersal predominates and (b) where both presaturation and saturation dispersal occur. See Figure 9.5b for situation in which presaturation dispersal predominates.

ture rather well. Most commonly reproductive strategies and dispersal strategies are treated as two separate fields of study (but see Bondrup-Nielsen, 1985). However, reproduction and dispersal are closely related (Iwao, 1977; Southwood, 1977; Solbreck, 1978; Chapter 2). Furthermore, dispersal may also proximally influence the reproductive performance of resident individuals, both those left behind and those in the new place entered (e.g. Charnov and Finerty, 1980). Very little data exist, however, to support these presumptions empirically.

Southwood (1977) has provided a general theoretical framework for studying the interaction between dispersal and reproduction (or demography) (Chapter 4). The central idea is that any organism's habitat is heterogeneous in time and space (e.g. Hansson, 1977; Stenseth, 1977, 1983; Anderson, 1980; Stenseth and Gustafsson, 1985; Chapter 4); hence, 'space' and 'time' become the axes in a reproduction matrix (Table 9.1). In the following, we suggest how to modify and extend Southwood's original idea.

Consider two instances in time (now and later) and two discrete habitat patches (here and elsewhere). In Table 9.1, D is the probability of dispersing; S is the probability of surviving from one instance in time to a later instance in time (given that the individual does not disperse); p_1 is the probability that a dispersing individual will both survive the dispersal phase and be able to enter successfully another patch; and r_1, r_2, r_3, and r_4 are the long-term expected specific rates of population growth in the

Table 9.1 The habitat template as defined by Southwood (1977) constituting the basis for analysis of the joint evolution of optimal reproductive and optimal dispersal rates. See main text for further discussion. (Modified from Southwood, 1977)

	Time	
	Now	*Later*
Space		
Here	$S\,(1 - D)\,(r_1 \pm V_1)$	$S\,(1 - D)\,(r_2 \pm V_2)$
Elsewhere	$S\,D\,p_1\,(r_3 \pm V_3)$	$S\,D\,p_1\,(r_4 \pm V_4)$

particular patches as given by the appropriate entry in Table 9.1. Finally, the corresponding V values express the variation in the specific population growth rates over time and space due to environmental variation. This will include non-predictable environmental changes that will require much detailed empirical information to assess (Stenseth and Gustafsson, 1985). Similarly, the effect of mating with close relatives or with unrelated individuals may be incorporated directly into the r values, as we have done earlier in this chapter and elsewhere in the book (Bengtsson, 1978).

The r_i values are density-dependent (i.e. regulating) rates. Utilizing the technique developed by Ugland and Stenseth (1985), we may then find the joint optimal reproductive and dispersal strategies given a certain (overall) pattern of density change in the habitat complex, a certain hostility of the transition habitat (p_1), and a certain level of environmental variation (V_i). Presumably, natural selection would in a fluctuating environment favour those strategies minimizing the time required to grow from a low density just after a crash to some critical density level, C.

Within this general conceptual framework discussed, it is obvious that what is the optimal strategy for any given individual will depend upon what other individuals do, a recurring theme in this discussion. For example, an individual may confront strangers through dispersal to a distant habitat patch or by dispersers arriving at its home patch. The optimal strategy may well be different in these two situations. Or, dispersal of a juvenile might be conditioned by how many siblings it has. Situations like this, where frequency dependencies are important, can ideally be analysed by game theory models (e.g. Maynard Smith, 1982).

Another parameter often discussed independently, but that is critically important to dispersal, is that of territoriality. Territorial behaviour varies interspecifically, intraspecifically in some cases, and even in the same population depending on density. If possession of a territory is essential for successful reproduction, as is generally the case in mammals, this must be viewed as an essential resource just like food, water, shelter, and an appropriate mate. Indeed, this suite of basic resources is often incorporated collectively in the guise of a suitable home range, which in turn often equates to a suitable territory. Dispersal behaviour must be intimately entwined, therefore, with the goal of achieving or maintaining a suitable home range, generally expressed as territorial behaviour.

If, for example, a population exhibits individual territoriality (so-called 'asocial' species), a juvenile has only the choices of dispersing or waiting for its mother to die or become so feeble she can be displaced along with any remaining siblings. Such populations should experience strong selection for dispersal behaviour. At the other extreme, a complexly social species may provide all the lifetime needs for a juvenile within the natal social group and group territory. Under these circumstances, dispersal would be motivated only by possible inbreeding costs, environmental catastrophe, or the need for successful groups to fission. A much lower level of more selective dispersal could be predicted to evolve in these circumstances. In between these two social extremes, we will find the mass of mammalian species all faced with a complex of social and environmental circumstances to which their dispersal strategy must adapt appropriately if they are to succeed.

The challenge for theorists is to address this complexity, incorporating all of the necessary simplifying assumptions without compromising reality to the point of uselessness. Presumably an extension of the Hamilton–May–Comins approach combined with the cost-benefit framework of Bengtsson could be a profitable avenue. In any case, it should be remembered that reproduction, survival, and dispersal rates are all likely to be density-dependent, as would be the probability of successful dispersal (p_1). Moreover, all of these may have different functions for increasing and decreasing densities, even at the same absolute density. One further suggestion is to defer development of more general models in favour of those addressing specific populations or groups of related species. In this way, some measure of simplification can be achieved without bending reality too much. If these are successful, attempts to make more general models can follow.

9.3 POPULATION DYNAMIC CONSEQUENCES OF DISPERSAL

As one of the basic parameters in the population growth equation, movements into and out of populations clearly have the potential for important demographic effects. In addition to effects on numbers, dispersal can be non-random with respect to age, sex, genetic composition, and social status. Hence, there are several routes by which dispersal can affect demographic features of a population. This aspect of dispersal biology is, in fact, one of the major themes of this book, and is discussed from different vantage points in most of the preceding chapters. This section focuses on the possible role of presaturation dispersal in explaining the multi-annual density cycles in microtine rodents (see also Chapter 7).

As recently pointed out by Krebs (1985), several microtine biologists have, over the past 10 years or so, argued that spacing behaviour and dispersal are at least components of the mechanism driving the density cycles in these small mammals (e.g. Lidicker, 1973, 1988; Krebs, 1978; Taitt and Krebs, 1985). The exact role of these factors and their causal relationships are, however, far from clear. Theoreticians generally claim that spacing behaviour such as territoriality and dispersal lead to stability (e.g. Stenseth, 1983, 1985a,b, 1986, 1988; Stenseth *et al.*, 1988; Stenseth and Sannes, 1988), not instability and cycles, as suggested by several microtine biologists. There are, however, a few exceptions to this general theoretical expectation, where authors (e.g. Stenseth, 1980; Abramsky and Van Dyne, 1980; Rosenzweig and Abramsky, 1980; Hestbeck, 1982, 1988; Morris, 1984), claim that dispersal may cause cycles. Two models explicitly incorporating presaturation dispersal as a possible mechanism for density cycles are discussed below.

9.3.1 The Charnov–Finerty hypothesis

Chitty (1960, 1967) proposed a much discussed theory saying that competitive interactions between genetically determined aggressive and docile morphs (see also Krebs, 1978, 1985), coupled with dispersal (Krebs *et al.*, 1973; Krebs, 1979; Chapter 7), would lead to regular density cycles of the kind commonly seen in small microtine rodents.

Charnov and Finerty (1980) proposed an alternative to the Chitty hypothesis, and Stenseth (1983) later expanded on this hypothesis. It is a phenotypic version of the genotypic Chitty hypothesis. Dispersal is essential in the Charnov–Finerty hypothesis, but not in the original Chitty hypothesis. In later versions, dispersal is postulated to drive the cycle because it causes qualitative changes in local populations. According to Charnov and Finerty, some individuals become aggressive because they are exposed repeatedly to strangers; according to Chitty, some individuals are born to be aggressive.

Warkowska-Dratnal and Stenseth (1985) analysed two models comparing these two hypotheses. They concluded that dispersal in fact could drive the cycle if the phenotypic mechanism of behavioural switching was assumed and if dispersal was of the presaturation type. However, further analysis of this model and the development of additional improved models (Stenseth and Łomnicki, 1990) demonstrated that the Charnov–Finerty model was implausible. Using reasonable demographic parameter values and assuming a patchy environment, but without seasonality, they found that only stable or damped oscillations could be generated. In particular, they discovered that the transition from aggressive to docile phenotypes needed to be synchronous among patches to achieve a reasonable level of destabilization, and this was not possible in their simulations. Thus, neither the Chitty nor the Charnov–Finerty approaches, assuming as they do a dominating role for intrinsic factors, seem to provide an adequate explanation for multi-annual cycles.

9.3.2 Stenseth–Antonsen model

Stenseth and Antonsen (1988) analysed a density regulation model incorporating a developmental delay. It takes a fairly fixed length of time (approximately 6–7 weeks in voles; e.g. Hasler, 1975) for a newly conceived female foetus before she herself can conceive her own foetus. It therefore makes sense to assume at least a 6-week delay in any reproductive response that a population might make to environmental perturbations.

This model predicts that cycles of the observed periodicity may result if density dependence is stronger on reproductive rates (including the fraction of the population reproducing) than on survival rates. This bias has been claimed to be the case for many cyclic microtines (Krebs and Myers, 1974; Bujalska, 1985; Gipps, 1985). According to this model, cycles of the appropriate periodicity will, however, only result if presaturation dispersal is stronger than saturation dispersal; in addition, dispersal changes must more than compensate for survival changes with density. Otherwise population densities will not change with a multi-annual periodicity (only annual cycles will result). That is, cycles are only expected to be caused by dispersal if the pattern depicted in Figure 9.5b applies. If the patterns depicted in Figure 9.6 apply, stability is expected. Notice, however, that this is neither a necessary nor a sufficient condition for cycles to occur. It is only one possible component. Notice in particular that presaturation dispersal may, in other models, also cause increased stability of the population (e.g. Stenseth *et al.*, 1988; Stenseth and Sannes, 1988).

We can conclude from the modelling efforts to date that: (1) if dispersal is to contribute to destabilization of density, it seems to have to be of the presaturation type; and (2) that it seems unlikely that even presaturation dispersal could be a sufficient explanation of cyclic behaviour. Future research should focus on the possible role of dispersal as one component in a multi-factorial explanation of density cycles as advocated by Lidicker (1988).

9.4 PRESATURATION AND SATURATION DISPERSAL 15 YEARS LATER

Over the last 15 years or so, there has been an explosion of studies on dispersal including investigations into both its causes and demographic effects (e.g. Swingland and Greenwood, 1983; Rankin, 1985; Chepko-Sade and Halpin, 1987; Chapter 1). Moreover, small mammals figure importantly in this literature (Krebs, 1979, 1985; Stenseth, 1983; Lidicker, 1985b). Why should that be?

In the case of small mammals, we believe that the discovery of the positive relation between dispersal rate and population growth rate (Figure 9.5a; Krebs *et al.*, 1973; Lidicker, 1975) was important. Before this, it had been generally thought that dispersal served as a safety valve for populations at high densities (Figure 9.6a), whereas the prevalence of presaturation dispersal (Figure 9.5), in fact, might be the most common pattern. Secondly, we believe that the coining of the terms 'saturation' and 'presaturation dispersal' has been of some value. Authors started to refer to them, and this in time led to the phenomena involved being studied in more detail. The importance of such 'catchy' words in science is often overlooked (but see preface to Greenwood *et al.*, 1985). Thirdly, there was the growing realization of the pivotal role that dispersal can play in bridging the disciplines of ecology, evolution, behaviour, and conservation biology (Chapter 1). The importance of this development should not be underestimated.

The most important new theoretical development supporting the empirical observations of presaturation dispersal is probably the paper by Hamilton and May (1977), although many models have been developed since then. Notice that even though we suggest that a positive relation between the specific dispersal rate and the specific growth rate will often be a convenient operational way of detecting presaturation dispersal, the phenomenon is by no means linked only to strongly fluctuating or cyclic species. As originally suggested (Lidicker, 1975), any population showing dispersal before carrying capacity is reached shows presaturation dispersal (Figure 9.1), and many papers support the validity of this idea today.

Reviewing available quantitative models, we feel that a clear consensus has developed concluding that presaturation dispersal will be particularly pronounced when:

1. habitat patches last for a short time relative to the organism's generation time;
2. transition habitats are not too hostile;
3. patch suitabilities are asynchronized;
4. patches are small; and
5. effects of inbreeding are pronounced.

These are generally conditions which have been assumed to favour *r*-selected (*sensu* MacArthur and Wilson, 1967; Pianka, 1970) or opportunistic strategies (Southwood, 1977).

Finally, we conclude that presaturation dispersal and rodent cycles relate to each other in various important ways. First, it is easier, we believe, to detect presaturation dispersal in cyclic rodents because they more often than other groups of organisms have densities far below saturation levels. However, presaturation dispersal, or voluntary dispersal (as used in Chapter 2) or adaptive dispersal (as suggested by Stenseth, 1983) is not restricted to cyclic small rodents, a fact which has been appreciated since the beginning by some (e.g. Lidicker, 1975), but unfortunately is often forgotten. Presaturation dispersal and cyclicity may also be related in a more profound and causal way. Although we have shown that presaturation dispersal by itself is not a sufficient cause of rodent cycles, it may still be an important component of the causal mechanism.

An interesting problem which now remains is to determine whether cyclic small rodents exhibit more extensive presaturation dispersal than non-cyclic organisms under otherwise similar conditions. If this is in fact the case, several important corollaries will emerge for study. Firstly, the ecologically changing conditions that cyclic rodents experience may provide the appropriate selective scenario for presaturation dispersal to evolve (Chapter 1; section 9.2; Stenseth, 1983). Secondly, a predominance of presaturation dispersal among cyclic species may help us untangle the mystery of why so many microtine rodents exhibit fairly regular multi-annual density cycles, even though dispersal tends to be a stabilizing demographic influence. Because of this, there is also the interesting possibility that presaturation dispersal may influence the patterns of density changes without itself generating instabilities.

Perhaps there is a closer connection than we had thought previously between the stories about lemmings marching to the sea and drowning themselves and the mystery of the lemming cycle? We are unable at this time to provide empirically based support for such a connection, but some of the available theory indicates such a possibility. Probably lemmings will continue to play a central role in both the study of dispersal and the study of the microtine multi-annual cycle (Stenseth, 1985a; Framstad *et al.*, 1991).

ACKNOWLEDGEMENTS

For helpful comments on earlier drafts of this manuscript, we warmly thank A. Cockburn.

REFERENCES

Abramsky, Z. and Van Dyne, G.M. (1980) Field studies and a simulation model of small mammals inhabiting a patchy environment. *Oikos*, **35**, 80–92.

Allee, W.C., Emerson, A.E., Park, O. *et al.* (1949) *Principles of Animal Ecology*, W.B. Saunders, Philadelphia, 837 pp.

Anderson, P.K. (1980) Evolutionary implications of mictrotine behavioral systems on the ecological stage. *The Biologist*, 62, 70–88.

Anderson, P.K. (1989) *Dispersal in Rodents: a resident fitness hypothesis*, Special Publication No. 9, American Society of Mammalogists, 141 pp.

Bengtsson, B.O. (1978) Avoiding inbreeding: at what cost? *Journal of Theoretical Biology*, **73**, 439–44.

Bondrup-Nielsen, S. (1985) An evaluation of the effects of space use and habitat patterns on dispersal in small mammals. *Annales Zoologica Fennici*, **22**, 373–85.

Bujalska, G. (1985) Regulation of female maturation in *Clethrionomys* spp., with special reference to an island population of *C. glareolus*. *Annales Zoologica Fennici*, **33**, 331–42.

Charnov, E.L. and Finerty, J.P. (1980) Vole population cycles: a case for kin selection? *Oecologia*, **45**, 1–2.

Chepko-Sade, S.D. and Halpin, Z.T. (eds) (1987) *Mammalian Dispersal Patterns: the effects of social structure on population genetics*, University of Chicago Press, Chicago, 342 pp.

Chitty, D. (1960) Population processes in the vole and their relevance to general theory. *Canadian Journal of Zoology*, **38**, 99–113.

Chitty, D. (1967) The natural selection of self-regulatory behaviour in natural populations. *Proceedings of the Ecological Society of Australia*, **2**, 51–78.

Comins, H.N. (1982) Evolutionarily stable strategies for localized dispersal in two dimensions. *Journal of Theoretical Biology*, **94**, 579–606.

Comins, H.N., Hamilton, W.D. and May, R.M. (1980) Evolutionarily stable dispersal strategies. *Journal of Theoretical Biology*, **82**, 205–30.

Engen, E. and Stenseth, N.C. (1989) Age-specific optimal diets and optimal foraging tactics: a life-historic approach. *Theoretical Population Biology*, **36**, 281–95.

Falconer, D.S. (1960) *Introduction to Quantitative Genetics*, Ronald Press, New York, 365 pp.

Framstad, E., Stenseth, N.C. and Østbye, E. (1991) Population fluctuations and variation in demographic characteristics of *Lemmus lemmus* in an alpine habitat. *Biological Journal of the Linnean Society*, in press.

Futuyma, D. (1986) *Evolutionary Biology*, Sinaur, Sunderland, Massachusetts, 600 pp.

Gadgil, M. (1971) Dispersal: population consequences and evolution. *Ecology*, **52**, 253–61.

Gadgil, M., Joshi, N.V. and Gadgil, S. (1983) On the moulding of population viscosity by natural selection. *Journal of Theoretical Biology*, **104**, 21–42.

Gaines, M.S. and McClenaghan, L.R. Jr (1980) Dispersal in small mammals. *Annual Review of Ecology and Systematics*, **11**, 163–96.

Gaines, M.S., Stenseth, N.C., Johnson, M.L. *et al.* (1991) A response to solving the enigma of population cycles with a multifactorial perspective. *Journal of Mammalogy*, 72, 627–31.

Getz, L.L. (1978) Speculation on social structure and population cycles of microtine rodents. *The Biologist*, **60**, 134–47.

Gipps, J.H.W. (1985) Spacing behaviour and male reproductive ecology in voles of the genus *Clethrionomys*. *Annales Zoologica Fennici*, **22**, 343–51.

Greenwood, P.J., Harvey, P.H. and Slatkin, M. (1985) *Evolution. Essays in Honour of John Maynard Smith*, Cambridge University Press, Cambridge, 328 pp.

Hamilton, W.D. (1964) The genetical evolution of social behaviour (I and II). *Journal of Theoretical Biology*, 7, 1–52.

Hamilton, W.D. and May, R.M. (1977) Dispersal in stable habitats. *Nature*, **269**, 578–81.

Hansson, L. (1977) Spatial dynamics of field voles *Microtus agrestis* in heterogeneous landscapes. *Oikos*, **29**, 539–44.

Hasler, J.F. (1975) A review of reproduction and sexual maturation in the microtine rodents. *The Biologist*, **57**, 52–86.

Hestbeck, J.B. (1982) Population regulation of cyclic mammals: the social fence hypothesis. *Oikos*, **39**, 157–63.

Hestbeck, J.B. (1988) Population regulation of cyclic mammals: a model of the social fence hypothesis. *Oikos*, **52**, 156–68.

Iwao, S. (1977) The m–m statistics as a comprehensive method for analyzing spatial patterns of biological populations and its application to sampling problems, in *Japan International Biology Programme Synthesis* (ed. M. Morisita), **17**, 21–46.

Krebs, C.J. (1978) A review of the Chitty hypothesis of population regulation. *Canadian Journal of Zoology*, **56**, 2463–80.

Krebs, C.J. (1979) Dispersal, spacing behaviour, and genetics in relation to population fluctuations in the vole *Microtus townsendii*. *Fortschritte der Zoologie*, **25**, 61–77.

Krebs, C.J. (1985) Do changes in spacing behaviour drive population cycles in small mammals?, in *Behavioural Ecology: ecological consequences of adaptive behaviour* (eds R.M. Sibly and R.H. Smith), Blackwell Scientific, Oxford, pp. 295–312.

Krebs, C.J., Gaines, M.S., Keller, B.L. *et al.* (1973) Population cycles in small rodents. *Science*, **179**, 35–41.

Krebs, C.J. and Myers, J.H. (1974) Population cycles in small mammals. *Advances in Ecological Research*, **8**, 267–399.

Lidicker, W.Z. Jr (1973) Regulation of numbers in an island population of the California vole, a problem in community dynamics. *Ecological Monographs*, **43**, 271–302.

Lidicker, W.Z. Jr (1975) The role of dispersal in the demography of small mammals, in *Small Mammals: their productivity and population dynamics* (eds F.B. Golley, K. Petrusewicz and L. Ryszkowski), Cambridge University Press, London, pp. 103–28.

Lidicker, W.Z. Jr. (1985a) Dispersal, in *Biology of New World* Microtus (ed. R.H. Tamarin), Special Publication No. 8, American Society of Mammalogists, pp. 420–54.

Lidicker, W.Z. Jr (1985b) An overview of dispersal in non-volant small mammals,

PART SIX

Invited commentaries

This section of the book is unique and quite special. Generally in modern science it is the case that specialists speak mainly to other specialists in their own discipline. Important as this is, and this book certainly contributes to such exchanges, we feel that in the long term it leads to fragmentation and consequently to intellectual stagnation. Just as the long-distance disperser hopes to achieve some successful new combination of genetic material, so too the scientist needs the intellectual heterosis of interdisciplinary exchange of ideas.

It is in this spirit of communication across a broad area of biology that we conceived this part of the book. We invited five well-known and respected experts on dispersal in groups of organisms other than small mammals to contribute commentaries on our basic chapters. We were anxious to know how they would react to the data presented by small mammal specialists and to their ways of conceptualizing that information. We asked them to respond to these chapters from their own quite different perspectives. To our pleasure and surprise, all of them accepted enthusiastically. We guessed that small mammal ecologists would profit immensely from reading these contributions, and we hoped, moreover, that non-small mammal researchers would likewise gain from this bridge to our subject.

The first two of these commentaries are by mammalogists, but who work on large species (Chapters 10 and 11). Anthony Sinclair is an expert on ungulates, and has worked extensively on African and North American species. Anne Pusey is a specialist on primates but has also done important research on lions. Both comment on the relatively better visibility of large mammals and the consequent advantage of being able to readily make direct observations of behaviour, including movements. Of course, this is coupled with the disadvantage that it takes a long time to follow many individuals through their life spans. Both groups of large mammals also are alike in exhibiting fairly stable population densities, and in being relatively K-selected. This leads in general to quite different dispersal patterns from the more wildly fluctuating and short-lived typical small mammal. Sinclair further emphasizes the important differences in spatial scale that characterize habitat heterogeneity in large versus small species, and these in turn lead to some semantic inconsistencies among researchers. One difference between the two groups of large mammals is that while primate researchers know a great deal about social factors leading to dispersal, ungulate biologists need much more information in this area before they can begin to see patterns. Finally, Pusey provides glimpses into the lives of the four species of primates which show predominately female dispersal, an unusual pattern for mammals. To this we can add a fifth species, North American humans (Koenig, 1989).

Adam Watson gives us insights from the perspective of an expert on grouse biology (Chapter 12). His counsel is especially valuable because he also has a long-standing familiarity with the small mammal literature. He provides direct comparisons on the two model systems over the whole gamut of issues from logistics and semantics to the role of dispersal in demographic processes. Since grouse and microtine rodents seemingly share the unusual feature of exhibiting multi-annual patterns of density fluctuations, these comparisons are particularly welcomed.

Chapter 13 has been written by two entomologists, both of whom have had extensive experience with research on insect movements. George Roderick and Roy Caldwell give us an impressive summary of the entomological perspective, and moreover connect this at every step with that of the small mammal ecologist. They give us a historical vignette of the separate paths taken in these two disciplines, explain the special features of insect dispersal and migration, point out where entomologists have made specially significant contributions to the subject, indicate where cross-fertilization is most likely to be productive, and finally give us the benefit of their thoughts on future directions.

After 'digesting' these commentaries, we experienced an infusion of hybrid vigour that will for sure enhance our future intellectual 'fitness'. We anticipate that other readers will share in these benefits.

REFERENCE

Koenig, W.D. (1989) Sex-biased dispersal in the contemporary United States. *Ethology and Sociobiology*, **10**, 263–77.

10

Do large mammals disperse like small mammals?

A.R.E. SINCLAIR

10.1 INTRODUCTION

This chapter asks the question: does dispersal in large mammals differ in principle from that described in small mammals? The subject has received limited treatment in carnivores (Pusey and Packer, 1987) and primates (Chapter 11), and with a few exceptions (McCullough, 1985) is largely neglected in ungulates. Since these species are visible and well described, one would have expected the dispersal phenomenon to be thoroughly documented; yet it is not. I define 'dispersal' as one-way movement with no predetermined direction in the population as a whole (Sinclair, 1986a). This is distinct from 'emigration', which is a directional one-way movement seen most commonly in birds breeding in temporary habitats; and also distinct from 'migration', which is a regular round trip movement within the life-span of the individual (see Chapter 1 for alternative definitions).

10.2 PATCHES AND DEMES

In the literature on small mammals it is implied that patches of suitable habitat are often separated by expanses of inferior habitat (Anderson, 1970; Wolff, 1980; Chapters 6 and 8). Animals disperse either by running the gauntlet across this inferior habitat to reach new optimal patches (Chapter 7), or along corridors (Fahrig and Merriam, 1985), or they remain in the suboptimal patches where they usually die before breeding (Dolbeer and Clark, 1975; Stenseth 1983; Chapter 4).

Is this scenario suitable for large mammals? Generally it is not. Large mammals certainly show well-developed habitat selection, this being particularly well studied in ungulates. However, these preferences are on a fine scale relative to the size of an individual's home range, or even average daily movement. Moose (*Alces alces*) can move from one willow patch to another several hundred metres away within a few minutes; impala

(*Aepiceros melampus*) from thicket to thicket; and water buffalo (*Bubalus bubalis*) from swamp to swamp in the same way (personal observations). An observer recording these movements on either a daily or seasonal basis would not consider them dispersal movements. Therefore, one distinction between the two groups (large and small mammals) is a matter of scale: large animals see their habitats as fine-grained and can average them over their home range. Consequently the population can spread more evenly over large areas, as moose do in the boreal forest and impala do in tropical savannah.

There is another scale problem that results from recording methods, a problem emphasized by Krebs (Chapter 7). Populations of small mammals are defined arbitrarily by those living within a study area (such as a trapping grid). In some cases dispersers are defined as those that leave the area, or those that disappear, confounding mortality with movement. In other cases (as in Chapter 7) dispersers are those that appear in an area previously cleared of animals. The reasons for these techniques and definitions are firstly that one cannot observe the extent of large populations, or demes, and secondly, that one cannot follow the movements of individuals for long periods. Since both of these are possible with large mammals, the small mammal techniques and definitions of dispersal are not employed by workers on large species.

For example, a study area of 10 km^2 set in a valley bottom with conifer forest in northern British Columbia, Canada, would contain many moose, since this is optimal winter habitat. This situation would be analogous to the small mammal study area. In spring a proportion of the moose population (mainly males) would disappear. By defining dispersers as those that disappear, these would be dispersers. Biologists know, however, that the animals are moving up the valley sides to reach new vegetation within their annual home range, and these seasonal movements are not regarded by large mammal biologists as dispersal (LeResche, 1974). With new techniques of radiotelemetry, smaller animals such as snowshoe hares and voles can be followed continuously. Boutin *et al.* (1985) have found that dispersal is either absent or very rare in snowshoe hares, contrary to the data derived from removal plots. Consequently some definitions of dispersal may be misleading (see also Chapter 1, Appendix 1, Chapter 4 and Chapter 7 for a critique of methods used to identify dispersal).

Given the greater visibility of large mammals, do we see dispersal occurring? Two situations are identified: those that leave the deme or whole population, and those that move within the deme. In the former case, few if any leave the population (Houston, 1982; Clutton-Brock *et al.*, 1985; Skogland, 1985). In the Serengeti National Park, Tanzania, ungulate populations are bounded by human or habitat barriers. Animals are capable of crossing the natural barriers, but at least in wildebeest (*Connochoetes taurinus*) and African buffalo (*Syncerus caffer*) none of them do so (personal observation). In northern Australia feral water buffalo inhabit riverine flood plains, and are mostly confined to them. But there are occasional reports of single males wandering several hundred kilometres across unsuitable dry habitat. A similar situation occurred with moose in British Columbia and elk (*Cervus elephas*) and American bison (*Bison bison*) in Yellowstone Park, USA (Houston, 1982; M. Meagher, personal communication). Thus in ungulates, at least, only a few interdemic dispersers have been recorded.

Within a large mammal population, however, movements between breeding groups (intrademic dispersal) are more frequently observed, especially in primates and carnivores (Bekoff, 1977; Packer, 1979, 1985; Dobson, 1982; Pusey and Packer, 1987). Amongst ungulates there is variation ranging from dispersal of all offspring in monogamous territorial species (Hendrichs, 1975; Bobek, 1977), to some dispersal of both sexes in those species with loose social organization (Bunnell and Harestad, 1983; Howard, 1986), to largely male dispersal in those that live in permanent herds (Sinclair, 1977; Murray, 1982). Thus in ungulates different social organization leads to different dispersal rates, a conclusion also reached by Bekoff (1977) for carnivores and Brandt (Chapter 5) for small mammals.

10.3 DISPERSAL IN INCREASING AND STATIONARY POPULATIONS

Presaturation dispersal is defined by Lidicker (1975, 1985; and used as such throughout this book) as that occurring when the carrying capacity of the habitat has not been reached. In contrast, saturation dispersal occurs when habitats are filled to capacity. These definitions imply that resources are superabundant in the former while limited in the latter case. Similar notions have also been labelled 'adaptive' and 'non-adaptive', respectively (Stenseth, 1983; Chapter 6). However, these definitions can be misleading because there may be a temptation to define them operationally according to whether the population is changing or stationary. In such cases, presaturation (adaptive) dispersal would be viewed as that occurring when the population is increasing or decreasing, while saturation dispersal is that occurring when the population is stationary. The implication would be that an increasing or decreasing population is not at carrying capacity, and of course this is ill-founded. The concept of carrying capacity has been reviewed more in the large mammal than small mammal literature (e.g. Caughley, 1976, 1979; Sinclair, 1981). Two important points are: (1) both increasing and decreasing populations can be above carrying capacity as well as at or below it, depending on whether the population is temporarily overshooting or not; and (2) a stationary population may be held below carrying capacity by other factors.

Since I identify presaturation and saturation dispersal operationally by the dynamic state of the population – increasing or stationary – they should be so labelled. Similarly, it is dangerous to say that a behaviour is either 'adaptive' or 'non-adaptive', since in most cases one does not know. Although Stenseth (1983) based his terminology on a theoretical argument, the 'non-adaptive' label was applied because most dispersers from stationary populations of small mammals die without breeding, which superficially appears non-adaptive. It is also argued that most of these 'non-adaptive' dispersers are forced out of an area and have no choice. Neither of these arguments hold, for if a disperser has a greater reproductive success (no matter how small) by leaving, then it should do so as an adaptive trait. Observations of aggression by territory holders tell us only about those animals, not about the recipients of aggression; in particular these observations do not tell us whether the recipient has no choice or is choosing between leaving and staying to fight it out. If the animal is making a choice then this is 'adaptive' behaviour. However, it is usually unknown whether dispersers are making a choice or not, and, therefore, it is premature to label dispersal in stationary populations as 'non-adaptive'. In short, general terms which imply undemonstrated general causes should be avoided (see Chapter 2 for further discussion of these issues).

Does 'increaser' and 'stationary' dispersal occur in large mammals? Since in my view increaser dispersal is peculiar to cyclic small mammal populations (Chapter 7), I would not expect to find it in large mammals except in the temporary situations of expanding populations. Even in these circumstances the proportion of the population dispersing is very small. For example, in elk it was around 3% (Houston, 1982).

The majority of dispersal in large mammals falls in the 'stationary' category. Ultimate and proximate causes for this dispersal are discussed below, but note that no author regards this dispersal as non-adaptive (see below). Thus large mammals are either very different from small mammals in the form and function of their social behaviour, or there is a major semantic difference between workers in these two areas. Future research could and should help to distinguish between these alternatives.

10.4 NATAL, BREEDING, AND TRANSFER DISPERSAL

Cockburn (Chapter 4) identifies and reviews three common forms of dispersal in small mammals defined largely by the stage in the life-cycle when they occur (see also Greenwood, 1980; Greenwood and Harvey, 1982). Thus natal dispersal occurs before breeding starts and is very largely male-biased. Subadult males leave the group in which they were born and, if they survive, enter a new group. In large mammals this is also the predominant form of dispersal within the deme, and is male-biased (Greenwood, 1980; Dobson, 1982). Detailed information is available from primates (e.g. Packer, 1979), carnivores (Pusey and Packer, 1987) and some ungulates (Murray, 1982). In lions (*Panthera leo*) all young males are evicted from their natal pride and become nomadic until they can take over a new pride (Packer *et al.*, 1988). With impala, only 2% of females leave their natal herd, while 73% of young males do so. These males form bachelor herds some distance away (>1 km) from their natal site (Murray, 1982). However, these bachelor herds associate with female herds; at least in the Serengeti each female herd has a satellite bachelor herd nearby which is kept away only by the aggressive behaviour of territorial males. This is an example of a complex social structure, but a similar male-biased regrouping occurs in the simpler social organization of wildebeest and topi (*Damaliscus korrigum*). In both species, close to the time when females give birth, yearling (subadult) males congregate together either on the periphery of the female herds or in separate groups. Because there is little obvious movement involved, merely a reassortment of age–sex groups, this has not been seen as dispersal. Clearly this regrouping behaviour is homologous with the formation of bachelor male groups in impala described above or in African buffalo (Sinclair, 1977), but because the distances moved are less (perhaps only a few tens of metres in topi) it is not seen as a form of dispersal.

Cockburn (Chapter 4) considers breeding dispersal in female small mammals to refer to females abandoning their home ranges to their weaned offspring. As these females move on to new habitats they breed rapidly, again abandoning their weaned litters and moving on to breed. Presumably some of these young return to the permanent habitats when conditions become unfavourable and the temporary habitats become uninhabitable. The original adult females are short-lived and will have died before the habitats become uninhabitable. Strictly this ought to be called 'emigration' rather than dispersal (Sinclair, 1986a; Chapter 1) and is similar to the breeding emigrations of crossbills (*Loxia curvirostra*) in holarctic boreal forests, red-billed queleas (*Quelea quelea*) in Africa, and budgerigars (*Melopsittacus undulatus*) in Australia (Sinclair, 1986a). To my knowledge this form of nomadic breeding movement is undescribed in large mammals, partly because the conditions where it might occur are difficult to study. It is possible that this type of dispersal occurs in lynx (*Lynx canadensis*) during the decline and low phases of the snowshoe hare cycle, for lynx perform extraordinary long distance movements (>500 km) at this time (Ward and Krebs, 1985). It is also possible that some gazelle species, in the arid areas of Africa and Asia, also perform nomadic movements, but little is known about these species (Sinclair, 1983a). Apart from these potential examples, large mammals do not show breeding dispersal. One reason is that, for large mammals, the habitat longevity/individual longevity ratio is either very high or very low. In the former case there is no need for animals to move; in the latter case (which describes seasonal changes) animals merely perform breeding migrations. Caribou (*Rangifer tarandus*) move to the tundra, and wildebeest move to short grass plains, both to give birth in ephemeral habitats, before returning to other more permanent habitats. In contrast, the same ratio for small mammals in temporary habitats is probably much closer to unity. A second reason is that large mammals, with their relatively low reproductive rates, have little to gain from emigrating to temporary habitats.

Transfer is a term used by Cockburn (Chapter 4) to describe movements of breeding animals from one social group to another. This might describe some male breeding groups in lions moving from pride to pride, but it does not happen often (Packer *et al.*, 1988). Transfer of males has been described in some primate species (Packer, 1985), and it may also occur at a low rate in ungulates with a well-defined herd structure (e.g. impala and African buffalo). However, in general, transfer is not a common process among large mammals.

10.5 BREEDING SYSTEMS AND THE FUNCTIONS OF DISPERSAL

The reviews of mating systems (Greenwood, 1980, 1983; Dobson, 1982) conclude that patterns of dispersal are closely related to the type of mating system. These reviewers suggest that in mammals, females are concerned with obtaining resources while males compete for mates. In general, males disperse in promiscuous and polygynous species because they are more likely to find new mates by doing so while females are philopatric because they are more likely to secure resources in areas known to them; but both sexes disperse in monogamous species and these generalities apply equally to small and large mammals (Dobson, 1982). Johnson (1986a) has made an alternative suggestion that the sex for which juvenile dispersal causes the greater delay in age at first breeding will be the least likely to disperse; and in mammals this is usually females.

If the ultimate evolutionary function of dispersal is to improve reproductive success, then dispersal is likely to be age-specific. Morris (1982) has predicted from a general evolutionary model that dispersal should occur either at young ages before reproduction or at old ages (see also Chapter 9). Old animals should disperse when their future individual contribution to the total fitness of a kin group becomes small relative to that of the offspring in the same area. Then it is better for the parent rather than the offspring to leave. Evidence for early (natal) dispersal is well supported in the literature but there are few data in small mammal studies to test the prediction for older age dispersal. However, in large mammals where age groups are easier to identify, there is some evidence to support Morris's predictions: for example, African buffalo males older than 10 years leave the breeding herds (Sinclair, 1977), and older male lions leave their prides (Packer *et al.*, 1988).

The proximate causes of dispersal are the subject of some debate, but they generally fall into three categories – competition for mates, avoidance of inbreeding, and competition for resources – of which the first two have received the most attention. These subjects have already been well reviewed (Greenwood, 1980; Greenwood and Harvey, 1982; Dobson, 1982; Moore and Ali, 1984; Dobson and Jones, 1985; Waser *et al.*, 1986). However, since much of the debate centres on the unusual dispersal patterns of zebra (*Equus burchelli*) and wild dogs (*Lycaon pictus*), it is worth considering these two more carefully.

Dobson (1982) states that the mate competition hypothesis predicts juvenile males as the predominant dispersers in polygynous and promiscuous mammals, while both sexes should disperse in monogamous species. The inbreeding avoidance and resource competition hypotheses do not predict such a relationship. Dobson claims that the large majority of case studies support the mate competition hypothesis. One of the exceptions to his conclusion is the wild dog, in which females disperse (Frame *et al.*, 1979), but they are classified as monogamous. A single female produces a large litter (as many as nine pups) and she needs many males to feed them (since males do most of the hunting). Two or more females with their litters could possibly overburden the males' ability to supply food. Therefore, females should compete to obtain as many males as they can, by excluding other females or killing their young. Johnson's (1986a) hypothesis above would predict female dispersal in wild dogs because males would gain few mating opportunities by dispersing whereas they can at least act as helpers by raising a sibling's pups; and in contrast subordinate females have low reproductive success if they are philopatric.

Moore and Ali (1984) extend the arguments for mate or resource competition and conclude that inbreeding avoidance is an unnecessary idea. They discuss female dispersal in zebra: subadult females attract other males, one of which eventually takes her from her natal group. Moore and Ali interpret this as male competition where the original owner loses because of the large number of males. However this is an inconsistent argument, for if mate competition is the relevant process, then harem males should be the dominant animals (by definition since they have acquired and maintained a harem). Hence one would expect that these males would win at least a significant proportion of contests over subadult females. Yet harem males (almost) never win such contests, a result inconsistent with the mate competition hypothesis. Furthermore, there seems to be a hidden assumption that if inbreeding avoidance is the process involved, females should leave without the need of a male consort. In fact, to do this would be dangerous indeed, for predation appears to be a significant mortality factor in zebra (Sinclair and Norton-Griffiths, 1982) and lone females would not survive long. It is possible, therefore, that females are reducing the cost of dispersal by advertising, and hence allowing new males to come to them instead. In short, inbreeding avoidance remains a plausible cause for dispersal in zebra as it does for some primates (Packer, 1985). In many cases mate competition and inbreeding avoidance predict the same dispersal phenomena and hence it would be unwise to suggest a single function for dispersal (Dobson and Jones, 1985).

10.6 DEMOGRAPHIC CONSEQUENCES OF DISPERSAL

If a small mammal species that shows population cycles is enclosed by a fence, there are marked changes in demography: the population reaches abnormally high density, the habitat is altered, fecundity and mortality schedules change and the population crashes. This is the 'fence effect' described by Krebs (Chapter 7), and from which he concludes that dispersal is necessary to maintain population cycles in microtine rodents. Furthermore, Krebs concludes that in these species dispersal is of the presaturation ('increaser') type, and is density-independent. Conversely, in non-cycling rodents, dispersal is of the saturation type and density–dependent. Krebs (1986) considers this type of dispersal to occur also in cycling snowshoe hare populations (but see Wolff, 1980). How do these ideas apply to large mammals?

McCullough (1985) has suggested that in solitary ungulates or those forming small groups such as white-tailed deer (*Odocoileus virginianus*), females disperse as individuals, and that dispersal is of the 'increaser' type. However, there is little evidence that dispersal takes place only during periods of population increase. In the black-tailed deer (*O. hemionus*) and southern reedbuck (*Redunca arundinum*) female dispersal occurs when the population is stationary (Bunnell and Harestad, 1983; Howard, 1986).

In ungulates living in moderately sized groups within home ranges, such as bighorn sheep (*Ovis canadensis*) and elk, females disperse in groups by splitting off from the parent group and setting up an adjacent home range (McCullough, 1985). This process has also been seen in American bison in Yellowstone (M. Meagher, personal communication) and in the MacKenzie Bison Sanctuary, NWT, Canada (Larter and Gates, 1990; Gates and Larter, 1990), and in muskox (*Ovibos moschatus*; Smith, 1989). Group fragmentation results in slow rates of population expansion. Caughley (1963) recorded dispersal rates of females from eight ungulate species introduced to New Zealand as being less than 1.5 km y^{-1}, and only the chamois (*Rupicapra rupicapra*) exceeded this (8.7 km y^{-1}). These examples are from increasing populations expanding into new habitat, and, therefore, they describe only transient situations. Dispersal in these expanding populations, however, is not necessarily of the 'increaser' type. Caughley (1970), for example, shows that dispersal in thar (*Hemitragus jemlahicus*) is from a central core population at or above carrying capacity, i.e. this is 'stationary' dispersal.

However, female dispersal may not occur once a population has inhabited all the available range and other regulatory mechanisms may take over (see below). Very large aggregations of ungulates such as wildebeest (personal observation) or caribou (Tyler, 1986) may expand their range occasionally by dispersing as large groups. This may not occur when the population is stationary, and hence is one of the few examples of pre-saturation dispersal in ungulates.

Clutton-Brock and Albon (1985) have suggested that where females live in groups, competition for food is common between group members. Dominant females have priority of access, survive better, and show superior breeding success. This effect is density–dependent, becoming more pronounced as the group size increases. Since group size increase in correlated with population increase, this means that density effects are disproportionately imposed on subordinate individuals via behavioural mechanisms (Sinclair, 1977, 1986b; Clutton-Brock and Albon, 1985).

Behavioural effects of density can result in greater mortality or lower reproduction as seen in red deer (Clutton-Brock and Albon, 1985) and red-necked wallaby (*Macropus rufogriseus*; Johnson, 1986b). Dominance behaviour may be a proximate cause of mortality in species with tight social groups, and dispersal may play little or no part in regulation. For example, dispersal plays little part in the regulation of red deer (*Cervus elephas*) and elk (Clutton-Brock *et al.*, 1985; Sauer and Boyce, 1983; Boyce, 1990), African buffalo and wildebeest (Sinclair, 1977, 1989; Sinclair *et al.*, 1985), and a number of other species including American bison, deer and sheep (Fowler, 1981) and red-necked wallaby (Johnson, 1986b). Some of these species such as red deer and African buffalo have well-defined social structure where female dominance hierarchies can affect competition for food by displacing subordinate animals from good food patches.

In contrast to strongly social species, monogamous or solitary species of large mammals often respond to declining resources through female dispersal (Clutton-Brock and Albon, 1985). Howard's (1986) study of southern reedbuck supports this idea. Bobek (1977) reports that roe deer (*Capreolus capreolus*) are regulated by summer food determining territory size. This results in yearlings of both sexes dispersing. Thus dispersal is the proximate regulating mechanism, but the role of the two sexes is not clear. Similarly, in the monogamous dikdik (*Madoqua kirki*) both sexes disperse. However, other examples do not fit the Clutton-Brock and Albon hypothesis: for example, moose, which are largely solitary, do not exhibit female dispersal when resources are limiting (Gasaway *et al.*, 1985; R.O. Peterson, personal communication). In general, therefore, the relationship between dispersal and social organization as regulating mechanisms suggested by Clutton-Brock and Albon, is a useful start but more data are needed.

The 'fence effect' has not been tested for large mammals although there is a widely held belief (without foundation) that fenced populations have to be culled to prevent substantial habitat alteration. One of the problems is that no fenced population has been allowed to remain undisturbed by man long enough to test whether habitat 'damage', 'overgrazing' and population collapse really do occur as a permanent effect, or whether habitat change is temporary, on the way to a stable equilibrium (Sinclair, 1983b). Island populations of reindeer which show eruptions and crashes (Scheffer, 1951; Klein, 1968) are difficult to interpret because they involve introduced species in habitats not adapted to grazing; one cannot conclude that lack of dispersal was the cause of the demographic changes. Leader-Williams (1988) does consider that island size (and hence the fence effect) at least partly determines whether reindeer populations crash or not. Other island populations such as red deer on Rhum (Clutton-Brock *et al.*, 1985) and moose on Isle Royale (R.O. Peterson, personal communication) do not show the fence effect.

10.7 CONCLUSIONS

Concepts of dispersal feature prominently in small mammal research. How do these apply to large mammals? First, some descriptions of dispersal may be artifacts of methodology (e.g. removal plots), and may only monitor movements within a home range; such movements are not recognized as dispersal in large mammals, as they are in some small mammal studies. Second, scale differences in the two mammal groups result in different types of movement: small mammals experience the habitats as coarse-grained, and disperse between optimal patches or into suboptimal patches, while large mammals include many patches within their home range (fine-grained habitats) and move among them on a daily basis. The latter has not been called dispersal in the large mammal literature (but see Chapter 1). Large mammals change habitats seasonally, and thus they differ from small mammals in showing habitat migrations rather than habitat dispersal. On the whole there is little interpopulation (interdeme) dispersal in large mammals. Third, the peculiarities of cycling microtine populations have led to the ideas of 'increaser' (presaturation) dispersal as distinct from 'stationary' (saturation) dispersal exhibited by non-cycling small mammals. Large mammals are similar to the non-cycling small mammals and there is little evidence of 'increaser' dispersal. In many small mammals 'stationary' dispersal involves high mortality before breeding (and is sometimes described as non-adaptive), whereas in large mammals a much lower proportion of dispersers die before breeding, and dispersal is considered adaptive. Fourth, natal dispersal is predominant in large mammals as it is in small mammals. Transfer is described particularly in primates, and breeding dispersal is relatively uncommon. Fifth, males are by far the most common dispersing sex, and the subadult age-groups predominate. In these respects large and small mammals are similar. There is evidence from large mammals for both mate competition and inbreeding avoidance as proximate causes of dispersal, and it is premature to conclude that there is a single function for dispersal. Finally, dispersal may be important in limiting density of cycling microtine populations (Chapter 7), whereas it is relatively less important in limiting large mammal populations. Caughley and Krebs (1983) conclude that small mammal populations are likely to be 'self-regulated' whereas those of large mammals are 'extrinsically regulated'. They suggest that 'self-regulation' involves behavioural mechanisms like dispersal and territoriality, and they conclude that behaviour plays a more conspicuous role in population regulation of small mammals compared to that of large mammals. At present more information is needed on the social regulation of large mammals.

Behaviour plays a part in the regulation of large mammals by determining who obtains the resources and who does not – the losers either die or disperse. Other aspects of 'self-regulation' mentioned by Caughley and Krebs are more dubious. The most parsimonious idea is that behaviour (such as that producing dominance hierarchies or dispersal) is an adaptation by individuals to avoid resource limitation, and that this has demographic consequences. Some consequences are discussed here, others are outlined by Clutton-Brock and Albon (1985).

REFERENCES

Anderson, P.K. (1970) Ecological structure and gene flow in small mammals. *Symposium of the Zoological Society of London*, **26**, 299–325.

Bekoff, M. (1977) Mammalian dispersal and the ontogeny of individual behavioral phenotypes. *American Naturalist*, **111**, 715–32.

Bobek, B. (1977) Summer food as the factor limiting roe deer population size. *Nature*, **268**, 47–9.

Boutin, S., Gilbert, B.S., Krebs, C.J. *et al.* (1985) The role of dispersal in the population dynamics of snowshoe hares. *Canadian Journal of Zoology*, **63**, 106–15.

Boyce, M.S. (1990) *The Jackson Elk Herd: intensive wildlife management in North America*, Cambridge University Press, Cambridge.

Bunnell, F. and Harestad, A.S. (1983) Dispersal and dispersion of black-tailed deer: models and observations. *Journal of Mammalogy*, **64**, 201–9.

Caughley, G. (1963) Dispersal rates of several ungulates introduced into New Zealand. *Nature*, **200**, 280–1.

Caughley, G. (1970) Eruption of ungulate populations, with emphasis on Himalayan thar in New Zealand. *Ecology*, **51**, 53–72.

Caughley, G. (1976) Plant–herbivore systems, in *Theoretical Ecology* (ed. R.M. May), W.B. Saunders, Philadelphia, pp. 94–113.

Caughley, G. (1979) What is this thing called carrying capacity?, in *North American Elk: ecology, behaviour and management* (eds M.S. Boyce and L.D. Haydon-Wing), University of Wyoming Press, Laramie, pp. 2–8.

Caughley, G. and Krebs, C.J. (1983) Are big mammals simply little mammals writ large? *Oecologia*, **59**, 7–17.

Clutton-Brock, T.H. and Albon, S.D. (1985) Competition and population regulation in social mammals, in *Behavioural Ecology – ecological consequences of adaptive behaviour* (eds R.M. Sibly and R.H. Smith), Blackwell Scientific, Oxford, pp. 557–75.

Clutton-Brock, T.H., Major, M. and Guiness, F.E. (1985) Population regulation in male and female red deer. *Journal of Animal Ecology*, **54**, 831–46.

Dobson, F.S. (1982) Competition for mates and predominant juvenile male dispersal in mammals. *Animal Behaviour*, **30**, 1183–92.

Dobson, F.S. and Jones, W.T. (1985) Multiple causes of dispersal. *American Naturalist*, **126**, 855–8.

Dolbeer, R.A. and Clark, W.R. (1975) Population ecology of snowshoe hares in the central Rocky Mountains. *Journal of Wildlife Management*, **39**, 535–49.

Fahrig, L. and Merriam, G. (1985) Habitat patch connectivity and population survival. *Ecology*, **66**, 1762–8.

Fowler, C.W. (1981) Density dependence as related to life history strategy. *Ecology*, **62**, 602–10.

Frame, L.H., Malcolm, J.R., Frame, G.W. and van Lawick, H. (1979) Social organization of African wild dogs (*Lycaon pictus*) on the Serengeti Plains, Tanzania 1967–1978. *Zeitschrift für Tierpsychologie*, **50**, 225–49.

Gasaway, W.C., Dubois, S.D., Preston, D.J. and Reed, D.J. (1985) Home range formation and dispersal of subadult moose in interior Alaska. Fed. Aid Wildl. Res. Proj. Final Report W-22-2, Job 1.26R, 26 pp. (mimeograph).

Gates, C.C. and Larter, N.C. (1990) Growth and dispersal of an erupting large herbivore population in northern Canada: the MacKenzie wood bison (*Bison bison athabascae*). *Arctic*, **43**, 231–8.

Greenwood, P.J. (1980) Mating systems, philopatry and dispersal in birds and mammals. *Animal Behaviour*, **28**, 1140–62.

Greenwood, P.J. (1983) Mating systems and the evolutionary consequences of dispersal, in *The Ecology of Animal Movement* (eds I.R. Swingland and P.J. Greenwood), Clarendon Press, Oxford, pp. 116–31.

Greenwood, P.J. and Harvey, P.H. (1982) The natal and breeding dispersal of birds. *Annual Review of Ecology and Systematics*, **13**, 1–21.

Hendrichs, H. (1975) Changes in a population of dikdik, *Madoqua (Rhynchotragus) kirki* (Gunther 1880). *Zeitschrift für Tierpsychologie*, **38**, 55–69.

Houston, D.B. (1982) *The Northern Yellowstone Elk*, Macmillan, New York.

Howard, P.C. (1986) Spatial organization of common reedbuck with special reference to the role of juvenile dispersal in population regulation. *African Journal of Ecology*, **24**, 155–71.

Johnson, C.M. (1986a) Sex-biased philopatry and dispersal in mammals. *Oecologia*, **69**, 626–7.

Johnson, C.M. (1986b) Philopatry, reproductive success of females, and maternal investment in the red-necked wallaby. *Behavioural Ecology and Sociobiology*, **19**, 143–50.

Klein, D.R. (1968) The introduction, increase, and crash of reindeer on St. Matthew Island. *Journal of Wildlife Management*, **32**, 350–67.

Krebs, C.J. (1986) Are lagomorphs similar to other small mammals in their population ecology? *Mammal Review*, **16**, 187–94.

Larter, N.C. and Gates, C.C. (1990) Home ranges of wood bison in an expanding population. *Journal of Mammalogy*, **71**, 604–7.

Leader-Williams, N. (1988) *Reindeer in South Georgia*, Cambridge University Press, Cambridge.

LeResche, R.E. (1974) Moose migrations in North America. *Naturaliste Canadien*, **101**, 393–415.

Lidicker, W.Z. (1975) The role of dispersal in the demography of small mammals, in *Small Mammals: productivity and dynamics of populations* (eds K. Petrusewicz, F.B. Golley and L. Ryszkowski), Cambridge University Press, Cambridge, pp. 103–28.

Lidicker, W.Z. (1985) An overview of dispersal in non-volant small mammals, in *Migration: mechanisms and adaptive significance* (ed. M.A. Rankin), Contributions in Marine Science, Suppl. 27, pp. 369–85.

McCullough, D.R. (1985) Long range movements of large terrestrial mammals, in *Migration: mechanisms and adaptive significance* (ed. M.A. Rankin), Contributions in Marine Science, Suppl. 27, pp. 444–65.

Moore, J. and Ali, R. (1984) Are dispersal and inbreeding avoidance related? *Animal Behaviour*, **32**, 94–112.

Morris, D.W. (1982) Age-specific dispersal strategies in iteroparous species: who leaves when? *Evolutionary Theory*, **6**, 53–65.

Murray, M.G. (1982) Home range, dispersal and the clan system of impala. *African Journal of Ecology*, **20**, 253–69.

Packer, C. (1979) Inter-troop transfer and inbreeding avoidance in *Papio anubis*. *Animal Behaviour*, **27**, 1–36.

Packer, C. (1985) Dispersal and inbreeding avoidance. *Animal Behaviour*, **33**, 676–7.

Packer, C., Herbst, L., Pusey, A.E. *et al.* (1988) Reproductive success of lions, in *Reproductive Success* (ed. T. Clutton-Brock), University of Chicago Press, Chicago, pp. 363–83.

Pusey, A.E. and Packer, C. (1987) Evolution of sex-biased dispersal in lions. *Behaviour*, **101**, 275–310.

Sauer, J.R. and Boyce, M.S. (1983) Density dependence and survival of elk in northwestern Wyoming. *Journal of Wildlife Management* **47**, 31–7.

Scheffer, V.B. (1951) The rise and fall of a reindeer herd. *Science Monthly*, **73**, 356–62.

Sinclair, A.R.E. (1977) *The African Buffalo*, The University of Chicago Press, Chicago.

Sinclair, A.R.E. (1981) Environmental carrying capacity and the evidence for overabundance, in *Problems in Management of Locally Abundant Wild Mammals* (eds P.A. Jewell and S. Holt), Academic Press, New York, pp. 247–57.

Sinclair, A.R.E. (1983a) The adaptations of African ungulates and their effects on community function, in *Tropical Savannas, Ecosystems of the World*, Vol. 13 (ed. F. Bourliere), Elsevier, Amsterdam, pp. 401–26.

Sinclair, A.R.E. (1983b) Management of conservation areas as ecological baseline controls, in *Management of Large Mammals in African Conservation Areas* (ed. N. Owen-Smith), Haum, Pretoria, pp. 13–22.

Sinclair, A.R.E. (1986a) The function of distance movements in vertebrates, in *The Ecology of Animal Movement*, 2nd edn (eds I.R. Swingland and P.J. Greenwood), Clarendon Press, Oxford, pp. 240–58.

Sinclair, A.R.E. (1986b) Testing multi-factor causes of population limitation: an illustration using snowshoe hares. *Oikos*, **47**, 360–4.

Sinclair, A.R.E. (1989) Population regulation in animals, in *Ecological Concepts* (ed. J.M. Cherrett), Blackwell Scientific, Oxford, pp. 197–241.

Sinclair, A.R.E., Dublin, H. and Borner, M. (1985) Population regulation of the Serengeti wildebeest: a test of the food hypothesis. *Oecologia*, **65**, 266–8.

Sinclair, A.R.E. and Norton-Griffiths, M. (1982) Does competition or facilitation regulate migrant ungulate populations in the Serengeti? A test of hypotheses. *Oecologia*, **53**, 364–9.

Skogland, T. (1985) The effects of density dependent resource limitations on the demography of wild reindeer. *Journal of Animal Ecology*, **54**, 359–74.

Smith, T.E. (1989) The role of bulls in pioneering new habitats in an expanding muskox population on the Seward Peninsula, Alaska. *Canadian Journal of Zoology*, **67**, 1096–101.

Stenseth, N.C. (1983) Causes and consequences of dispersal in small mammals, in *The Ecology of Animal Movement* (eds I.R. Swingland and P.J. Greenwood), Clarendon Press, Oxford, pp. 63–101.

Tyler, N.J.C. (1986) Natural limitation of the abundance of wild reindeer on Spitsbergen. Ph.D. Thesis, University of Cambridge, England.

Ward, R.M.P. and Krebs, C.J. (1985) Behavioural responses of lynx to declining snowshoe hare abundance. *Canadian Journal of Zoology*, **63**, 2817–24.

Waser, P.M., Austad, S.N. and Keane, B. (1986) When should animals tolerate inbreeding? *American Naturalist*, **128**, 529–37.

Wolff, J.O. (1980) The role of habitat patchiness in the population dynamics of snowshoes hares. *Ecological Monographs*, **50**, 111–30.

– 11

The primate perspective on dispersal

A.E. PUSEY

11.1 INTRODUCTION

There are two major differences between primates and most of the small mammals discussed in this book that are particularly pertinent to the subject of dispersal. First, many species of small mammals have generation lengths of much less than one year and give birth to large litters, with the result that most show strong seasonal fluctuations in numbers and some show longer cyclical fluctuations. Much dispersal in these species is connected with these cyclic increases in numbers. In contrast, primates are long-lived animals with long generation lengths and slow reproductive rates (Harvey *et al.*, 1987). Their populations tend to remain stable over long periods, with their densities determined by the long-term carrying capacity of the environment. Nevertheless, although they do not show strong fluctuations in numbers, they still show high rates of dispersal. Thus a comparison between overall patterns of dispersal in primates and small mammals should reveal features of dispersal that may be universal to mammals.

A second major difference between primates and small mammals is that most primates are relatively large, slow-moving and diurnal. Individuals are easily recognized, and their social interactions and movements can be monitored over long periods [see Smuts *et al.* (1987) for a comprehensive review]. This makes it feasible to determine the rates of dispersal by each age–sex class by direct observation instead of having to resort to such indirect methods as trapping, or the creation of a dispersal sink. Also, by observing the behaviour of individuals it is possible to examine the proximate as well as infer the ultimate causes of dispersal. Such direct observation of individuals in most species of small mammals is virtually impossible, and has greatly hampered the determination of proximate causes of dispersal (Chapters 2 and 4). A notable exception is the ground squirrel group, and the abundant data on dispersal in this group are discussed briefly by several authors in this book (e.g. Chapters 4 and 5).

This chapter concentrates on one pattern of dispersal that has been identified by several authors in this book as being widespread among

Table 11.1 The direction of sex bias in dispersal for the 31 species of primates (number of genera in parentheses) for which sufficient data are available (data from Pusey and Packer, 1987a)

Social system	*Male*	*Female*	*Both*
Gregarious	15(8)	2(2)	4(3)
Monogamous			5(2)
Solitary	5(3)		

mammals. This is sex-biased dispersal; in particular the tendency in most species of mammals for most males, rather than females, to disperse from the site of their birth before breeding (Greenwood, 1980).

11.2 PATTERNS OF DISPERSAL IN PRIMATES

Unlike studies of small mammals, which often involve the trapping and marking of large numbers of individuals over fairly short periods, field studies of primates have tended to involve the identification and long-term observation of the individuals of one or a few social groups. Sample sizes are thus small, but many species have now been studied long enough for overall dispersal patterns to be determined.

Primates exhibit a variety of social organizations (Smuts *et al.*, 1987). Most species are gregarious and live in permanent social groups containing several breeding females and one to several adult males. In some species the adults are solitary, and in others they form monogamous pairs. Although individuals of most age–sex classes have been observed to disperse in primates, the predominant pattern in all species is for a proportion of one sex, or more rarely both sexes, to emigrate from their natal group or area before breeding. This is known as natal dispersal. Table 11.1 gives the direction of the sex bias in dispersal in those species for which there are sufficient data (Pusey and Packer, 1987a). Most species show an overwhelming male bias in dispersal, a few show a female bias, and in only the monogamous species and four gregarious species is dispersal by both sexes common. In two of these gregarious species – gorillas (*Gorilla gorilla*) and red colobus (*Colobus badius*) – a proportion of both sexes leave their natal group, but only females normally join other breeding groups, so in this sense dispersal is female-biased. In most gregarious species most or all members of the dispersing sex leave their natal group at or before attaining full size, while all members of the opposite sex remain in their natal group (Figure 11.1). The result of this pattern is that social groups tend to be composed of a kin group of one sex, while all the adults of the other sex are immigrants.

In many gregarious species, emigrating animals directly join a new group, usually adjacent to their original group (Pusey and Packer, 1987a). In others, particularly those that live in social groups with only one adult male, young males usually emigrate to become solitary, or join all-male bands, and only later join bisexual groups. In most cases little is known of their subsequent fates, but at least some eventually join non-adjacent groups.

In most species, members of the predominantly dispersing sex often disperse again as adults, a phenomenon termed secondary or breeding dispersal. This is more common when the predominantly dispersing sex is male. For example, male olive baboons (*Papio anubis*) have been known to reside in at least five different social groups during their lifetime (Packer, 1979). In all species for which extensive secondary dispersal by males has been documented, return to the natal group is extremely rare (Pusey and Packer, 1987a). Because of these patterns of dispersal, close relatives of the opposite sex rarely reside in the same group, and many animals eventually breed in groups far from their natal group.

11.3 PROXIMATE CAUSES OF DISPERSAL

11.3.1 Natal dispersal

When considering the proximate causes of primate dispersal it is useful to distinguish between intrinsic and extrinsic factors (Chapter 2). The ubiquity of natal dispersal by one sex, and the fact that it usually occurs at a particular developmental stage (Pusey and Packer, 1987a) strongly implicate the operation of intrinsic genetic and hormonal influences as has been demonstrated in the natal emigration of Belding's ground squirrels (*Spermophilus beldingi*; Holekamp, 1986). Two important extrinsic factors which also appear to influence natal dispersal are aggression from conspecifics, and sexual attraction to unfamiliar mates.

It is often supposed that much dispersal is the result of aggression from conspecifics and the eviction of subordinate individuals (Wilson, 1975; Moore and Ali, 1984; Chapter 5). There are certainly some species (e.g. lions; Hanby and Bygott, 1987; Pusey and Packer, 1987b) where aggression does play an important part in natal emigration. However, several recent reviews have questioned the general importance of aggression in natal dispersal. Direct observations of sciurids, lagomorphs and hyracoids suggest that aggression by parents or siblings plays little part in most emigration (Bekoff, 1977; Dobson, 1979; Holekamp, 1984; Chapter 5) although the importance of subtle or non-visual threats such as odours cannot always be ruled out. Brandt (Chapter 5) also concludes from a review of more indirect studies of the effects of aggression on dispersal

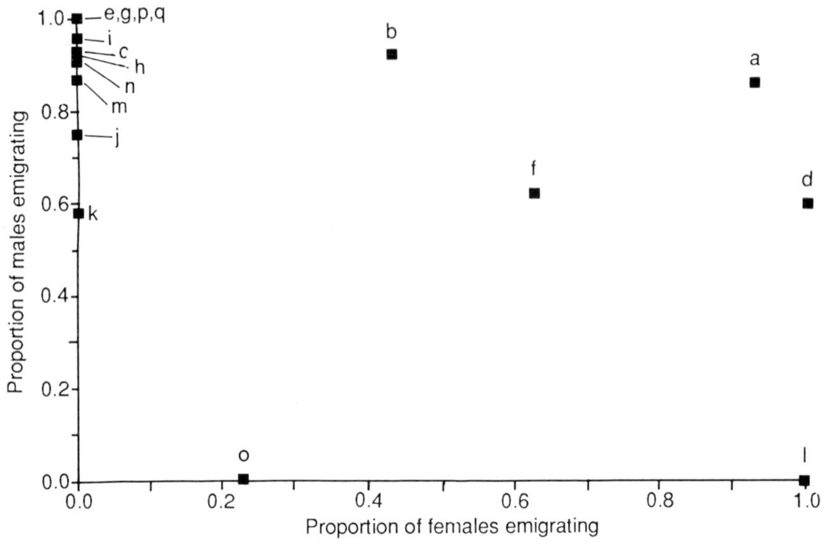

Proportion of males emigrating
Proportion of females emigrating
e,g,p,q
i
c
h
n
m
j
k
b
a
f
d
o
l

(a)

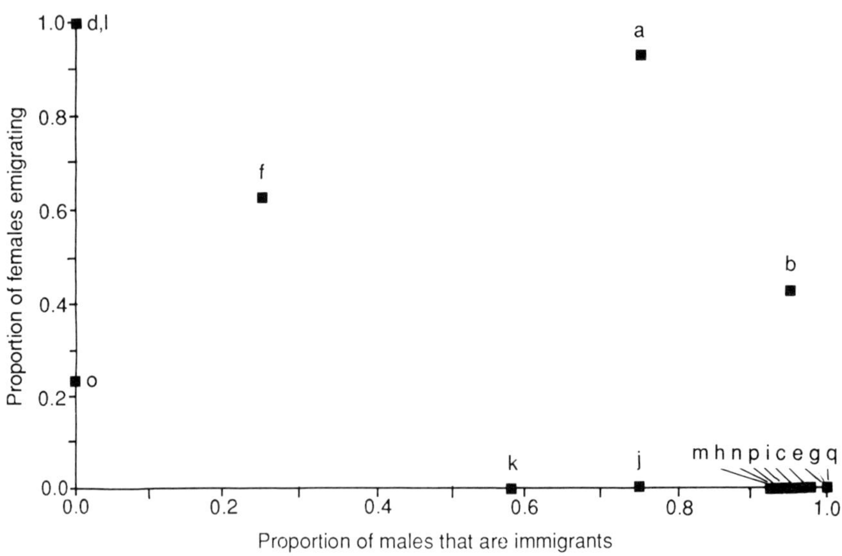

Proportion of females emigrating
Proportion of males that are immigrants
d,l
a
f
b
o
k
j
m h n p i c e g q

(b)

among murids and cricetids that there are no good data directly implicating dominance or aggression in the emigration of any but a small proportion of emigrants in these groups. Because of the natural history of two *Antechinus* species, Cockburn *et al.* (1985) could rule out any effect of male competition on natal emigration, although they suggest that mothers might force their sons to emigrate (see also Chapter 4).

Among primates also, there are only a few species in which natal dispersal appears to be forced by obvious aggression from conspecifics (Pusey and Packer, 1987a; Table 11.2). In some species that live in groups consisting of several breeding females and one adult male, aggressive male replacements occur where the group is taken over by a new adult male who evicts the previous breeding male. In two species of langurs (*Presbytis* spp.) there is good evidence that the new male also forces the emigration of the juvenile males (and sometimes juvenile females) from the group. Such data on the effects of immigrants on the recipient population are rarely available for other mammals, but similar observations have been made in lions (Pusey and Packer, 1987b). Among the monogamous gibbons (*Hylobates* spp.), aggression from the like-sexed parent hastens the emigration of subadults. In many other species, however, the natal emigration of the predominantly dispersing sex appears to occur voluntarily, in

Figure 11.1(a) The extent of male versus female natal emigration in gregarious species of primates. Each letter represents one species. The proportion of individuals emigrating is the proportion of individuals of that sex that were known to emigrate at or before attaining full size. There is a significant inverse relationship between the extent of natal emigration by each sex (across species: $r_s = -0.613$, $N = 17$, $p < 0.02$; across genera: $r_s = -0.836$, $N = 9$, $p < 0.01$). a, *Alouatta palliata*; b, *A. seniculus*; c, *Cercopithecus aethiops*; d, *Colobus badius*, e, *Erythrocebus patas*; f, *Gorilla gorilla*; g, *Macaca fascicularis*; h, *M. fuscata*; i, *M. mulatta*; j, *M. sinica*; k, *M. sylvanus*; l, *Pan troglodytes*; m, *Papio anubis*; n, *P. cynocephalus*; o, *P. hamadryas*; p, *P. ursinus*; q, *Presbytis entellus*. Correlations are also calculated from generic averages since most species of *Macaca* and *Papio* showed similar patterns. Redrawn from Figure 21–3 (data in Table 21–1) of Pusey and Packer (1987a), with permission from the University of Chicago Press. (b) Female emigration versus availability of unrelated mates in natal group. The probability that females emigrate from their natal group is highest where the proportion of immigrant males in that group is lowest (across species: $N = 17$, $r_s = -0.70$, $p < 0.005$; across genera: $N = 10$, $r = -0.86$, $p < 0.005$). The proportion of male immigrants in a female's natal group is highly correlated with the extent of male emigration ($r_s > 0.95$). However, in species where emigration is not followed by transfer into another group (e.g. male gorillas and red colobus), the extent of emigration overestimates the proportion of immigrant males. See (a) for key and sources. Redrawn from Figure 21–4 of Pusey and Packer (1987a), with permission from the University of Chicago Press.

Table 11.2 Features of natal dispersal in 22 species of primates. Summarized from Pusey and Packer (1987a)

Species	*Sex*	*Natal emigration forced*	*Natal emigration voluntary*	*More aggression in new group*	*Attracted to unfamiliar mates*	*Reference*
Presbytis entellus	M	Evicted by new males				Sugiyama (1967)
P. senex	M and F*	Evicted by new males				Rudran (1973)
Hylobates syndactylus	M†	Evicted by father				Chivers and Raemakers (1980)
H. lar	M†	Evicted by father				Chivers and Raemakers (1980)
H. hoolock	M†	Evicted by father				Tilson (1981)
H. klossi	M and F	Evicted by like-sexed parent				Tilson (1981)
Colobus badius	M*	Occasionally evicted by older male relatives				Struhsaker and Leland (1979); personal communication
	F		+			
Alouatta seniculus	F†	Sometimes evicted by adult females (not mother)				Crockett (1984)
A. palliata	F†	Sometimes evicted by adult females (not mother)				Jones (1980)

Papio hamadryas	F	Abducted by adult males				Abegglen (1984)
Gorilla gorilla	F	Occasional abduction by adult males				Fossey (1983)
			Usually			Harcourt (1978)
					+	Stewart and Harcourt (1987)
Pan troglodytes	F	Possible abduction by adult males				Goodall (1986)
			Usually	+	+	Pusey (1980)
Macaca mulatta	M		+			Colvin (1983)
				+		Lindburg (1969)
M. fuscata	M		+			Sugiyama (1976)
M. fascicularis	M		+		+	van Noordwijk and van Schaik (1985)
M. sinica	M			+	+	Dittus (1977)
Cercopithecus aethiops	M		+	+	+	Cheney (1983, 1987); Henzi and Lucas (1980); Harrison (1983)
C. campbelli	M		+			Bourlière *et al.* (1970)
C. ascanius	M		+			Struhsaker and Leland (personal communication)
Papio anubis	M		+	+	+	Packer (1979)
Erythrocebus patas	M		+			Olson (personal communication
Lemur catta	M		+	+		Jones (1983)

* This sex is not the predominantly dispersing sex.
† Both sexes disperse.

the absence of any obvious aggression from conspecifics. Indeed, in those species in which dispersing individuals directly enter new groups, immigrants often receive considerably more aggression from group members than do natal individuals of the same age and sex (Table 11.2).

A common cause of dispersal in primates seems to be sexual attraction to animals in other groups (Pusey and Packer, 1987a). In many seasonally breeding species, transfer of males from one group to another occurs primarily during the mating season, suggesting that it is motivated by sexual activity. Also, extra-group males often join groups during this time. Sexual attraction also appears to be an important cause of transfer in non-seasonally breeding species. In chimpanzees (*Pan troglodytes*), where females but not males transfer between groups, the first visits of adolescent females to other groups occur when they are in oestrus. In baboons (*Papio* spp.) male immigration is most frequent when the group contains many oestrous females.

Natal emigration in primates often appears to be motivated in particular by attraction to unfamiliar mates. As in many other mammals (Ralls *et al.*, 1986) there is good evidence from several species of primates that individuals that were closely familiar during immaturity (usually close relatives) show reduced sexual attraction as adults (Pusey and Packer, 1987a). For example, male chimpanzees often reside as adults in the same group as their mothers, but most have never been observed to court or mate with their mothers (Goodall, 1986). Female chimpanzees with older male siblings associate with them frequently until they begin sexual cycling, at which time the association drops markedly, and the females eventually move to other groups (Pusey, 1980). Nevertheless, several females have been observed in proximity with their siblings when they were in oestrus, and the males showed almost no sexual interest in them although they readily courted other females of the same age, nor did the females solicit their brothers although they solicited other males in the group. In addition, females showed more interest in and mated more readily with males of unfamiliar groups than with males of their natal group (Pusey, 1980). In a detailed study of olive baboons, Packer (1979) found that natal males mate surreptitiously with females of their group, but the degree to which they compete for these females with other males of their group actually declines as their competitive ability increases. All natal males transfer as young adults. Their first interactions in the new group are directed primarily at females, and they subsequently compete at far higher levels for access to these females than they did in their natal group. Similar greater attraction to mates outside their natal group has also been recorded for natal males in several other species (Table 11.2).

Because in polygynous species females invest more than males in their joint offspring, it might be expected that females should be more reluctant than males to mate with close relatives because their reproductive success would be more greatly affected by costs associated with close inbreeding (Waser *et al.*, 1986; Chapter 5). This has led to the suggestion that males might emigrate simply because females in their group refuse to mate with them (e.g. Chapter 5). In primates, as in other mammalian groups (Chapter 5), it appears that sexual attraction is often reduced between relatives of both sexes. However, it is possible that males might learn to ignore their female relatives if their early attempts to mate with them are rebuffed. Primates would provide particularly good subjects in which this question could be investigated by detailed observations of the first mating attempts of young males.

11.3.2 Secondary or breeding dispersal

Few primate studies have separated natal from secondary dispersal in their discussion of proximate factors influencing dispersal, although, as Brandt (Chapter 5) points out, the reasons for dispersal in each age group are likely to be different. There is no doubt that some secondary dispersal in primates and other mammals such as lions results from aggression. Where male takeovers take place, the emigration of the defeated male or males results directly from the aggression of the invader(s). Another prevalent factor in secondary dispersal seems to be an attraction to groups or areas which provide increased breeding opportunities. Male olive baboons transferring for a second or subsequent time show a significant tendency to move to groups containing more cycling females, whereas natal males transferring for the first time do not (Packer, 1979). Packer suggests that for males dispersing for the first time, any other group will contain more attractive females than the natal group, while for males transferring a second time, the relative number of receptive females is important. Most other studies have lumped natal and secondary transfer when considering whether males move to groups with more female-biased sex ratios or not. Several have found that they do, while others have found no effect of this factor (Pusey and Packer, 1987a). The relative intensity of male–male competition in different neighbouring groups can also influence transfer patterns. Ten of 13 adult male vervets (*Cercopithecus aethiops*) (many of which were presumably transferring for a second time) improved their rank and probably their reproductive success when they transferred (Cheney and Seyfarth, 1983). Playback experiments in this population show that the monkeys have a detailed knowledge of individuals in neighbouring groups (Cheney and Seyfarth, 1982). It therefore seems likely that these males may have monitored male dominance relations in other groups and timed their transfers accordingly.

In black-tailed prairie dogs (*Cynomys ludovicianus*), males are more likely to move to a new breeding group if they have adult daughters present in their own group (Hoogland, 1982). Similarly, the only times that competitively superior groups of male lions voluntarily abandon large prides for smaller prides is when they have maturing daughters in the former group (Pusey and Packer, 1987b). It has been suggested that male primates also may move on from social groups or areas containing daughters when their daughters reach maturity (e.g. Itani, 1972; Henzi and Lucas, 1980). Henzi and Lucas showed that in several species, average male tenure in groups is shorter than the time it takes females to mature. However, there is as yet no direct evidence from primates that avoidance of mature daughters is a proximate factor in secondary dispersal. In olive baboons, fathers sometimes continue to reside in the same troop as their adult daughters, but matings between them are rare (Packer, 1979).

11.4 EVOLUTIONARY EXPLANATIONS OF DISPERSAL

Three major explanations for dispersal have been proposed: (1) that individuals disperse to seek better or unexploited environmental resources (Howard, 1960; Lidicker, 1962); (2) that dispersal is a consequence of intrasexual competition or the search for better access to mates (Dobson, 1982; Waser and Jones, 1983; Moore and Ali, 1984); and (3) that individuals disperse to avoid the deleterious genetic consequences of close inbreeding (Darling, 1937; Itani, 1972; Bengtsson, 1978; Packer, 1979; Greenwood 1980; Shields, 1982; Ralls *et al.*, 1986; see also Chapter 1). Because most primate populations appear to be at or near carrying capacity, it is unlikely that dispersing primates can regularly find sufficient unexploited resources to offset the cost of emigration. Indeed, there are a few examples of natal male primates voluntarily leaving groups with high quality resources for those with low (Pusey and Packer, 1987a). Similarly, the probability of dispersal by the predominantly dispersing sex does not appear to vary with population density or resource availability in a number of species of ground squirrels (*Spermophilus* spp.) (Dobson, 1979; Holekamp, 1984).

Dispersal resulting from attraction to more mates, or to groups or areas where competition for mates is less intense, appears to be more common and may account for a considerable proportion of secondary dispersal. Nevertheless, intrasexual competition alone is insufficient to account for the ubiquity of natal dispersal by one sex across primate species. If dispersal were caused by this factor alone, individuals might be expected to forgo emigration whenever their natal group contained larger numbers of animals of the opposite sex than surrounding groups, or lower levels of intrasexual competition. There are some rare examples of individual primates remaining in their natal group under such circumstances (Pusey and Packer, 1987a), but in the best-studied species it appears to be the rule that the great majority of individuals of one sex leave their natal group at or before reaching full size virtually regardless of its composition. Indeed, in one group of Japanese macaques (*Macaca fuscata*), young males continued to leave the group even in the total absence of adult males (Sugiyama, 1976). Similar observations have been made in California ground squirrels (*Spermophilus beecheyi*) (Dobson, 1979) and *Antechinus* species [Cockburn *et al.*, 1985; Chapter 4; see also Brandt (Chapter 5), who suggests that rates of dispersal may depend on the degree of polygyny and of local mate competition].

The overall pattern of sex-biased dispersal in primates, the rarity of return to the natal group, and the fact that sexual attraction to unfamiliar mates is sometimes a motivating factor in natal dispersal, are all consistent with the inbreeding avoidance hypothesis. If all members of one sex leave and never return, the chance of breeding between close relatives is greatly reduced. As long as all the members of one sex leave, there is no reason for members of the other sex to pay the costs of dispersal. Across gregarious primate species there is a significant negative correlation between the proportion of males and the proportion of females that disperse (Figure 11.1a). In those few species where males do not disperse, most or all females do. In most species, therefore, the tendency for females to disperse depends on the likelihood that breeding males in their natal group are close relatives (Figure 11.1b; Pusey and Packer, 1987a).

If sex-biased dispersal has evolved in part as an inbreeding avoidance mechanism, it still remains to explain which sex disperses. Inbreeding can be avoided by the dispersal of all members of either sex. Why then do mammals show predominantly male-biased dispersal in contrast to birds where dispersal is female-biased? Current theories to explain these patterns are discussed in Chapters 4 and 5 and a number of other recent reviews (e.g. Packer, 1979; Greenwood, 1980; Dobson, 1982; Liberg and von Schantz, 1985; Pusey, 1987). The identity of the sex that disperses will depend on the relative costs and benefits of dispersal and philopatry to each sex. Even if the costs of close inbreeding to females are higher than to males, as is likely to be the case in polygynous mammals (see above), the costs of dispersal might be higher and the benefits lower for females than for males, with the result that males and not females disperse. Male-biased dispersal appears to be closely linked to polygyny, the mating system most prevalent among mammals (cf. birds where most species are monogamous). In polygynous species, females generally invest more in each offspring than males (Trivers, 1972). Female reproductive success is therefore limited primarily by nutritional constraints, while male reproductive success is limited more by the number of mates that they can inseminate (Bradbury and Vehrencamp, 1977; Wrangham, 1980; Ostfeld, 1985). There is evidence that female reproductive success is higher for philopatric than dispersing females in some species, e.g. brown bears (*Ursus arctos*; Waser and Jones, 1983) and lions (Pusey and Packer, 1987b), probably because of the advantages of familiarity with the area, the phenology of food sources and the location of refuges or denning sites (Waser and Jones, 1983) or the presence of kin (Wrangham, 1980). Therefore, the benefits of philopatry might be particularly strong for females. In addition, males might be more likely to disperse than females because their reproductive success depends more than female reproductive success on access to mates, and therefore they may be more likely to benefit from moving to groups or areas where there are large numbers of mates (Clutton-Brock and Harvey, 1976; Packer, 1979). Finally, intrasexual competition for mates is probably more intense among males than among females in polygynous species (Trivers, 1972) with the result that males are more likely to be evicted and have to disperse to find mates (Clutton-Brock and Harvey, 1976; Dobson, 1982).

11.5 SPECIES WITH FEMALE-BIASED DISPERSAL

Although male-biased dispersal is the rule among polygynous mammals, there are some interesting exceptions that show female-biased dispersal. There seem as yet to be no general explanations for these exceptions. Brandt (Chapter 5) and Gliwicz (Chapter 6) suggest a relationship in microtines but not other groups between female dispersal and territoriality by females rather than males. Among primates there are four species in which female-biased dispersal has been well documented: chimpanzees, gorillas, hamadryas baboons (*Papio hamadryas*), and some populations of red colobus (Pusey and Packer, 1987a). A corollary of sex-biased dispersal is that individuals of the philopatric sex are surrounded by relatives and there is consequently scope for the evolution of co-operation by kin selection. Brandt (Chapter 5) suggests that sex differences in co-operation are probably the result rather than the cause of sex differences in emigration, but a case has been made for the opposite view in primates (Wrangham, 1980). Wrangham argues that primates generally live in female-bonded kin groups because of the advantages of co-operative defence of patchy food sources. One primate species in which the advantages of co-operation by males may have forced dispersal by females is chimpanzees (Pusey and Packer, 1987a). Female chimpanzees and female orang-utans (*Pongo pygmaeus*) have very similar diets and occupy the same system of overlapping core areas. Male orang-utans are solitary and occupy larger home ranges overlapping those of females. As in most other primates, orang-utans show higher levels of male dispersal. In contrast, male chimpanzees co-operate with their kin to defend a large group range encompassing the ranges of many females. Wrangham (1979) has argued that because the costs of locomotion are much lower in chimpanzees than orang-utans, each male can gain access to greater numbers of females by jointly defending a large range than by defending an individual range. The advantages to males from this co-operation may exceed the advantages to females of philopatry with the result that females are 'forced' to emigrate to avoid close inbreeding (Clutton-Brock and Harvey, 1976; Pusey, 1979). Although both male hamadryas baboons and male red colobus also show high levels of co-operation, the differences between these species and others with similar feeding habits and social structure that show male-biased dispersal are not clear (Pusey and Packer, 1987a).

11.6 CONCLUSIONS

Primates share with most other mammals, including small mammals, the pattern of male-biased dispersal. The primate evidence is consistent with the views expressed by Cockburn (Chapter 4) and Brandt (Chapter 5) that sex-biased natal dispersal is strongly connected with the avoidance of inbreeding, and that male bias in dispersal in most mammals is linked to the prevalence of polygyny among mammals, and the consequent sex differences in intrasexual competition and reproductive strategies.

Cockburn (Chapter 4) discusses two other patterns of dispersal also involving interaction between close relatives which are common to many species of small mammals, but do not appear to occur among primates: breeding dispersal by females involving the abandonment of the nest to her offspring, and transfer between winter nesting aggregations. The former pattern appears to be closely tied to fluctuating environmental conditions and relatively short generation lengths, conditions that do not pertain in primate populations. In the latter case, Cockburn proposes that a moderate dispersal of daughters by a mother to surrounding groups might result in the 'capture' of adjacent habitat for her genotype. Such advantages might also be expected to accrue to primates, but the dispersal of females to adjacent groups is very rare (Pusey and Packer, 1987a), and even in those species in which it occurs (e.g. howlers; *Alouatta* spp.), dispersal is not forced by the mother, but by other females (Table 11.2). Possibly among primates the advantages of co-operation among female kin, and the hostility of other groups to emigrants, overcome this potential advantage of female dispersal.

REFERENCES

Abegglen, J.J. (1984) *On Socialisation in Hamadryas Baboons*, Associated University Presses, Cranbury, New Jersey.

Bekoff, M. (1977) Mammalian dispersal and the ontogeny of individual behavioral phenotypes. *American Naturalist,* **111**, 715–32.

Bengtsson, B.O. (1978) Avoiding inbreeding: at what cost? *Journal of Theoretical Biology*, 73, 439–44.

Bourlière, F., Hunkeler, C. and Bertrand, M. (1970) Ecology and behaviour of Lowe's guenon (*Cercopithecus campbelli lowei*) in the Ivory Coast, in *Old World Monkeys* (eds J.R. Napier and P.H. Napier), Academic Press, New York, pp. 297–350.

Bradbury, J.W. and Vehrencamp, S.L. (1977) Social organization and foraging in Emballonurid bats, III. Mating systems. *Behavioral Ecology and Sociobiology*, 2, 1–17.

Cheney, D.L. (1983) Proximate and ultimate factors relating to the distribution of male emigration, in *Primate Social Relationships: an integrated approach* (ed. R.A. Hinde), Blackwell, Oxford, pp. 241–9.

Cheney, D.L. (1987) Interactions and relationships between groups, in *Primate Societies* (eds B.B. Smuts, D.L. Cheney, R.M. Seyfarth *et al.*), University of Chicago Press, Chicago, pp. 267–81.

Cheney, D.L. and Seyfarth, R.M. (1982) Recognition of individuals within and between groups of free-ranging vervet monkeys. *American Zoologist*, **22**, 519–29.

Cheney, D.L. and Seyfarth, R.M. (1983) Non-random dispersal in free-ranging vervet monkeys: social and genetic consequences. *American Naturalist*, **122**, 392–412.

Chivers, D.J. and Raemakers, J.J. (1980) Long-term changes in behaviour, in *Malayan Forest Primates: ten years' study in a tropical rain forest* (ed. D.J. Chivers), Plenum Press, New York, pp. 209–60.

Clutton-Brock, T.H. and Harvey, P.H. (1976) Evolutionary rules and primate societies, in *Growing Points in Ethology* (eds P.P.G. Bateson and R.A. Hinde), Cambridge University Press, Cambridge, pp. 195–237.

Cockburn, A., Scott, M.P. and Scotts, D. (1985) Inbreeding avoidance and male-biased natal dispersal in *Antechinus* spp. *Animal Behaviour*, **33**, 908–15.

Colvin, J. (1983) Influences of the social situation on male emigration, in *Primate Social Relationships: an integrated approach* (ed. R.A. Hinde). Blackwell, Oxford, pp. 160–71.

Crockett, C.M. (1984) Emigration by female red howler monkeys and the case for female competition, in *Female Primates: studies by women primatologists* (ed. M.F. Small), pp. 159–73. Alan R. Liss, New York.

Darling, F.F. (1937) *A Herd of Red Deer*. Oxford University Press, London.

Dittus, W. (1977) The social regulation of population density and age–sex distribution in the toque monkey. *Behaviour*, **63**, 281–322.

Dobson, F.S. (1979) An experimental study of dispersal in the California ground squirrel. *Ecology*, **60**, 1103–9.

Dobson, F.S. (1982) Competition for mates and predominant juvenile male dispersal in mammals. *Animal Behaviour*, **30**, 1183–92.

Fossey, D. (1983) *Gorillas in the Mist*, Houghton Mifflin, Boston.

Goodall, J. (1986) *The Chimpanzees of Gombe: patterns of behavior*, Harvard University Press, Cambridge, Massachusetts.

Greenwood, P.J. (1980) Mating systems, philopatry and dispersal in birds and mammals. *Animal Behaviour*, **28**, 1140–62.

Hanby, J.P. and Bygott, J.D. (1987) Emigration of subadult lions. *Animal Behaviour*, **35**, 161–9.

Harcourt, A.H. (1978) Strategies of emigration and transfer by primates, with particular reference to gorillas. *Zeitschrift für Tierpsychologie*, **48**, 401–20.

Harrison, M.J.S. (1983) Territorial behaviour in the green monkey, *Cercopithecus sabaeus*: seasonal defence of local food supplies. *Behavioural Ecology and Sociobiology*, **12**, 85–94.

Harvey, P.H., Martin, R.D. and Clutton-Brock, T.H. (1987) Life histories in comparative perspective, in *Primate Societies* (eds B.B. Smuts, D.L. Cheney, R.M. Seyfarth *et al.*), University of Chicago Press, Chicago, pp. 181–96.

Henzi, S.P. and Lucas, J.W. (1980) Observations on the inter-troop movement of adult vervet monkeys (*Cercopithecus aethiops*). *Folia Primat.*, **33**, 220–35.

Holecamp, K.E. (1984) Dispersal in ground-dwelling sciurids, in *Biology of Ground Squirrels: annual cycles, behavioural ecology, and sociality* (eds J.O. Murie and G.R. Michener), University of Nebraska Press, Lincoln, Nebraska, pp. 297–320.

Holekamp, K.E. (1986) Proximal causes of natal dispersal in Belding's ground squirrels (*Spermophilus beldingi*). *Ecological Monographs*, **56**, 365–91.

Hoogland, J.L. (1982) Prairie dogs avoid extreme inbreeding. *Science*, **215**, 1639–41.

Howard, W.E. (1960) Innate and environmental dispersal of individual vertebrates. *American Midland Naturalist*, **63**, 152–61.

Itani, J. (1972) A preliminary essay on the relationship between social organisation and incest avoidance in non-human primates, in *Primate Socialization* (ed. F.E. Poirier), Random House, New York, pp. 165–71.

Jones, C.B. (1980) The functions of status in the mantled howler monkey, *Alouatta palliata* Gray: intraspecific competition for group membership in a folivorous Neotropical primate. *Primates*, **21**, 389–405.

Jones, K.C. (1983) inter-troop transfer of *Lemur catta* males at Berenty, Madagascar. *Folia Primatol.*, **40**, 145–60.

Liberg, O. and von Schantz, T. (1985) Sex-biased philopatry and dispersal in birds and mammals: the Oedipus hypothesis. *American Naturalist*, **126**, 129–35.

Lidicker, W.Z. Jr (1962) Emigration as a possible mechanism permitting the regulation of population density below carrying capacity. *American Naturalist*, **96**, 29–33.

Lindburg, D.G. (1969) Rhesus monkeys: mating season mobility of adult males. *Science*, **166**, 1176–8.

Moore, J. and Ali, R. (1984) Are dispersal and inbreeding avoidance related? *Animal Behaviour*, **32**, 94–112.

Ostfeld, R.S. (1985) Limiting resources and territoriality in microtine rodents. *American Naturalist*, **126**, 1–15.

Packer, C. (1979) Inter-troop transfer and inbreeding avoidance in *Papio anubis*. *Animal Behaviour*, **27**, 1–36.

Pusey, A.E. (1979) Intercommunity transfer of chimpanzees in Gombe national park, in *The Great Apes* (eds D.A. Hamburg and E.R. McCown), Benjamin-Cummings, Menlo Park, pp. 465–79.

Pusey, A.E. (1980) Inbreeding avoidance in chimpanzees, *Animal Behaviour*, **28**, 543–52.

Pusey, A.E. (1987) Sex-biased dispersal and inbreeding avoidance in birds and mammals. *Trends in Ecology and Evolution*, **2**, 295–9.

Pusey, A.E. and Packer, C. (1987a) Dispersal and philopatry, in *Primate Societies* (eds B.B. Smuts, D.L. Cheney, R.M. Seyfarth *et al.*), University of Chicago Press, Chicago, pp. 250–66.

Pusey, A.E. and Packer, C. (1987b) The evolution of sex-biased dispersal in lions. *Behaviour*, **101**, 275–310.

Ralls, K., Harvey, P.H. and Lyles, A.M. (1986) Inbreeding in natural populations of birds and mammals, in *Conservation Biology* (ed. M.E. Soulé), Sinauer, Sunderland, Massachusetts, pp. 35–56.

Rudran, R. (1973) Adult male replacement in one-male troops of purple-faced langurs (*Presbytis senex senex*) and its effect on population structure. *Folia Primatol.*, **19**, 166–92.

Shields, W.M. (1982) *Philopatry, Inbreeding, and the Evolution of Sex*, State University of New York Press, Albany.

Smuts, B.B., Cheney, D.L. Seyfarth, R.M. *et al.* (eds) (1987) *Primate Societies*, University of Chicago Press, Chicago.

Stewart, K. and Harcourt, A.H. (1987); Gorillas: variation in female relationships, in *Primate Societies* (eds B.B. Smuts, D.L. Cheney, R.M. Seyfarth *et al.*), University of Chicago Press, Chicago, pp. 155–64.

Struhsaker, T.T. and Leland, L. (1979) Socioecology of five sympatric monkey species in the Kibale forest, Uganda, in *Advances in the Study of Behavior*, vol. 9 (eds J.S. Rosenblatt, R.A. Hinde, C. Beer and M.C. Busnel), Academic Press, New York, pp. 159–228.

Sugiyama, Y. (1967) Social organization of hanuman langurs, in *Social Communication among Primates* (ed. S.A. Altmann), University of Chicago Press, Chicago, pp. 221–36.

Sugiyama, Y. (1976) Life history of male Japanese monkeys. *Advances in the Study of Behavior*, 7, 255–84.

Tilson, R. (1981) Family formation strategies of Kloss's gibbons. *Folia Primatol.*, **35**, 259–87.

Trivers, R.L. (1972) Parental investment and sexual selection, in *Sexual Selection and the Descent of Man* (ed. B. Campbell), Aldine, Chicago, pp. 136–79.

van Noordwijk, M.A. and van Schaik, C.P. (1985) Male migration and rank acquisition in wild long-tailed macaques *Macaca fascicularis*. *Animal Behaviour*, 33, 849–61.

Waser, P.M., Austad, S.N. and Keane, B. (1986) When should animals tolerate inbreeding? *American Naturalist*, **128**, 529–37.

Waser, P.M. and Jones, W.T. (1983) Natal philopatry among solitary mammals. *Quarterly Review of Biology*, **58**, 355–90.

Wilson, E.O. (1975) *Sociobiology: the new synthesis*, Harvard University Press, Cambridge, Massachusetts.

Wrangham, R.W. (1979) On the evolution of ape social systems. *Social Science Information*, **18**, 335–68.

Wrangham, R.W. (1980) An ecological model of female-bonded primate groups. *Behaviour*, 75, 262–99.

12

A red grouse perspective on dispersal in small mammals

A. WATSON

12.1 INTRODUCTION

This chapter includes a brief comparison of red grouse (*Lagopus lagopus scoticus*) and small mammals as study animals. I write more on how I, as a worker who has used red grouse to tackle general questions on population ecology, view the literature on dispersal in small mammals. Occasionally comparisons are drawn with red grouse and it is noted where work on small mammals has been of particular benefit. Cycles are mentioned often because several species of small mammals and grouse (Tetraonidae) are two of the notable groups of animals which show population cycles, and because advances in understanding with each group have been useful to workers on the other group.

The present state of understanding (and controversy) on population limitation in red grouse, including dispersal, is briefly reviewed. Such reviews have been published (Moss and Watson, 1985, 1990; Watson and Moss, 1987, 1990). They integrate the roles of extrinsic factors (food, cover, weather, predation, parasitism and disease) and self-regulation. They represent a school of thought which emphasizes the role of self-regulation in population limitation, and which attributes cycles in numbers to changes in spacing behaviour (e.g. Watson *et al.*, 1988; Mountford *et al.*, 1990 for red grouse). Another school of thought (Hudson and Dobson, 1989; Hudson, 1990) gives little credence to spacing behaviour in limiting numbers, and emphasizes parasites and other extrinsic factors to the extent of stating for cyclic fluctuations of red grouse in northern England that (p. 34) 'without the parasite they could not occur' (Potts *et al.*, 1984, criticized by Moss and Watson, 1985; Moss *et al.*, 1990). Bergerud *et al.* (1985) and Bergerud (1988) challenged our main findings on population limitation by territorial behaviour, claiming that change in numbers of *Lagopus* from one spring to the next change solely in response to variations in breeding success in the intervening summer. After stating that a correlation between breeding success and change in spring numbers is not a test between their two hypotheses, they used such correlations as their main evidence. In fact we have long incorporated both hypotheses in one: breeding success influences spring numbers by varying the number of competitors for territory. We have now produced a mathematical model which integrates these, along with spacing behaviour, dispersal, change in resources and extrinsic factors (Moss and Watson, 1990).

Wynne-Edwards (1986) used our work on red grouse as the main example to back his case on evolution through group selection. Unfortunately his account of the birds' population ecology contains many errors and is out of date, for example understating the frequency of emigration as an important demographic factor, and his use of evidence on red grouse to argue for group selection is overstated (Watson, 1987).

Small mammals and red grouse differ greatly, as do practical problems for workers studying them, but the questions asked are often the same. Research on small mammals has greatly benefited understanding of population dynamics in general, and work on red grouse. To me, the most valuable studies in decades past were Elton's (1942) classic work on population fluctuations, and Chitty's (1967) philosophical analysis of the problems. The most exciting research nowadays is the ingenious experiments that are being done. However, so much work has been done in recent years and there have been so many different accounts on rodent dispersal and demography that the advance and clarity which come from simplifying, unifying and generalizing in science still seem elusive.

12.2 RED GROUSE AND SMALL MAMMALS AS STUDY ANIMALS

How do red grouse and small mammals compare for study? Wynne-Edwards (1986) recently wrote of our work on red grouse (p. 177) that lemmings and voles are 'incomparably more difficult animals than grouse to study in the wild ... the study of a single red-grouse cycle, though it had many shortcomings, has provided insights that have never been fully appreciated in the innumerable investigations of microtine cycles, which have involved more than 20 species', and (p. 90) 'the red grouse is an exceptionally amenable animal for experimental study in the wild because it poses so few insurmountable difficulties.' Those who have actually done field work on red grouse will regard these statements as unrealistic exaggerations. Like any other animals, red grouse have their good points as study animals, as do small mammals, but both also pose some serious difficulties. Workers on small mammals can usually get big samples easily, and mark many animals easily, and with 3–4 year fluctuations they soon see differences between population declines and increases. With fluctuations of 6–7 years and even up to 10 years for red grouse in east Scotland, one has to wait longer. Fences can be erected that keep voles in, or out, or let them in, or out. Red grouse would fly over!

The main disadvantage in studying dispersal of small mammals is that they are hard to observe behaving naturally in the wild, though determined efforts are now being made to tackle this (Chapter 5). It is often difficult to define density and to compare it with dispersal, reproduction, and mortality as causes of demographic change, because different generations overlap and because breeding seasons are so variable. It is hard to define populations, as a 'population' may expand and contract its range during fluctuations. The work usually rests on trapping, with its well-known biases.

An advantage of red grouse is that at high densities one can watch territorial and certain other aspects of spacing behaviour in the wild, and can often find whether spacing behaviour is followed by net immigration or emigration in numbers. It is usually possible to count the total population, and in some very intensive studies it has been possible to find most individuals that die on a given area and to follow family trees and land-ownership throughout a cycle. Some of these procedures, particularly the observations on spacing behaviour, are extremely time-consuming, but they can be done. A disadvantage is that it is very difficult to find all nests in an area, but here the small-mammal man and grouse man are in the same boat.

12.3 DISPERSAL OR EMIGRATION?

Dispersal is a vague term because it means different things to different workers (Taylor, 1986; see also Chapter 1). It can even mean different things to the same workers (e.g. Desy and Thompson, 1983, pp. 135 and 136, with voles). If demographic change is occurring due to animals moving in or out of a population, it would be better to speak of immigration or emigration. This would follow Lidicker's (1975, p. 104) useful early statement that, when dispersal movements result in individuals leaving or entering the population under study, this becomes emigration and immigration, respectively.

In recent years I have been using the terms emigration and immigration more, and the ambiguous 'dispersal' less. Taylor (1986) stated that what most researchers on small mammals call 'dispersal' is really emigration or immigration (for a discussion or this, see Chapter 1). This is followed below, but sometimes 'dispersal' is used more loosely, when citing what was written in papers.

12.4 MODELS ON EMIGRATION AND IMMIGRATION AS DIRECT CAUSES OF DEMOGRAPHIC CHANGE IN RED GROUSE AND VOLES

Consider an area with good habitat, and an adjoining, generally bigger area with poorer habitat. In model A_1, animals occupy the poor area only after immigration from the rich one, a well-known feature in small mammals (Chapter 4). Alternatively, in model A_2 they occupy most or all of the poor area in all years, but generally at lower density. Model A_2 is usually the more realistic of the two in red grouse, willow ptarmigan (*Lagopus lagopus*) and rock ptarmigan (*Lagopus mutus*). There is no absolute division between it and A_1, as small patches of habitat that are either poorer or are relatively isolated may not be occupied in years of low numbers. There is an obvious difference of degree, however. I know of no case in the literature on *Lagopus* where the great majority of the habitat that is occupied in years of high numbers is unoccupied in other years, but this often seems to be the case with small mammals.

Workers on any animal species tend to choose good habitats because these support high densities that provide good samples for statistical analysis. Hence most study areas may have denser populations than average, and results should be treated with caution accordingly. Perhaps partly for this reason, studies of voles and red grouse have found an exodus in most years, and seldom an influx. Furthermore, evidence on the movements of individuals rests largely or entirely on marking, usually on one area. Since the area covered by a marked population is so much smaller than that into which individuals may emigrate, sampling effort usually falls off with distance from the source population. With such data it is very difficult to get an unbiased estimate of the rate of emigration. Moreover, single population studies using marking and recapture do not necessarily provide any evidence of immigration, so most animal studies consider only emigration. Workers seldom mark animals and count numbers on two nearby areas where there is some emigration or immigration from at least one to the other. This drawback applies to studies on red grouse, other birds and insects as well as voles.

Model B involves direct density-dependence; higher percentages emigrate at high density than at low, and higher percentages emigrate at low density than at high. By contrast, in model C, the proportion emigrating, or immigrating, shows strong, delayed density-dependence, being highly correlated with density in some previous year.

Any, or all, of models A–C may apply in a given population or area at different times, or even at the same time. For instance, in a recent fluctuation of red grouse the proportion emigrating in spring before egg-laying was correlated with density that spring, whereas the proportion emigrating after their chicks hatched showed delayed density-dependence (Watson *et al.*, 1984). Models B and C can apply in the same summer, for instance in red grouse (Watson *et al.*, 1984), but whether B or C has the overriding effect on numbers will depend on which relationship is stronger and on which involves the higher proportion of individuals. However, B will tend to cancel out C and to lead to irregular fluctuations of low amplitude (i.e.

more stable numbers), and C to more regular fluctuations, if not cycles (i.e. fluctuations significantly more regular than random). There are other models but only a few are discussed here. It may be hard to apply B and C to small mammals, where much or most of the range is unoccupied during troughs in numbers, and where consequently it is hard to define density and to draw boundaries to occupied, studied areas. After a trough, dispersal may be associated with increase in density on a large scale, but this dispersal may lead to emigration from a small part of the area and immigration into another part.

12.5 EMIGRATION OR IMMIGRATION AS A DIRECT CAUSE OF DEMOGRAPHIC CHANGE

Is emigration a direct cause of big population declines, or immigration of big increases? It is generally agreed that immigration of small mammals can cause large increases in less favourable areas, and this is well known in some insects such as aphids (Taylor, 1986). But what is the demographic role of emigration on favourable areas? For emigration to be a direct demographic cause of a population decline, two points are important to consider, and, if fulfilled, allow one to call emigration a 'key factor' (Varley and Gradwell, 1970). Is there a high, negative correlation between the change in numbers from years $i-1$ to i, and the proportion emigrating during the intervening period? Do emigrants form a high proportion of total losses during the decline? If these two points both apply, this strongly suggests that emigration is a major cause of population decline, although in the absence of extra biological evidence this numerical evidence is not in itself conclusive.

Both the above points were found in a recent fluctuation of red grouse (Watson *et al.*, 1984). To judge from the summary by Krebs (1985), this is not usually so in small mammals, because emigration tends to be higher during population increases than during declines, but the point has not been conclusively shown by data analysis. Moreover there is some evidence of a direct demographic role for emigration in declines of voles (Desy and Thompson, 1983; Mihok *et al.*, 1985). A difficulty in many studies of small mammals is that 'dispersal' is measured as immigration into a trapping grid other than the main reference grid, so workers cannot usually tell how the measured rates relate to actual emigration rates from the reference grid.

Watson and Moss (1970) proposed conditions which will show that behaviour, via socially induced mortality (or depression of recruitment), limits a breeding population. In the present context of a book on dispersal, it may be useful to note that the term 'socially induced depression of recruitment' covers net, socially induced emigration. As used for small mammals (e.g. Tamarin, 1983; Taitt, 1985), these conditions have often been phrased as 'spacing behaviour'. Strictly speaking, the conditions do not in themselves necessarily entail 'behaviour' that limits numbers, unless extra behavioural evidence is available. What they do show, if they are met, is at least that the presence of socially dominant individuals limits a breeding population.

12.6 DIFFERENT TYPES OF EMIGRATION AND THEIR DEMOGRAPHIC CONSEQUENCES

Lidicker (1975) suggested that 'pre-saturation dispersal' in small mammals occurs at low density or in the early increase phase before resources become limiting, whereas saturation dispersal involves surplus animals emigrating from a population at or near its carrying capacity. These terms have been widely used since (Chapter 9). A drawback is that they imply abstract concepts which are hard to measure and may not have been demonstrated (such as saturation and carrying capacity). Gaines and McClenaghan (1980) wrote that there is no operational way of distinguishing the two types, because there is no way of defining, let alone measuring, carrying capacity (but see Chapters 7 and 9). This is controversial, and many workers find 'carrying capacity' a useful term, but others think it suffers from drawbacks and so is better avoided. One reason for the difficulty of defining and measuring carrying capacity in animal studies in general is that the limiting resource (assuming that there is one) is so seldom identified. If it is identified, as in some insect studies, it can usually be measured. Or, an area's carrying capacity can be defined differently as the density above which the population growth rate becomes negative, but this would be a circular definition.

Nevertheless, it is worth discussing different types of emigration behaviour. One can think broadly of two types of behaviour. Type A involves a subordinate individual, which is unlikely to breed unless it finds a vacancy due to a more dominant individual dying, and which otherwise is likely to die itself. Type B involves an individual, not necessarily subordinate, which moves to places where density is lower, or perhaps more importantly where the increase phase of the population fluctuation is about to begin, and which may thereby be more likely to produce recruits. [Types A and B may be the same as, but are not necessarily the same as, Lidicker's (1975) suggestion that saturation dispersers in small mammals would be social outcasts in poor condition, whereas presaturation dispersers would survive well in newly colonized habitats (see also Chapters 2 and 10).]

We have found examples of A and B in red grouse. With A, aggressiveness and breakup of families in autumn were followed by the formation of a non-territorial social class. Most of its members died of socially induced mortality and emigration over winter, though a few replaced the relatively few territorial birds that died (Watson, 1985; Watson and Moss, 1990). Red grouse with type B behaviour showed emigration of two kinds during years of high and declining density, and a third kind at the end of population troughs (Watson *et al.*, 1984). In the first kind, the proportion of adult pairs emigrating shortly before egg-laying was zero at low density and high at high density. In the second kind, the proportion that stayed in the area until their chicks hatched, but then emigrated with their chicks and returned in autumn without them, was highest in years of declining densities. In such years, very few of the young reared in the area were successful in obtaining territories and thus being recruited into the breeding population. The third kind of type B behaviour involved birds immigrating to areas at the end of population troughs in these areas, when the increase phase was about to begin there. Moss and Watson (1985, 1990) suggested in evolutionary terms why such emigrant red grouse should leave. It is important to realize that animals showing behaviour of type A might show type B in a different social setting or in a different phase of the population fluctuation.

Lidicker's (1975) comparisons of numbers with density of small mammals did not involve the percentage emigrating, but he came near it by writing that if emigration were to stop population growth at carrying capacity, it necessarily 'makes up a very large fraction of the losses.' Unfortunately, many workers on small mammals ignored this lead until recently, by paying no attention to the proportion emigrating. Lidicker (1975) also wrote that if losses from presaturation emigration occur during periods of population growth, the resulting growth rate will be 'correspondingly less', and 'if a population accepts immigrants they add directly to the growth rate by increasing the additions per unit of time' (p. 114). The assumption here, which has often been made with birds as well as small mammals, is that losses (or gains) are additive and that none shows complete compensation. It is an assumption that needs testing. Krebs (1986) wrote, 'Whether mortality factors are additive or compensatory is one of the most pressing and most difficult questions in mammalian ecology'. Yes, and add 'bird ecology' to this for good measure.

Krebs *et al.* (1973) found that the dispersal 'rate' of small mammals, which in this case was the number dispersing per two-week trapping period, was inversely correlated with the rate of change in numbers. The use of 'rate' instead of number in that paper and in other papers on small mammals has led other workers to publish misleading accounts, including myself in Moss *et al.* (1982, Figure 4.7), and also Morris (1984, summary and Figure 1). Although Dekker (1975) wrote (p. 65) of 'the proportionality between dispersal and the rate of change of the total population', what he actually meant was that the number of dispersers per acre per week was correlated with the rate of population change. In their review of the literature on 'dispersal' in small mammals, Gaines and McClenaghan (1980) mentioned (p. 169) a positive correlation between 'dispersal rate' and density, but their Table 2 reveals that for nine species with adequate data, only four showed a statistically significant positive correlation.

Gaines and McClenaghan pointed to studies where percentage losses due to dispersal were correlated with the population's rate of increase. But does this mean that dispersal is one limiting factor preventing continued population increase though not causing declines, or is it simply a compensatory form of loss? They also found evidence in the literature that percentage losses from dispersal during the increase phase exceeded those during declines. Krebs *et al.* (1973) had already reached this conclusion from their observation that the number dispersing per two weeks was highest during the increase phase rather than during the peak or the decline. They were probably right, though their observation on its own is not sufficient to be sure about this conclusion. If one takes it at face value, the direction of the trend is the opposite of that needed if emigration is to be a direct, important demographic cause of population declines [see Warkowska-Dratnal and Stenseth (1985) and Stenseth and Łomnicki (1990) for further discussion and references].

Gaines and McClenaghan concluded (p. 189) that considering all the evidence 'the lack of density-dependence suggests that dispersal may not operate as a mechanism of population regulation', and that dispersal is clearly density-independent. Rates of loss that are not positively density-dependent obviously cannot 'regulate' numbers, if one uses 'regulate' in the sense of density-dependent returns of numbers to a supposed 'equilibrium density' after departures from it (Varley *et al.*, 1973). Such losses that are not positively density-dependent will, however, tend to lengthen the time taken for peak density to be reached, or will reduce the peak density, provided that there is not complete compensation for these losses. A different, related point is that it might be useful with small mammals and birds to consider models other than the equilibrium one, for example the ceiling model as used by some entomologists (Milne, 1957; Dempster, 1983).

Krebs (Chapter 7) has made a useful advance and cleared up some confusion by doing statistical analysis on whether dispersal 'rate' (this time an index of per capita dispersal rate, and so equivalent to a percentage) was correlated with breeding density in a non-cyclic population of *Microtus townsendii*. The two were correlated. However, what about cyclic populations of small mammals? Before we accept the statement that dispersal rate is not correlated with density in such cyclic populations, the lack of this correlation requires equally rigorous statistical analysis. And, although 'dispersal' rate may seem unlikely to be a strong, delayed density-dependent factor in small mammals, this question too does not appear to have been answered clearly by statistical analysis.

12.7 EMIGRATION, IMMIGRATION, CYCLES, AND IRREGULAR FLUCTUATIONS

There is interesting anecdotal evidence that ruffed grouse (*Bonasa umbellus*) in small isolated fragments of forest remaining after clearance for agriculture do not show cycles and are fairly stable in numbers, unlike those in large tracts of boreal forest (Leopold, 1931; Graham and Hunt, 1958; Ammann, 1972). Watson and Moss (1979) suggested that populations need not show cycles when 'dispersal' is continuously high, as strong competition for space or other resources at home may thereby be avoided. They gave as a possible example the blue grouse (*Dendragapus obscurus*), where new habitat continually becomes available due to forest succession after fire (Zwickel and Bendell, 1972).

Red grouse in marginal habitat in Mayo, western Ireland (Watson and O'Hare, 1979; Watson and Moss, 1979) and on Mull in western Scotland (Watson *et al.*, 1987) is another example. Shooting bags from former decades fluctuated greatly, as in east Scotland today. Declines, and increases, often lasted several years each. Densities now are lower, and on unmanipulated moorland on Mull they fluctuated little during a recent study. Habitat in Mayo and Mull has deteriorated, due to careless burning of vegetation and overgrazing by sheep. Areas of favourable habitat now adjoin large areas of poorer habitat, and temporarily vacant ground is common. Emigration to poorer areas may be preventing longer population increases and higher peaks on the better habitat. Large-amplitude fluctuations of red grouse may be associated with large areas of good habitat from which it is possible to emigrate to poorer areas (Moss and Watson, 1990), but perhaps not where there is too big a 'dispersal sink' in small-mammal terms (Lidicker, 1973; Tamarin, 1978). This is compatible with Lidicker's (1985a,b, 1988) views on dispersal and cycles in microtines, namely that the configuration of habitat patches may be important.

Pearson (1966) stated that predation is necessary for vole cycles. Others argued that the associated inability to emigrate from islands is the reason why small mammals there do not show 3–4 year cycles. When fencing prevented emigration, voles inside fences increased to abnormally high density, overgrazed, and eventually starved to death even though predators were present (Krebs *et al.*, 1969). Hence space to emigrate into may be necessary for cycles.

Or, preventing immigration to an area, not preventing emigration from it, may be what prevents cycles. This is evidently refuted by experiments where voles were within fences which allowed different extents of egress and ingress (Tamarin *et al.*, 1984). This result is of such theoretical importance for research on animals in general, let alone voles, that the experiment needs to be repeated. Meanwhile, it suggests that preventing large-scale emigration from an area of favourable habitat, or completely preventing immigration into it, may not be necessary for cycles, but that allowing a small number to emigrate may be necessary. Such advances with voles are of great importance to some of us working with grouse.

Krebs (1985, p. 309) wrote that he had 'in the past emphasized the role of dispersal in cycles', but as dispersal predominates in the increase phase (Gaines and McClenaghan, 1980), he had to explain 'the obviously increased mortality that occurs in the peak and decline phases. Dispersal losses do not seem to be the delayed density-dependent factor that generates cycles.' He added that the evidence from enclosures is that '*in situ* mortality is the cause of cyclic declines in voles.' Boonstra and Krebs (1977) gave an example, although mortality here was inferred from differences in numbers at different times, rather than being studied directly. I wonder whether heavy mortality occurs outside enclosures. And, even though the immediate cause of loss may be death, are such animals more vulnerable to mortality because of earlier social events?

12.8 EMIGRATION AS AN INDIRECT CAUSE OF DECLINES

Emigration from one area entails immigration to another. Such animals may be an important cause of increase in numbers on areas with a net influx. However, the consensus seems to be that emigration is not a major, direct, demographic cause of population declines in small mammals. Hence, if it is an important mechanism for demographic change on favourable habitat where there is a net exodus, one must postulate some indirect effect. Perhaps emigration changes the population's genetic structure in such a way that strife is raised, and a decline in numbers then follows due to more deaths, more emigration or poorer reproduction (a variant of the hypothesis of Chitty, 1967). Workers on small mammals have long known that changes in animal quality characterize different phases of the fluctuation (Chitty, 1967), and workers on red grouse have benefited greatly from their ideas and findings on this (Moss *et al.*, 1984). Animals which move through a population while emigrating, or which immigrate into it, may increase strife and hostility simply because they are strangers which are unknown or less well known to residents (Abramsky and Tracy, 1979; Charnov and Finerty, 1980). Mountford *et al.* (1990) produced a mathematical model showing that cycles in red grouse can be explained by the joint action of two factors: territory size and kin selection as expressed in differential behaviour towards kin and non-kin.

12.9 FINAL REMARKS

There is still much confusion on dispersal and demography in small mammals, and in *Lagopus* and other birds, let me hasten to add. The question arises whether demographic data on small mammals in recent studies could sometimes have been better analysed and modelled, and whether this could still be done with these data or in future studies. An example is the frequent usc of 'number emigrating', and not the more important 'percentage emigrating'. Although re-analysis may not solve the main problems, it might clarify them and suggest useful new experiments. However, I do not wish to underestimate the difficulties (perhaps some would say the drawbacks!) of applying conventional population concepts to small mammal 'populations' which rapidly expand and contract the range they occupy during a fluctuation in numbers.

Some studies are more influenced by interest in the species than in questions of general biological interest, so we have lagomorph conferences, grouse conferences, and even a woodland grouse conference. However, I think workers who tackle population problems using small mammals have on the whole taken a wider view than workers using grouse. The editors of that worthy periodical *Mouse News* once asked me for an article on grouse, but I cannot imagine the editors of a grouse periodical asking a small-mammal worker for an article on mice.

ACKNOWLEDGEMENTS

Thanks to R. Moss, J.P. Dempster and the editors all of whom gave useful comments.

REFERENCES

Abramsky, Z. and Tracy, C.R. (1979) Population biology of a 'noncycling' population of prairie voles and a hypothesis on the role of migration in regulating microtine cycles. *Ecology*, **60**, 349–61.

Ammann, G.A. (1972) Discussion on tetraonid symposium. *Proceedings of the International Ornithological Congress*, **15**, 179.

Bergerud, A.T. (1988) Population ecology of North American grouse, in *Adaptive Strategies and Population Ecology of Northern Grouse* (eds A.T. Bergerud and M.W. Gratson), University of Minnesota Press, Minneapolis, pp. 578–685.

Bergerud, A.T., Mossop, D.H. and Myrberget, S. (1985) A critique of the mechanics of annual changes in ptarmigan numbers. *Canadian Journal of Zoology*, **63**, 2240–8.

Boonstra, R. and Krebs, C.J. (1977) A fencing experiment on a high-density population of *Microtus townsendii*. *Canadian Journal of Zoology*, **55**, 1166–75.

Charnov, E.L. and Finerty, J. (1980) Vole population cycles: a case for kin selection? *Oecologia*, **45**, 1–2.

Chitty, D. (1967) The natural selection of self-regulatory behaviour in animal populations. *Proceedings of the Ecological Society of Australia*, **2**, 51–78.

Dekker, H. (1975) A simple mathematical model of rodent population cycles. *Journal of Mathematical Biology*, **2**, 57–67.

Dempster, J.P. (1983) The natural control of populations of butterflies and moths. *Biological Reviews*, **58**, 461–81.

Desy, E.A. and Thompson, C.F. (1983) Effects of supplemental food on a *Microtus pennsylvanicus* population in central Illinois. *Journal of Animal Ecology*, **52**, 127–40.

Elton, C.S. (1942) *Voles, Mice and Lemmings*, Clarendon Press, Oxford.

Gaines, M.S. and McClenaghan, L.R. (1980) Dispersal in small mammals. *Annual Review of Ecology and Systematics*, **11**, 163–96.

Graham, S.A. and Hunt, G.S. (1958) A non-cyclic ruffed grouse population near Ann Arbor, Michigan. *Journal of Wildlife Management*, **22**, 427–32.

Hudson, P.J. (1990) Territorial status and survival in a low density grouse population, in *Red Grouse Population Processes* (eds A.N. Lance and J.H. Lawton), Royal Society for the Protection of Birds, Sandy, Bedfordshire, pp. 20–28.

Hudson, P.J. and Dobson, A.P. (1989) Population biology of *Trichostrongylus tenuis*, a parasite of economic importance for red grouse management. *Parasitology Today*, **5**, 283–91.

Krebs, C.J. (1985) Do changes in spacing behaviour drive population cycles in small mammals? *Symposia of the British Ecological Society*, **25**, 295–312.

Krebs, C.J. (1986) Are lagomorphs similar to other small mammals in their population ecology? *Mammal Review*, **16**, 187–94.

Krebs, C.J., Gaines, M.S. Keller, B.L. *et al.* (1973) Population cycles in small rodents. *Science*, **179**, 35–41.

Krebs, C.J., Keller, B.L. and Tamarin, R.H. (1969) *Microtus* population biology: demographic changes in fluctuating populations of *M. ochrogaster* and *M. pennsylvanicus* in southern Indiana. *Ecology*, **50**, 587–607.

Leopold, A. (1931) *Report on a Game Survey of the North Central States*, Sporting Arms Ammunition Manufacturing Institute, Madison, Wisconsin.

Lidicker, W.Z. Jr (1973) Regulation of numbers in an island population of the California vole, a problem in community dynamics. *Ecological Monographs*, **43**, 271–302.

Lidicker, W.Z. Jr (1975) The role of dispersal in the demography of small mammals, in *Small Mammals: their productivity and population dynamics* (eds F.B. Golley, K. Petrusewicz and L. Ryszkowski), Cambridge University Press, Cambridge, pp. 103–28.

Lidicker, W.Z. Jr (1985a) Dispersal, in *Biology of New World Microtus* (ed. R.H. Tamarin), Special Publication No. 8, American Society of Mammalogists, pp. 420–54.

Lidicker, W.Z. Jr (1985b) Population structuring as a factor in understanding microtine cycles. *Acta Zoologica Fennica*, **173**, 23–7.

Lidicker, W.Z. Jr (1988) Solving the enigma of microtine 'cycles.' *Journal of Mammalogy*, **659**, 225–35.

Mihok, S., Turner, B.N. and Iverson, S.L. (1985) The characterization of vole population dynamics. *Ecological Monographs*, **55**, 399–420.

Milne, A. (1957) The natural control of insect populations. *Canadian Entomologist*, **89**, 193–213.

Morris, D.W. (1984) Rodent population cycles: life history adjustments to age-specific dispersal strategies and intrinsic time lags. *Oecologia*, **64**, 8–13.

Moss, R., Shaw, J.L. Watson, A. and Trenholm, I.B. (1990) Role of the caecal threadworm *Trichostrongylus tenuis* in the population dynamics of red grouse, in *Red Grouse Population Processes* (eds A.N. Lance and J.H. Lawton), Royal Society for the Protection of Birds, Sandy, Bedfordshire, pp. 62–71.

Moss, R. and Watson, A. (1985) Adaptive value of spacing behaviour in population cycles of red grouse and other animals. *Symposia of the British Ecological Society*, **25**, 275–94.

Moss, R. and Watson, A. (1990) Population regulation in red grouse – a theoretical outline, in *Red Grouse Population Processes* (eds A.N. Lance and J.H. Lawton), Royal Society for the Protection of Birds, Sandy, Bedfordshire, pp. 29–34.

Moss, R., Watson, A. and Ollason, J. (1982) *Animal Population Dynamics*, Chapman & Hall, London.

Moss, R., Watson, A. and Rothery, P. (1984) Inherent changes in the body size, viability and behaviour of a fluctuating red grouse (*Lagopus lagopus scoticus*) population. *Journal of Animal Ecology*, **53**, 171–89.

Mountford, M.D., Watson, A., Moss, R. *et al.* (1990) Land inheritance and population cycles of red grouse, in *Red Grouse Population Processes* (eds A.N. Lance and J.H. Lawton), Royal Society for the Protection of Birds, Sandy, Bedfordshire, pp. 78–83.

Pearson, O.P. (1966) The prey of carnivores during one cycle of mouse abundance. *Journal of Animal Ecology*, **35**, 217–33.

Potts, G.R., Tapper, S.C. and Hudson, P.J. (1984) Population fluctuations in red grouse: analysis of bag records and a simulation model. *Journal of Animal Ecology*, **53**, 21–36.

Stenseth, N.C. and Łomnicki, A. (1990) On the Charnov-Finerty hypothesis: the unproblematic transition from docile to aggressive and the problematic transition from aggressive to docile. *Oikos*, **58**, 234–8.

Taitt, M.J. (1985) Experimental analysis of spacing behaviour in the vole, *Microtus townsendii, Symposia of the British Ecological Society*, **25**, 313–30.

Tamarin, R.H. (1978) Dispersal, population regulation, and K-selection in field mice. *American Naturalist*, **112**, 545–55.

Tamarin, R.H. (1983) Animal population regulation through behavioural interactions, in *Advances in the Study of Mammalian Behaviour* (eds J.F. Eisenberg and D.V. Kleiman), Special Publication No. 7, American Society of Mammalogists, pp. 698–720.

Tamarin, R.H., Reich, L.M. and Moyer, C.A. (1984) Meadow vole cycles within fences. *Canadian Journal of Zoology*, **62**, 1796–804.

Taylor, L.R. (1986) Synoptic dynamics, migration and the Rothamsted insect survey. *Journal of Animal Ecology*, **55**, 1–38.

Varley, G.C. and Gradwell, G.R. (1970) Recent advances in insect population dynamics. *Annual Review of Entomology*, **15**, 1–24.

Varley, G.C., Gradwell, G.R. and Hassell, M.P. (1973) *Insect Population Ecology: an analytical approach*, Blackwell Scientific, Oxford.

Warkowska-Dratnal, H. and Stenseth, N.C. (1985) Dispersal and the microtine cycle: comparison of two hypotheses. *Oecologia*, **65**, 468–77.

Watson, A. (1985) Social class, socially-induced loss, recruitment and breeding of red grouse. *Oecologia*, **67**, 493–8.

Watson, A. (1987) Review of Wynne-Edwards, V.C., 'Evolution through Group Selection,' Blackwell Scientific, Oxford, 1986, 386 pp. *Journal of Animal Ecology*, **56**, 1089–90.

Watson, A. and Moss, R. (1970) Dominance, spacing behaviour and aggression in relation to population limitation in vertebrates. *Symposia of the British Ecological Society*, **10**, 167–218.

Watson, A. and Moss, R. (1979) Population cycles in the Tetraonidae. *Ornis Fennica*, **56**, 87–109.

Watson, A. and Moss, R. (1987) The mechanics of annual changes in ptarmigan numbers: a reply to Bergerud, Mossop and Myrberget. *Canadian Journal of Zoology*, **65**, 1043–7.

Watson, A. and Moss, R. (1990) Spacing behaviour and winter loss in red grouse, in *Red Grouse Population Processes* (eds A.N. Lance and J.H. Lawton), Royal Society for the Protection of Birds, Sandy, Bedfordshire, pp. 35–52.

Watson, A. and O'Hare, P.J. (1979) Red grouse populations on experimentally treated and untreated Irish bog. *Journal of Applied Ecology*, **16**, 433–52.

Watson, A., Moss, R. and Parr, R. (1987) Grouse increase on Mull. Landowning in Scotland, 207:6.

Watson, A., Moss, R., Parr, R. *et al.* (1988) Preventing a population decline of red grouse (*Lagopus lagopus scoticus*) by manipulating density. *Experientia*, **44**, 274–5.

Watson, A., Moss, R. Rothery, P. and Parr, R. (1984) Demographic causes and predictive models of population fluctuations in red grouse. *Journal of Animal Ecology*, **53**, 639–62.

Wynne-Edwards, V.C. (1986) *Evolution through Group Selection*, Blackwell Scientific, Oxford.

Zwickel, F.C. and Bendell, J.F. (1972) Blue grouse, habitat, and populations. *Proceedings of the International Ornithological Congress*, **15**, 150–69.

– 13

An entomological perspective on animal dispersal

G.K. RODERICK AND R.L. CALDWELL

13.1 INTRODUCTION

Mammalian, avian, entomological, and botanical views of the movements of organisms have developed simultaneously during the past century (Rankin, 1985b). Although the paths have often been separate, links among the taxonomic viewpoints have provided new insights and the impetus for studies in other taxa. Here, we first sketch historical differences and similarities in the approaches of studies of movements of different taxa. We concentrate on why divergent views have developed in the studies of birds, mammals, and insects. We then provide a summary of what studies of insects have contributed to the understanding of the ecology and evolution of the movements of organisms. Finally, we point to areas where interactions among students of different taxa may be especially rewarding.

13.2 SEPARATE PATHS

The most dramatic of animal movements are usually the best studied, such is the case for both birds and mammals. Ornithologists have been preoccupied with long-distance, round-trip migration and, not surprisingly, questions concerning avian dispersal have concentrated on the adaptive value of movement, especially with respect to food and reproduction (Dingle, 1972, 1980; Baker, 1978; Gauthreaux, 1980; Lidicker and Caldwell, 1982; Swingland and Greenwood, 1983). However, several early avian studies emphasized movements that were not of the long-distance, round-trip type. For example, in his studies of marked crows, Rowan (see Baker, 1978) documented within-range north–south movements for distances of 200–300 miles. Likewise, the phenomenon of nomadism has been extensively studied in parakeets and East African ploceid finches (*Quelea*). A second, and later focus of avian movements incorporated both proximate and ultimate questions in the search for an understanding of homing and navigation (Baker, 1978). Homing studies emphasized the mechanisms of behaviour, and less so the evolution and ecology of movements (Gauthreaux, 1980). An interesting entomological parallel to the study of avian movements was the early study of butterfly migration, especially that of monarchs (*Danaus plexippus*) (Williams, 1930; Urquhart, 1960). Yet, these efforts seemed to be prompted more out of 'avian envy' and an interest in particular species than by a desire to understand general movement patterns in insects.

The development of an understanding of mammalian movements has arisen through several perspectives (Lidicker and Caldwell, 1982; Swingland and Greenwood, 1983; Chapter 2). As in the study of avian movements, conspicuous movements have received much attention. For example, studies of cyclic and irruptive populations of small rodents have given focus to much of the understanding of mammalian dispersal and have produced several mechanistic hypotheses involving both environmental and genetic factors (Chapters 4 and 8). In another example, mass migrations of wildebeests and caribou demonstrated the role of the environment, especially weather and forage, in mammalian movements (McCullough, 1985). Other work in an applied context has concerned conspicuous mammals as, for example, studies of migration in cetaceans as a predictive harvesting tool and studies of the role of dispersal in pest management and conservation (Chapter 8). Less well studied have been patterns and causes of dispersal in non-cyclic populations of small rodents (Chapter 6) and in animals showing less dramatic migrations. A final perspective to the mammalian view of dispersal stems from important aspects of mammalian biology, such as the importance of home range and social factors in animal movements (Swingland and Greenwood, 1983; Chapter 5). By contrast, understanding home range and social interactions has contributed far less to the understanding of insect movements (Lidicker and Caldwell, 1982).

13.3 INSECT SPATIAL MOVEMENTS

13.3.1 Natural variation

The insects have provided a different, yet complementary view of animal movement. Of great importance have been the vast natural variations in behaviour and morphology related to dispersal that are available for study both within and among species of insects. For example, a dramatic, yet common, form of variation in dispersal is that of flight polymorphism, which can be manifest as morphological, behavioural, and/or physiological variation. Flight polymorphisms are widespread in insects, occurring most notably in the Orthoptera, Homoptera, Hemiptera, and Coleoptera (Harrison, 1980; Dingle, 1984b; Roff, 1986, 1990; Roff and Fairbairn, in press). In some species, such as planthoppers and aphids, wing dimorphism can be extreme; within a species both fully winged and flightless or reduced-flight forms can exist (Lees, 1966, 1967; Denno, 1983, 1985; Dingle, 1984b; Dixon, 1985; Denno *et al.*, 1989). Because flight polymorphisms represent easily quantified variation that directly affects migration and dispersal, they have been important historically in the understanding of the ecology and evolution of insect migration and dispersal. In a well-known example, natural variation in locusts, expressed as migratory and gregarious phases, were studied for economic reasons as early as the 1930s (Uvarov, 1928; Kennedy, 1956; Waloff, 1966).

During the same period, Dobzhansky, Wright, and co-workers were performing field releases of marked *Drosophila* in an effort to measure natural levels of dispersal of individual flies to estimate gene flow (Dobzhansky and Wright, 1943). Here, small size and rapid generation times allowed the production of experimental material. Although this early work provided a series of actual measurements of rates of dispersal, the results were somewhat limited in generality. Only more recently have population geneticists and ecologists taken a combined effort to examine dispersal and to estimate population genetic parameters (Carson *et al.*, 1970; Ehrlich *et al.*, 1975; McCauley *et al.*, 1981; Coyne *et al.*, 1982, 1987; McCauley, 1983; Taylor *et al.*, 1984; Bull *et al.*, 1987; Coyne and Milstead, 1987; Chapter 3).

Guided by earlier work on locusts and butterflies, insect spatial movements have been divided into two categories — migration and 'vegetative' reflexes (Kennedy, 1961). In this sense, migration has been described as a distinct physiological, behavioural, and ecological syndrome (Kennedy, 1961; Johnson, 1963; Southwood, 1962). 'Vegetative' reflexes, or 'trivial' movements (Heape, 1931) were activities involving feeding, mating, and oviposition (Kennedy, 1961). Indeed, in at least some species (e.g. the giant milkweed bug *Oncopeltus*), behaviours involving feeding, mating, and oviposition can be separated readily from migratory behaviours (Caldwell and Rankin, 1974).

Migration (Kennedy, 1975; Solbreck, 1978; Dingle, 1984a), along with diapause (Tauber and Tauber, 1976, 1981; Tauber *et al.*, 1986), have been considered to be as important to life histories of insects as other components of fitness, such as fecundity schedules and survivorship. This suggests that the sets of traits above have not evolved independently (Lees, 1966, 1967; Dingle, 1974, 1979, 1980, 1984a; Denno, 1978; Istock, 1978; Solbreck, 1978; Vepsäläinen, 1978). For example, among insects that are colonizers, one might predict a suite of correlated life-history traits representing a migration syndrome (Dingle, 1972, 1986). If colonizers are also to survive and reproduce (MacArthur and Wilson, 1967), they should increase fecundity, possibly including a decrease in the age at first reproduction (Cole, 1954; Lewontin, 1965), and maintain a high reproductive value (Fisher, 1958; Johnson, 1969; Dingle, 1972; Safriel and Ritte, 1980, 1983). For a number of migratory species, migratory syndromes do occur as parts of their life-history patterns (Dingle, 1979, 1984b; Palmer, 1985). However, for other insects, particularly those that are shorter-lived or spend much of their lives in one habitat, migratory syndromes are not obvious, and it is difficult to adequately predict insect movement or life-history variation (Davis, 1984). For this reason, some authors view insect dispersal as a continuum of displacement movements, from those related to 'vegetative' processes of feeding and reproduction to those involved in migration (Lidicker and Caldwell, 1982; Chapter 1).

In a variety of species, short-distance movements have been described and modelled with considerable success (e.g. Begon, 1976; Jones, 1977; McCauley *et al.*, 1981; Kareiva, 1983b; Rausher, 1983; Davis, 1984; Wetzler and Risch, 1984; Roderick, 1987; and numerous others). Documentation of longer-distance movements has been more difficult (Dingle, 1984b). Noted exceptions include studies of the monarch (Brower, 1985), some *Drosophila* species (see above), coccinellid beetles (Hagen, 1962), and several species of economic importance, including rice planthoppers (Kisimoto, 1976), orthoptera locusts (Rainey, 1951), Colorado potato beetle (Johnson, 1967), pine weevils (Solbreck, 1980), gypsy moth (Mason and McManus, 1981), and cabbage white butterfly (Jones *et al.*, 1980).

13.3.2 Causal mechanisms of migration and dispersal in insects

Traditionally, studies of causal mechanisms of migration or dispersal in insects have tended to focus on the effects of either environmental or genetic factors, but seldom on their interaction. Numerous studies have shown that in many orders of insects environmental factors can influence characters or behaviours related to emigration, migration, and/or dispersal (Dingle, 1984b). Southwood (1977), when suggesting a link between generation time and duration of favourable habitats, recognized the important effects on migration and dispersal of environmental inputs such as photoperiod and food plant physiology, Likewise, wing or flight polmorphisms in a variety of species can be strongly influenced by environmental factors, including temperature, photoperiod, and crowding (Harrison, 1980; Strong and Stiling, 1983; Dingle, 1984b; Denno *et al.*, 1985; Roff, 1986, 1990; Roderick, 1987; Denno and Roderick, 1990). Only a few studies have examined the effects of local sex ratio. In the milkweed bug, *Lygaeus kalmii*, if mates are not available, and depending on temperature and food conditions, virgin females will fly before mating (Caldwell, 1974). Lawrence (1987) showed that sex ratio affected the rate of emigration of male, but not female, milkweed beetles (*Tetraopes*). Finally, dispersal in a variety of insects has been shown to be influenced by a vast array of habitat variables, including patchiness (Kareiva, 1982, 1985; Denno, 1983; Turchin, 1986; Antolin and Strong, 1987), habitat structure (Kareiva, 1983a; Denno and Roderick, 1991), host-plant quality (Rausher, 1983; Denno, 1983), and prey abundance (Kareiva, 1987).

Less is known about the effects of genetic variation in the expression of characters or behaviours related to insect migration and dispersal. Models by Roff (1975) suggest that the proportion of migrants maintained in a population depends not only on habitat parameters, but also on the mode of inheritance of traits affecting dispersal. While heritable components of variation in dispersal and migration were documented in a few species some time ago (e.g. Caldwell and Hegmann, 1969), only recently have more detailed studies begun to elucidate the mechanisms of genetic control. Genetic variation in dispersal in insects appears to be polygenic in most systems studies, with a few exceptions (Dingle, 1984b, 1986; Roff, 1986; Gatehouse, 1986, 1991). Polygenic effects are common in a variety of wing-dimorphic systems, e.g. gerrids (Zera *et al.*, 1983), a pyrrhocorid bug (Honek, 1976, 1979), *Gryllus* crickets (Harrison, 1979; Roff, 1984), and planthoppers (Ammar, 1973; Mochida, 1973; Mahmud, 1980; Roderick, 1987).

The roles of genotype–environment interactions in the expression of characters affecting migration and dispersal are poorly studied, and few, if any, consistent patterns emerge at present. In a study of aphids, field-caught clones (genotypes) differed in their production of alates in response to specific environmental conditions (Lamb and MacKay, 1979, 1983). Other studies of aphids show clonal differences in tendency to produce alates in response to crowding or host plant quality (Sutherland, 1969, 1970). In gerrid waterstriders, crosses of long × long and short × short-winged forms differed in responses to several photoperiods. The percentage macroptery increased with less daylight, but was always less for short × short crosses (Zera *et al.*, 1983; Zera, 1985). In a salt marsh planthopper, wing form is determined largely by environmental factors in females and by genetic factors in males (Roderick, 1987). In the cowpea weevil, Messina (1987) documented a complicated genotype by crowding interaction for a dispersal polymorphism in crosses of monomorphic laboratory populations with polymorphic wild populations. Population by environment interactions in migration behaviour have been measured in *Oncopeltus* (Dingle *et al.*, 1980) and wild strains of *Drosophila* (Sakai *et al.*, 1957). Although, as these examples show, the presence of interactions between environmental and genetic factors have been documented in insects, broad underlying principles connecting environmental and genetic variation with dispersal and life-history variation are still being formulated.

13.3.3 Dispersal and migration as an adaptation

Entomologists have given considerable attention to dispersal and migration as an adaptation, especially with respect to colonization. For example, dispersal and migration among patches in a subdivided population can affect the dynamics (Andrewartha and Birch, 1954) and/or persistence (termed 'spreading the risk'; den Boer, 1968) of the population as a whole. Studies on dispersal of aquatic insects by Brown (1951) and Jackson (1952) showed that in certain species migrants were more likely to inhabit small or temporary ponds. Migration seems to occur often for insects in 'temporary' habitats, where variation in carrying capacity is large, while less migration and even winglessness occurs often in stable or isolated habitats (Southwood, 1962; Vepsäläinen, 1978; Dingle, 1984b; Andrewartha and Birch, 1984; Denno *et al.*, in press). As many studies show, migration and dispersal can have important consequences for population dynamics, especially for fluctuations in space and time; yet, for relatively few insects is the connection between habitat variables, individual variation in migration or dispersal, and population dynamics well understood (Taylor and Taylor, 1977; Taylor *et al.*, 1983; Kareiva, 1983a; Rausher, 1983).

As a first step in understanding the adaptive significance of characters influencing migration and dispersal, one can assess the 'fitness costs' (Roff, 1986) associated with the ability to fly, migrate, or disperse, by examining phenotypic correlations between traits affecting migration or dispersal and other components of fitness. Recent reviews show that within a particular environment individuals of different flight abilities or wing forms can differ in other components of fitness (Harrison, 1980; Dingle, 1984b; Roff, 1986; Roderick, 1987; Denno *et al.*, 1989). For example, fecundity of macropterous forms is less than that of flightless forms in crickets (Tanaka, 1976; Roff, 1984), aphids (Dixon, 1972; Wratten, 1977), waterstriders (Andersen, 1973; Zera, 1984), corixids (Young, 1965), and pea weevils (Utida, 1972). Reproduction is delayed in migratory forms of crickets (Roff, 1984), waterstriders (Andersen, 1973), locusts (Norris, 1950), milkweed bugs (Dingle, 1978, 1981, 1982), and *Dysdercus* bugs (Dingle and Arora, 1973). In planthoppers (Delphacidae), brachypterous (short-winged) individuals, relative to macropterous (long-winged) individuals, have a shorter pre-oviposition period and reach peak reproduction earlier (Denno, 1985; Roderick, 1987; Denno and Roderick, 1990).

Many evolutionary predictions concerning dispersal and other life-history characters require knowledge or assumptions about the genetic variance and co-variance structure of dispersal and life-history traits. Such information represents the so-called 'substructure of fitness' (Istock, 1978) and can be used to calculate genetic correlations among components of fitness (Falconer, 1981). Genetic correlations, which indicate to what extent selection of one character will cause a change in another character, may represent the ultimate constraints in life-history evolution (Rose and Charlesworth, 1981; Lande, 1982; Hegmann and Dingle, 1982; Via, 1984; Service and Rose, 1985). Unfortunately, genetic correlations cannot be predicted from phenotypic correlations alone, because strong environmental correlations, especially if in the opposing direction, may mask the effects of genetic correlations.

Genetic correlations among dispersal or migratory characters and other components of fitness have been little studied. With a halfsib/fullsib design with *Oncopeltus fasciatus*, Hegmann and Dingle (1982) examined a suite of traits including morphological characters related to migration. Longer-winged bugs (large and potential colonizers) developed rapidly and had large clutches. In another set of experiments, selection for longer wings in *O. fasciatus* increased tethered flight duration and fecundity, but had no effect on age of first reproduction (Palmer, 1985; Dingle *et al.*, 1988), while selection for delayed flight also delayed oviposition and altered levels of juvenile hormone (Rankin, 1978). There is additional evidence that the genetic correlations between flight and reproductive characteristics in *O. fasciatus* differs among populations (Dingle *et al.*, 1988). In *Tribolium castaneum* a line selected for high 'dispersal' also showed increased rates of development, (egg production in the first 4 days) and rates of egg fertility (Lavie and Ritte, 1978). Wu (1981) confirmed these results for *T. castaneum*, but found so such correlations in a less dispersive species, *T. confusum*. In one of a few studies to look at both dispersal and life-history variation in more than one environment, Honek (1979) selected for macroptery in a phyrrhocorid bug cultured in two different photoperiods and found that although selection did affect the proportion of macropterous individuals, there was no resulting change in diapause.

Theoretical work suggests that the reliability of environmental cues may be important in the evolution of migratory and dispersal responses to environmental variables. Summaries of modelling efforts to examine the evolution of dispersal can be found in Horn (1983), Bull *et al.* (1987), and Chapters 3, 4, and 9. Cohen (1967), Ziegler (1976) and Giesel (1976) have argued that, with respect to dispersal, unreliable cues will favour genotypic stereotypy and reliable cues will favour phenotypic plasticity (see also Dingle, 1986). However, the relationships between environmental cues and the effects of environmental and genetic components of variation on migration and dispersal have been little studied experimentally in natural populations (Dingle, 1984b; Gatehouse, 1991).

13.3.4 Entomological breakthroughs and failures

Two major successes in the study of insect dispersal and migration have been the role in insect movement of genetics (reviewed above) and of behavioural and physiological substrates underlying dispersal, especially hormonal control and the use and switching of flight fuels. Hormonal control has been studied extensively, for example, in *Drosophila*, the giant milkweed bug *Oncopeltus*, various waterstriders, and the Colorado potato beetle (deKort, 1969; Rankin, 1978, 1985a). Use and switching of flight fuels has been documented, for example, in grasshoppers (Jutsam and Goldsworthy, 1976) and planthoppers (Padgham, 1983).

Entomologists have also been very successful in describing and modelling short-distance movements of insects (reviewed above), but documenting longer-distance movement patterns has been less successful. The same attributes that make insects useful for genetic studies – small size and short generation times – make them difficult subjects for long-distance mark and recapture methods (see also methods in Appendices 1–3). However, as noted above, there are well-known exceptions to the lack of understanding of long-distance movements, for example, the work on the monarch butterfly (Brower, 1985), and the yearly re-invasion of the little brown rice planthopper (*Nilaparvata lugens*) to Japan from the mainland (Kisimoto, 1976).

13.4 CROSS-CONTRIBUTIONS AND FUTURE DIRECTIONS

Here, we note several areas which might benefit from an interaction between students of dispersal who examine different taxa. For example, understanding behavioural interactions and mating systems has been fruitful in assessing the function of dispersal in some mammals (see papers in Swingland and Greenwood, 1983), but has received little attention in insects. Likewise, methods and understanding of genetic and physiological influences on dispersal in insects may help guide mammalian research. One could ask, are reproductive hormones as important in dispersal and migration in mammals as they appear to be in some insects and birds? Or, are there fundamental differences among taxa in the underlying proximate mechanisms of dispersal, or even in the ultimate functions of dispersal?

We suspect that a particularly productive interaction among students of different taxa will be at the interface of insect (or arthropod) and vertebrate life histories. Understanding dispersal and migration will be important in the growing numbers of studies of mutualism and parasitism involving these two groups. For example, what role do ectoparasites play in determining the movements of mammals (for a consideration of this hypothesis in ground squirrels, see Holekamp, 1986)? Conversely, mammalian movements necessarily determine the habitats and movements of their arthropod parasites and vectors (May and Anderson, 1983). Undoubtedly, studies of epidemiology will benefit from such an interaction in the future.

Finally, success in the study of insect movements has been achieved only by a multidisciplinary approach of physiology, behaviour, ecology, and evolution. As this book will attest, this approach is not new to the study of dispersal of small mammals. Yet, we hope the questions posed by entomologists and the types of answers available will enlighten and benefit mammalogists and students of other taxa.

ACKNOWLEDGEMENTS

We thank the following people for helpful discussions concerning the dispersal and migration of insects, R.F. Denno, H. Dingle, R.F. Chapman, R.F. Colwell, P.M. Kareiva, S. Naeem, J.R. Ott, V.H. Resh, D. Roff, W.P. Sousa, D.R. Strong and S. Via. F.A. Pitelka and W.Z. Lidicker read and improved the manuscript. GKR was supported by NSF Dissertation Grant BSR-8413593.

REFERENCES

Ammar, E.D. (1973) Factors related to the two wing forms in *Javesella pellucida* (Fab.) (Homoptera: Delphacidae). *Zeitschrift für Angiwanch Entimologie*, **74**, 211–6.

Andersen, N.M. (1973) Seasonal polymorphism and developmental changes in bivoltine pondskaters (Hem. Gerridae) *Entomological Scandinavia*, **4**, 1–20.

Andrewartha, H.G. and Birch, L.C. (1954) *The Distribution and Abundance of Animals*, University of Chicago Press, Chicago.

Andrewartha, H.G. and Birch, L.C. (1984) *The Ecological Web*, University of Chicago Press, Chicago.

Antolin, M.F. and Strong, D.R. (1987) Long-distance dispersal by a parasitoid (*Anagrus delicatus*, Mymaridae) and its host. *Oecologia*, **73**, 288–92.

Baker, R.R. (1978) *The Evolutionary Ecology of Animal Migration*, Holmes & Meier, New York.

Begon, M. (1976) Dispersal, density, and microdistribution in *Drosophila subobscura* Collin. *Journal of Animal Ecology*, **45**, 441–56.

Brower, L.P. (1985) New perspectives on the migration biology of the monarch butterfly, *Danaus plexippus* L, in *Migration: mechanisms and adaptive significance* (ed. M.A. Rankin), Contributions in Marine Science, Suppl. **27**, pp. 748–85.

Brown, E.S. (1951) The relation between migration-rate and type of habitat in aquatic insects, with special reference to certain species of Corixidae. *Proceedings of the Zoological Society of London*, **121**, 539–45.

Bull, J.J., Thompson, C., Ng, D. and Moore, R. (1987) A model for natural selection of genetic migration. *American Naturalist*, **129**, 143–57.

Caldwell, R.L. (1974) A comparison of the migratory strategies of two milkweed bugs, in *The Experimental Analysis of Insect Behavior* (ed. L.B. Browne), Springer, Heidelberg, pp. 304–16.

Caldwell, R.L. and Hegmann, J.P. (1969) Heritability of flight duration in the milkweed bug, *Lygaeus kalmii*. *Nature*, **223**, 91–2.

Caldwell, R.L. and Rankin, M.A. (1974) Separation of migratory from feeding and reproductive behaviour in *Oncopeltus fasciatus*. *Journal of Comparative Physiology*, **88**, 383–94.

Carson, H.L., Hardy, D.E., Spieth, H.T. and Stone, W.S. (1970) The evolutionary biology of the Hawaiian Drosophilidae, in *Essays in Evolution and Genetics in Honor of Theodosius Dobzhansky* (eds M.K. Hecht and W.C. Steere), Appleton-Century-Crofts, New York, pp. 437–543.

Cohen, D. (1967) Optimization of seasonal migratory behavior. *American Naturalist*, **101**, 5–17.

Cole, L.C. (1954) The population consequences of life history phenomena. *Quarterly Review of Biology*, **29**, 103–37.

Coyne, J.A., Boussy, I.A., Prout, T. *et al.* (1982) Long-distance migration of *Drosophila*. *American Naturalist*, **119**, 589–95.

Coyne, J.A., Bryant, S.H. and Turelli, M. (1987) Long-distance migration of *Drosophila*. 2. Presence in desolate sites and dispersal near a desert oasis. *American Naturalist*, **129**, 847–61.

Coyne, J.A. and Milstead, B. (1987) Long-distance migration of *Drosophila*. 3. Dispersal of *D. melanogaster* alleles from a Maryland orchard. *American Naturalist*, **130**, 70–82.

Davis, M.A. (1984) The flight and migration ecology of the red milkweed beetle (*Tetraopes tetraopthalmus*). *Ecology*, **65**, 230–4.

deKort, C.A.D. (1969) Hormones and the structural and biochemical properties of the flight muscle of the Colorado beetle. *Mededelingen Landboushogeshool Wageningen* **69**(2), 1–63.

den Boer, P.J. (1968) Spreading the risk and stabilization of animal numbers. *Acta Biotheoretica*, **18**, 165–94.

Denno, R.F. (1978) The optimum population strategy for planthoppers (Homoptera: Delphacidae) in stable marsh habitats. *Canadian Entomologist*, **110**, 135 42.

Denno, R.F. (1983) Tracking variable host plants in space and time, in *Variable Plants and Herbivores in Natural and Managed Systems* (eds R.F. Denno and M.S. McClure), Academic Press, New York, pp. 291–341.

Denno, R.F. (1985) Fitness, population dynamics, and migration in planthoppers: the role of host plants, in *Migration: mechanics and adaptive significance* (ed. M.A. Rankin), Contributions in Marine Science, Vol. 27, Suppl., pp. 623–40.

Denno, R.F., Douglass, L.W. and Jacobs, D. (1985) Crowding and host plant nutrition: environmental determinants of wing-form in *Prokelisia marginata*. *Ecology*, **66**, 1588–96.

Denno, R.F., Olmstead, K.L. and McCloud, E.S. (1989) Reproductive cost of flight capability: A comparison of life history traits in wing dimorphic planthoppers. *Ecology and Entomology*, **14**, 31–44.

Denno, R.F. and Roderick, G.K. (1990) Population biology of planthoppers. *Annual Review of Entomology*, **35**, 489–520.

Denno, R.F. and Roderick, G.K. (1991) Influence of patch size, vegetation texture, and host plant architecture on the diversity, abundance, and life history styles of sap-feeding herbivores, in *Habitat Structure: the physical arrangement of objects in time and space* in (eds H.R. Mushinsky, S. Bell and E.D. McCoy), Chapman & Hall, London, in press.

Denno, R.F., Roderick, G.K., Olmstead, K.L. and Döbel, H.G. (1991) Density-related migration in planthoppers (Homoptera: Delphacidae): The role of habitat persistence. *American Naturalist*, in press.

Dingle, H. (1972) Migration strategies of insects. *Science*, **175**, 1327–35.

Dingle, H. (1974) The experimental analysis of migration and life history strategies in insects, in *The Experimental Analysis of Insect Behaviour* (ed. L. Barton Browne), Springer-Verlag, Heidelberg, pp. 329–42.

Dingle, H. (1978) Migration and diapause in tropical, temperate, and island milkweed bugs, in *The Evolution of Insect Migration and Diapause* (ed. H. Dingle), Springer-Verlag, New York, pp. 329–42.

Dingle, H. (1979) Adaptive variation in the evolution of insect migration, in *Movement of Highly Mobile Insects: concepts and methodology in research* (eds R.J. Rabb and G.G. Kennedy), North Carolina State University Press, Raleigh, North Carolina, pp. 64–87.

Dingle, H. (1980) Ecology and evolution of migration, in *Animal Migration, Orientation, and Navigation* (ed. S.A. Gauthreaux Jr), Academic Press, New York, pp. 1–101.

Dingle, H. (1981) Geographical variation and behavioral flexibility in milkweed bug life histories, in *Insect Life History Patterns* (eds R.F. Denno and H. Dingle), Springer-Verlag, New York, pp. 57–73.

Dingle, H. (1982) Function of migration in the seasonal synchronization of insects. *Entomological Experimental Applications*, **31**, 36–48.

Dingle, H. (1984a) Behavior, genes, and life histories: Complex adaptations in uncertain environments, in *A New Ecology: novel approaches to interactive systems* (eds P.W. Price, C.N. Slobodchikoff and W.S. Gaud), Wiley, New York, pp. 169–94.

Dingle, H. (1984b) Migration, in *Comprehensive Insect Physiology, Biochemistry, and Pharmacology* (eds G.A. Kerkut and L.I. Gilbert), Pergamon Press, Oxford, pp. 375–415.

Dingle, H. (1986) The evolution and genetics of insect migration, in *Insect Flight, Dispersal, and Migration* (ed. W. Danthanarayana), Springer-Verlag, New York, pp. 11–26.

Dingle, H. and Arora, G. (1973) Experimental studies of migration in the bugs of the genus *Dysdercus*. *Oecologia*, **12**, 119–40.

Dingle, H., Blakley, N.R. and Miller, E.R. (1980) Variation in body size and flight performance in milkweed bugs (Oncopeltus). *Evolution*, **34**, 371–85.

Dingle, H., Evans, K.E. and Palmer, J.O. (1988) Responses to selection among life-history traits in a nonmigratory population of milkweed bugs (*Oncopeltus fasciatus*). *Evolution*, **42**, 79–92.

Dixon, A.F.G. (1972) Fecundity of brachypterous and macropterous alatae in *Drepanosiphum dixoni* (Callaphididae, Aphididae). *Entomological Experimental Applications*, **15**, 335–40.

Dixon, A.F.G. (1985) *Aphid Ecology*, Blackie and Sons, London.

Dobzhansky, T. and Wright, S. (1943) Genetics of natural populations. X. Dispersion rates in *Drosophila pseudoobscura*. *Genetics*, **28**, 304–40.

Ehrlich, P.R., White, R.R., Singer, M.C. *et al.* (1975) Checkerspot butterflies: a historical perspective. *Science*, **188**, 221–8.

Falconer, D.S. (1981) *Introduction to Quantitative Genetics*, 2nd edn, Longman, London.
Fisher, R.A. (1958) *The Genetical Theory of Natural Selection*, Dover, New York.
Gatehouse, A.G. (1986) Migration in the African armyworm *Spodoptera exempta*: Genetic determination of migratory capacity and a new synthesis, in *Insect flight: dispersal and migration* (ed. W. Danthanarayana) Springer-Verlag, Berlin, pp. 128–44.
Gatehouse, A.G. (1991) *Genes, Environment and Insect Flight*, CRC Handbook.
Gauthreaux, S.A. Jr (ed.) (1980) *Animal Migration, Orientation, and Navigation*, Academic Press, New York, 387 pp.
Giesel, J.T. (1976) Reproductive strategies as adaptations to life in temporally heterogeneous environments. *Annual Review of Ecology and Systematics*, 7, 57–79.
Hagen, K.S. (1962) Biology and ecology of predaceous Coccinellidae. *Annual Review of Entomology*, 7, 289–326.
Harrison, R.G. (1979) Flight polymorphism in the field cricket *Gryllus pennsylvanicus*. *Oecologia*, **40**, 125–32.
Harrison, R.G. (1980) Dispersal polymorphism in insects. *Annual Review of Ecology and Systematics*, **11**, 95–118.
Heape, W. (1931) *Emigration, Migration and Nomadism*, Heffer, Cambridge.
Hegmann, J.P. and Dingle, H. (1982) Phenotypic and genetic covariance structure in milkweed bug life history traits, in *Genetics and Evolution of Life Histories*, (eds H. Dingle and J.P. Hegmann), Springer-Verlag, New York, pp. 177–84.
Holekamp, K.E. (1986) Proximal causes of natal dispersal in Belding's ground squirrels (*Spermophilus beldingi*) *Ecological Monographs*, **56**, 365–91.
Honek, A. (1976) The regulation of wing polymorphism in natural populations of *Phyrrhocoris apterus* (Heteroptera, Pyrrhocoridae) *Zoologische Jahrbuch der Systematic*, **103**, 547–70.
Honek, A. (1979) Independent response of 2 characters to selection for insensitivity to photoperiod in *Pyrrhocoris apterus*. *Experienta*, **35**, 762–3.
Horn, H.S. (1983) Some theories about dispersal, in *Animal Movements* (eds I. Swingland and P.J. Greenwood), Oxford University Press, Oxford, pp. 54–62.
Istock, C.A. (1978) Fitness variation in a natural population, in *Evolution of Insect Migration and Diapause* (ed. H. Dingle), Springer-Verlag, New York, pp. 171–90.
Jackson, D.J. (1952) Observations on the capacity for flight of water beetles. *Proceedings of the Royal Society of London (A)*, **27**, 57–70.
Jarvinen, O. and Vepsäläinen, K. (1976) Wing dimorphism as an adaptive strategy in water-striders. *Hereditas*, **84**, 61–8.
Johnson, C.G. (1963) Physiological factors in insect migration by flight. *Nature*, **198**, 423–27.
Johnson, C.G. (1967) International dispersal of insects and insect-borne viruses. *Netherlands Journal of Plant Pathology*, **73**(Suppl. 1), 21–43.
Johnson, C.G. (1969) *The Migration and Dispersal of Insects by Flight*, Methuen, London.
Jones, R.E. (1977) Movement patterns and egg distribution in cabbage butterflies. *Journal of Animal Ecology*, **46**, 195–212.

Jones, R.E., Gilbert, N., Guppy, M. and Nealis, V. (1980) Long-distance movement of *Pieris rapae. Journal of Animal Ecology*, **49**, 629–42.

Jutsam, A.J. and Goldsworthy, G.J. (1976) Fuels for flight in *Locusta. Journal of Insect Physiology*, **22**, 243–9.

Kareiva, P. (1982) Experimental and mathematical analyses of herbivore movement: quantifying the influence of plant spacing and quality on foraging discrimination. *Ecological Monographs*, **52**, 261–82.

Kareiva, P. (1983a) Influence of vegetation texture on herbivore populations: resource concentration and herbivore movement, in *Variable Plants and Herbivores in Natural and Managed Systems* (eds R.F. Denno and M.S. McClure), Academic Press, New York, pp. 259–89.

Kareiva, P.M. (1983b) Local movement in herbivorous insects: applying a passive diffusion model to mark-recapture field experiments. *Oecologia,* **57**, 322–7.

Kareiva, P. (1985) Finding and losing host plants by *Phyllotreta*: patch size and surrounding habitat. *Ecology*, **66**, 1809–16.

Kareiva, P. (1987) Habitat fragmentation and the stability of predator–prey interactions. *Nature*, **326**, 388–90.

Kennedy, J.S. (1956) Phase transformation in locust biology. *Biological Review*, **31**, 349–70.

Kennedy, J.S. (1961) A turning point in the study of insect migration. *Nature*, **189**, 785–91.

Kennedy, J.S. (1975) Insect dispersal, in *Insects, Science, and Society* (ed. D. Pimentel), Academic Press, New York, pp. 103–19.

Kisimoto, R. (1956) Factors determining the wing-form of adult, with special reference to the effect of crowding during the larval period of the brown planthopper, *Nilaparvata lugens* Stal. Studies on the polymorphism in the planthoppers (Homoptera, Araeopidae). I. *Oyo-Kontyu*, **12**, 105–11.

Kisimoto, R. (1976) Synoptic weather conditions inducing long distance immigration of planthoppers *Sogatella furcifera* Horvath and Nilaparvata lugens Stal. *Ecology and Entomology*, **1**, 95–109.

Lamb, R.J. and MacKay, P.A. (1979) Variability in migratory tendency within and among natural populations of the pea aphid, *Acyrthosiphon pisum. Oecologia*, **39**, 289–99.

Lamb, R.J. and MacKay, P.A. (1983) Micro-evolution of the migratory tendency, photoperiodic response and developmental threshold of the pea aphid, *Acrythosiphon pisum*, in *Diapause and Life Cycle Strategies in Insects* (eds V.K. Brown and I. Hodek), Junk, The Hague, pp. 209–17.

Lande, R. (1982) A quantitative genetic theory of life history evolution. *Ecology*, **63**, 607–15.

Lavie, B. and Ritte, U. (1978) The relation between dispersal behavior and reproductive fitness in the flour beetle *Tribolium castaneum. Canadian Journal of Genetic Cytology*, **20**, 589–95.

Lawrence, W.S. (1987) Effects of sex ratio on milkweed beetle emigration from host plant patches. *Ecology*, **68**, 539–46.

Lees, A.D. (1966) The control of polymorphism in aphids. *Advances in Insect Physiology*, **3**, 207–77.

Lees, A.D. (1967) The production of the apterous and alate forms in the aphid,

Megoura viciae Buckton, with special reference to the role of crowding. *Journal of Insect Physiology*, **13**, 289–318.

Lewontin, R.C. (1965) Selection for colonizing ability, in *The Genetics of Colonizing Species* (eds H.G. Baker and G.L. Stebbins), Academic Press, New York, pp. 77–94.

Lidicker, W.Z. Jr and Caldwell, R.L. (1982) *Dispersal and Migration. Benchmark Papers in Ecology*, Vol. 11, Hutchinson Ross, Stroudsburg, Pennsylvania.

MacArthur, R.H. and Wilson, E.O. (1967) *The Theory of Island Biogeography*, University of Princeton Press, Princeton, New Jersey.

Mahmud, F.S. (1980) Alary polymorphism in the small brown planthopper *Laodelphax striatellus* (Homoptera: Delphacidae) *Entomological Experimental Applications*, **28**, 47–53.

Mason, C.J. and McManus, M.L. (1981) Larval dispersal of the gypsy moth, in *The Gypsy Moth: research toward integrated pest management* (eds C.C. Doane and M.L. McManus), Forest Service Science and Education Agency, Tech. Bull. 1584, Washington, DC.

May, R.M. and Anderson, R.M. (1983) Parasite–host coevolution, in *Coevolution* (eds D. Futuyma and M. Slatkin), Sinauer, Sunderland, Massachusetts.

McCauley, D.E. (1983) Gene flow distances in natural populations of *Tetraopes tetraopthalmus*. *Evolution*, **37**, 1239–46.

McCauley, D.E., Ott, J.R., Stine, S. and McGrath, S. (1981) Limited dispersal as one determinant of population structure in the milkweed beetle (*Tetraopes tetraophthalmus*). *Oecologia*, **51**, 145–50.

McCullough, D.R. (1985) Long range movements of large terrestrial mammals, in *Migration: mechanisms and adaptive significance* (ed. M.A. Rankin), Contributions in Marine Science, Suppl. 27, pp. 444–65.

Messina, F.J. (1987) Genetic contribution to the dispersal polymorphism of the cowpea weevil (Coleoptera: Bruchidae). *Annals of the Entomology Society of America*, **80**, 12–6.

Mochida, O. (1973) The characters of the two wing-forms of *Javesella pellucida* (F.) (Homoptera: Delphacidae), with special reference to reproduction. *Transactions of the Royal Entomology Society of London*, **125**, 177–225.

Norris, M.J. (1950) Reproduction in the African migratory locust (*Locusta migratoria migratorioides* R&F) in relation to density and phase. *Anti-Locust Bulletin*, **6**, 1–50.

Padgham, D.E. (1983) Flight fuels in the brown planthopper *Nilaparvata lugens*. Journal of Insect Physiology, **29**, 95–9.

Palmer, J.O. (1985) Ecological genetics of wing length, flight propensity, and early fecundity in a migratory insect, in *Migration: mechanisms and adaptive significance* (ed. M.A. Rankin), Contributions in Marine Science, Suppl. **27**, pp. 663–73.

Rainey, R.C. (1951) Weather and the movement of locust swarms: a new hypothesis. *Nature*, **168**, 1057–60.

Rankin, M.A. (1978) Hormonal control of insect migratory behavior, in *Evolution of Insect Migration and Diapause* (ed. H. Dingle), Springer-Verlag, New York, pp. 5–32.

Rankin, M.A. (1985a) Endocrine influences on insect migratory behavior, *Migra-*

tion: mechanisms and adaptive significance (ed. M.A. Rankin), Contributions in Marine Science, Suppl. 27, pp. 817–41.

Rankin, M.A. (ed.) (1985b) *Migration: mechanisms and adaptive significance*, Contributions in Marine Science, Suppl. 27, 868 pp.

Rausher, M.D. (1983) Ecology of host-selection behavior of phytophagous insects, in *Variable Plants and Herbivores in Natural and Managed Systems* (eds R.F. Denno and M.S. McClure), Academic Press, New York, pp. 223–57.

Roderick, G.K. (1987) Ecology and evolution of dispersal of Californian populations of a salt marsh insect. Ph.D. Thesis, University of California, Berkeley.

Roff, D.A. (1975) Population stability and the evolution of dispersal in a heterogeneous environment. *Oecologia*, **19**, 217–37.

Roff, D.A. (1984) The cost of being able to fly: a study of wing polymorphism in two species of crickets. *Oecologia*, **63**, 30–7.

Roff, D.A. (1986) The genetic basis and evolution of wing dimorphism in insects. *Evolution*, **40**, 1009–20.

Roff, D.A. (1990) The evolution of flightlessness in insects. *Ecological Monographs*, **60**, 389–421.

Roff, D.A. and Fairbairn, D.J. (1990) Wing dimorphisms and the evolution of migratory polymorphisms among the Insecta. *American Zoologist*, in press.

Rose, M.R. and Charlesworth, B. (1981) Genetics of life history in *Drosophila melanogaster*. II. Exploratory selection experiments. *Genetics*, **97**, 187–96.

Safriel, U.N. and Ritte, U. (1980) Criteria for the identification of potential colonizers. *Biological Journal of the Linnean Society*, **13**, 287–97

Safriel, U.N and Ritte, U. (1983) Universal correlates of colonizing ability, in *The Ecology of Animal Movement* (eds I.R. Swingland and P.J. Greenwood), Clarendon Press, Oxford, pp. 215–39.

Sakai, K., Narise, T. and Iyama, S. (1957) Migration studies in several wild strains of *Drosophila melanogaster*. National Institute of Genetics, Japan Annual Report No. 7, pp. 73–5.

Service, P.M. and Rose, M.R. (1985) Genetic covariation among life-history components: the effect of novel environments. *Evolution*, **39**, 943–5.

Solbreck, C. (1978) Migration, diapause, and direct development as alternative life histories in a seed bug, *Neacoryphus bicrucis*, in *The Evolution of Insect Migration and Diapause* (ed. H. Dingle), Springer-Verlag, New York, pp. 195–217.

Solbreck, C. (1980) Dispersal distances of migrating pine weevils, *Hylobius abietis*, Coleoptera: Curculionidae. *Entomological Experimental Applications*, **28**, 123–31.

Southwood, T.R.E. (1962) Migration of terrestrial arthropods in relation to habitat. *Biological Review*, **37**, 171–214.

Southwood, T.R.E. (1977) Habitat, the templet for ecological strategies? *Journal of Animal Ecology*, **46**, 337–65.

Strong, D.R. and Stiling, P.D. (1983) Wing dimorphism changed by experimental density manipulation in a planthopper (*Prokelisia marginata*, Homoptera, Delphacidae). *Ecology*, **64**, 206–9.

Sutherland, O.R.W. (1969) The role of crowding in the production of winged forms by two strains of the pea aphid, *Acyrthosiphon pisum*. *Journal of Insect Physiology*, **15**, 1385–410.

Sutherland, O.R.W. (1970) The role of host plant in the production of winged forms by two strains of the pea aphid, *Acyrthosiphon pisum*. *Journal of Insect Physiology*, **15**, 2179–201.

Swingland, I.R. and Greenwood, P.J. (1983) *The Ecology of Animal Movement*, Clarendon Press, Oxford, 311 pp.

Tanaka, S. (1976) Wing polymorphism, egg production and adult longevity in *Pteronemobious taprobanensis* Walker (Orthoptera, Gryllidae). *Kontyu*, **44**, 327–33.

Tauber, C.A. and Tauber, M.J. (1981) Insect seasonal cycles: genetics and evolution. *Annual Review of Ecology and Systematics*, **12**, 281–308.

Tauber, M.J. and Tauber, C.A. (1976) Insect seasonality: diapause maintenance, termination, and postdiapause development. *Annual Review of Entomology*, **21**, 81–107.

Tauber, M.J., Tauber, C.A. and Masaki, S. (1986) *Seasonal Adaptations of Insects*, Oxford University Press, New York.

Taylor, C.E., Powell, J.R., Kekic, V. *et al.* (1984) Dispersal rates of species of the *Drosophila obscura* group: implications for population structure. *Evolution*, **38**, 1397–401.

Taylor, L.R. and Taylor, R.A.J. (1977) Aggregation, migration and population mechanics. *Nature*, **265**, 415–21.

Taylor, L.R., Taylor, R.A.J., Woiwod, I.P. and Perry, J.N. (1983) Behavioral dynamics. *Nature*, **303**, 801–4.

Turchin, P.B. (1986) Modelling the effect of host patch size on Mexican bean beetle emigration. *Ecology*, **67**, 124–32.

Urquhart, F.A. (1960) *The Monarch Butterfly*, University of Toronto Press, Toronto.

Utida, S. (1972). Density dependent polymorphism in the adult of *Callosabruchus maculatus* (Coleoptera: Bruchidae). *Journal of Stored Products Research*, **8**, 111–26.

Uvarov, B.P. (1928) *Grasshoppers and Locusts*, Imperial Bureau of Entomology, London.

Vepsäläinen, K. (1978) Wing dimorphism and diapause in *Gerris*: Determination and adaptive significance, in *Evolution of Insect Migration and Diapause* (ed. H. Dingle), Springer-Verlag, New York, pp. 218–53.

Via, S. (1984) The quantitative genetics of polyphagy in an insect herbivore. II. Genetic correlations in larval performance within and among host plants. *Evolution*, **38**, 896–905.

Waloff, Z. (1966) The upsurges and recessions of the desert locust plague: a historical survey. *Anti-Locust Memoirs*, **8**, 1–111.

Wetzler, R.E. and Risch, S.J. (1984) Experimental studies of beetle diffusion in simple and complex crop habitats. *Journal of Animal Ecology*, **53**, 1–19.

Williams, C.B. (1930) *The Migration of Butterflies*, Edinburgh.

Wratten, S.D. (1977) Reproductive strategy of winged and wingless morphs of the aphids *Sitobon avenae* and *Metopolophium dirhodum*. *Annals of Applied Biology*, **85**, 319–31.

Wu, A.C. (1981) Life history traits correlated with emigration in flour beetle populations. Ph.D. Thesis, University of Illinois at Chicago Circle.

Young, E.C. (1965) Flight muscle polymorphism in British Corixidae: Ecological observations. *Journal of Animal Ecology*, **34**, 353–90.

Zera, A.J. (1984) Differences in survivorship, development rate, and fertility between the longwinged and wingless morphs of the waterstrider, *Limnoporus canaliculatus*. *Evolution*, **38**, 1023–32.

Zera, A.J. (1985) Wing polymorphism in waterstriders (Gerridae: Hemiptera): Mechanisms of morph determination and fitness differences between morphs, in *Migration: mechanisms and adaptive significance* (ed. M.A. Rankin), Contributions in Marine Science, Suppl. 27, pp. 674–703.

Zera, A.J., Innes, D.J. and Saks, M.E. (1983) Genetic and environmental determinants of wing polymorphism in the waterstrider *Limnoporus canaliculatus*. *Evolution*, **37**, 513–22.

Ziegler, J.R. (1976) Evolution of the migration response: emigration by *Tribolium* and the influence of age. *Evolution*, **30**, 579–92.

PART SEVEN

Appendices

The three appendices primarily concern techniques. While each focuses on a particular class of techniques for investigating dispersal, new data in the form of illustrative material are also included. We intend that these three appendices collectively will: (1) provide an introduction to the methodology of research on dispersal in small mammals for those beginning research in this area, and (2) serve as a launching pad for those specialists in the field clever enough to devise ever more imaginative, innovative, and effective ways to increase our understanding of dispersal.

In Appendix 1, we as editors provide a state-of-the-art overview of the techniques, experimental designs, and methods of data analysis used in the study of dispersal in small mammals and insects. Special attention is paid to the problem of deriving null models for testing hypotheses about dispersal distances.

One of the most powerful tools in the study of dispersal is radiotelemetry. The idea has been around for several decades, but it has been only recently that radio transmitters and especially their accompanying batteries have become small enough to be used on small mammals. Appendix 2 reviews this approach. It is written by Dale Madison and William McShea, who have been among the pioneers in the application of radiotelemetry to small mammal ecology. They have worked intensively on movement patterns in the North American meadow vole, *Microtus pennsylvanicus*, and their results have been, and continue to be, an inspiration and a model for other researchers.

Finally, Appendix 3 summarizes the use of radioactive isotopes in dispersal investigations. This technique has been used to measure dispersal distances and per capita rates as well as to mark litters of young before they are weaned. In the latter case, young are tagged *in utero* or during lactation by transfer of radioactive isotopes from their mothers. Dispersal can then be monitored in juveniles with known birth sites, mothers, and siblings. We are pleased also to include in this review a report on a major, but unpublished, project by Arvo Myllymäki and Lennart Hansson on the field vole, *Microtus agrestis*, carried out in southern Finland. Their data from this effort were subsequently analysed by Jann Morten Hoff (formerly of the University of Oslo). We are grateful to these three for giving us permission to summarize their results. They add an important case history to our knowledge of microtine rodent dispersal biology.

Appendix 1. Where do we stand methodologically about experimental design and methods of analysis in the study of dispersal?

N.C. STENSETH AND W.Z. LIDICKER, JR.

A1.1 INTRODUCTION

The oldest and most reliable method of investigating dispersal is to directly observe individual movements from one place to another without influencing them in any way. However, for this to be a useful method, the species studied must be large in size, visible without being disturbed, and its individuals must be distinguishable. Moreover, population densities in the study area must not be too high: if there are too many individuals in the field of view, confusion may easily prevent focal animal observations. In such situations we will at best be able to observe dispersal if there is directional and synchronized movement like that seen in several of the large herbivores in Africa (Chapter 10) or in migrating fish and birds.

For most species it is almost impossible to distinguish individuals that are dispersing or have recently dispersed from animals making routine exploratory movements. In some relatively rare cases dispersers define themselves through particular morphological attributes (e.g. winged versus non-winged aphids; Chapter 13) or some particular behavioural trait (e.g. migratory restlessness in birds). Unfortunately, small mammals do not provide such identifying features. In fact, the rather secretive life of small mammals makes the study of their dispersal difficult. It is far easier to observe wildebeests migrating across the African plains than to see the seasonal habitat shifts in lemmings (Kalela, 1961, 1971; Henttonen and Järvinen, 1981). Nevertheless, small mammals offer special advantages and profitable opportunities of their own, as this book attests (see especially Chapter 1).

Recently it has been increasingly common to manipulate organisms or their environments in one way or another in order to obtain data about dispersal. These manipulations include: (1) marking of individuals (e.g.

tagging, fluorescent powder, radio-collaring, and radioisotope marking); (2) manipulations of the population under study (e.g. experimental removal of all or a certain fraction of the population) in order to see who disperses under given conditions; and (3) building various sorts of enclosures in order to determine which individuals are motivated to leave.

This chapter reviews some of the experimental techniques most commonly applied for studying dispersal in small mammals. Rather than providing an exhaustive account of all possible methods available, this chapter is intended primarily to be a guide to those techniques and methods that are available and to where detailed guidelines on their use can be found.

Radiotelemetry and radioisotope marking are singled out for treatment in separate chapters (Appendices 2 and 3) because of their special importance and potential usefulness.

This chapter also includes some useful methods for analysing data on dispersal. However, standard statistical methods are not covered because these can easily be found in general biometry textbooks (e.g. Snedecor and Cochran, 1967; Sokal and Rohlf, 1981) and in standard texts on ecological methods (Caughley, 1977; Southwood, 1978; Krebs, 1989). Besides, little can be said about statistics that applies to dispersal specifically, and dispersal biology offers even less that would be generally interesting in this context without reference to the details of particular models being studied. Our objective is rather to draw attention to some particular methods of analysis as well as point to some potential pitfalls frequently encountered in the handling of data on dispersal. We also give references to some useful computer software packages.

The conceptual framework we have in mind in this chapter is a source–sink type of landscape (*sensu* Pulliam, 1988; see also Hansson, 1977; Stenseth, 1983).

A1.2 TECHNIQUES FOR OBTAINING DATA

A1.2.1 Obtaining data on dispersal distances

In some species, such as migratory birds and locusts, dispersal movements can often be inferred directly from the observation that at one point in time the species was absent from an area, whereas at a later point in time the species is present there. Given that the 'increase' in number in such a particular locality cannot be attributed to reproduction on the site, immigration must have occurred. If we know the closest possible donor habitat (*sensu* Hansson, 1977), minimal dispersal distances can be inferred.

The basic problem associated with this method is that it may be difficult to be sure that the area was in fact empty initially. There would then be the possibility that the inferred new arrivals into an area were the result of reproduction *in situ*. If, however, the presumed dispersers were marked elsewhere, then we can exclude this possible error. Moreover, knowing the source of immigrants will give a measure of minimal dispersal distances. Among ornithologists, marking (or banding) is of course a very common technique for observing minimum dispersal distances. A similar technique used by biologists working with small mammals is the use of nest-boxes combined with various types of marking of young. Then, capturing of marked individuals at breeding sites would provide estimates of natal dispersal distances. Such methods have been used in small mammal studies by Howard (1949), Dice and Howard (1951), Bondrup-Nielsen (1983, 1984), and others. In Appendix 3, this approach is discussed in reference to the study of Wolff and Holleman (1978).

A similar technique is the method of trapping at various distances from a marked source population (Stoddart, 1970; Wolff and Lidicker, 1980; see in this connection the study by Myllymäki and co-workers discussed in Appendix 3). This can be done with assessment lines extending from the source population or by sampling at various distances away from the source. This technique does, of course, provide data on minimum dispersal distances. However, both this and the nest-box method described above are very inefficient and unlikely to provide quantitatively useful data on the frequency distribution of dispersal distances because of the dilution effect stemming from the geometry of area-distance covered. We return to this in section A1.3.2.

Radiotelemetry may also provide data on minimum dispersal distances. With this method, however, there are great problems in obtaining sufficiently large samples for adequately describing dispersal distances. The method is therefore of most use when deducing how individuals use their home range and in relation to seasonal shifts in habitats. See Appendix 2, Kenwood (1987), Harris *et al.* (1990) for further discussion of this method.

A1.2.2 Obtaining experimental data on dispersal rates and demographic effects

(a) Conceptual preamble

Per capita (or specific) dispersal rates are defined as the number of individuals dispersing per time unit relative to how many could have dispersed. Unfortunately, there is much confusion in the literature on this point (e.g. Stenseth, 1983; Chapter 9). As explained in Chapter 9, what we in most cases need to know is how many individuals dispersed relative to how many could have dispersed. This is particularly an important issue when working on species with cyclic density changes in which the ratio of peak to low densities may vary more than 1000-fold (e.g. Henttonen *et al.*, 1987). Then, if the specific dispersal rate is 10 times less at high densities than at low densities (something which would be possible if the idea of presaturation dispersal applies; Chapters 2 and 9), we could still observe 100 times as many dispersers at high densities. Under low density conditions there will be so few individuals present that necessarily very few dispersers will be recorded regardless of the prevailing dispersal rates.

It is essential therefore to know not only how many individuals actually disperse but the size of the source population as well (i.e. what is the size of the potential disperser pool). Moreover, it is critical to have a control population if any experimental manipulations are done. This may be difficult to arrange because small mammal populations tend to be very patchily distributed (e.g. Lidicker, 1985a, 1988; Ostfeld *et al.*, 1985). Even though most experimental studies designed to provide estimates on dispersal rates use control treatments, few researchers try to ascertain whether in fact their controls really are true controls, allowing them to compare legitimately numbers from one experimental population to the other.

Density estimates for source populations can be provided by standard methods (Seber, 1982, 1986; Smith *et al.*, 1975; Otis *et al.*, 1978; White *et al.*, 1982; Cormack, 1989; Pollock *et al.*, 1990) given that a suitable reference population is located. Replication of experiments over both space and time is of course also necessary in order to permit appropriate statistical treatment of data and improve the generality of results. As in other ecological studies, many experimental studies on dispersal are troubled by the lack of treatment replicates or of pseudo-replication (*sensu* Hurlbert, 1984).

Experimental approaches in which a population is partially or completely fenced, usually in order to improve control over conditions or to more easily identify dispersing individuals, risk the development of the 'fence effect'. This syndrome is characterized by unusually rapid population growth often leading to overexploitation of resources and density crashes (e.g. Krebs *et al.*, 1969; MacArthur, 1972; Boonstra and Krebs, 1977; Lidicker, 1975, 1985a,b; Gipps and Jewell, 1979; Chapter 7). The researcher using enclosures must take these complications into account.

(b) Removal grids

A widely used technique for obtaining data on dispersal is to trap out an area and then assume that individuals caught subsequently on the vacated area are dispersers (e.g. Stickel, 1946; Calhoun and Webb, 1953; Myers and Krebs, 1971; Hilborn and Krebs, 1976; Krebs *et al.*, 1976; Fairbairn, 1977, 1978; Staples and Terman, 1977; Sullivan, 1977; Tamarin, 1977; Gaines *et al.*, 1979a,b; Keith and Tamarin, 1981; Baird and Birney, 1982;

Johnson and Gaines, 1985; Jensen *et al.*, 1991a). However, this method poses several sources of error (e.g. Lidicker, 1985b). Because small mammals tend to have relatively low trappabilities (Hilborn and Krebs, 1976; Boonstra and Krebs, 1978; Stoddart, 1982; Jensen *et al.*, 1991b), it is extremely difficult to eradicate them from an area. Thus, some residents will almost surely remain in the area, possibly to be caught later and then be classified incorrectly as 'dispersers'. Additionally, residents on the edge of such low-density areas are likely to expand their home ranges into it (e.g. Stickel, 1946; Stenseth and Hansson, 1979; Bondrup-Nielsen, 1983, 1984). Thirdly, low-density areas may induce some individuals to become colonists, such as those on an excursion, who otherwise might not become dispersers [see e.g. Verner (1979) for discussion]. Finally, the subsequent history of colonizers to a vacated area may be significantly different from those that disperse into undisturbed areas; below we return to this last point.

Some workers (Beacham, 1979, 1980a,b, 1981; Hestbeck, 1986; Tamarin, 1980) have attempted to minimize these biases by providing barriers of poor habitat strips around their trapped-out areas (i.e. creating dispersal sinks; Lidicker, 1975; Chapters 4 and 7). Boundary strips are intended to discourage nearby residents and others not highly motivated to move from reaching the removal grids. Hestbeck's (1986) experiments were designed to include social as well as physical barriers between the source and sink areas. A variant of this experimental design was the one by Tamarin *et al.* (1984) using partial enclosures with openings only into non-optimal areas (forest in this case).

To minimize the 'magnetic' effect of low-density patches, Krebs *et al.* (1978) introduced the technique of pulsed removals. [For discussion of this technique, see Chapter 7 and Johnson and Gaines (1985).] By regular but infrequent trapping episodes, more normal densities were maintained on experimental plots. This procedure produces plots with relatively high frequencies of dispersers, but further complicates the problem of distinguishing them from residents.

Dueser *et al.* (1981, 1984) adopted a single trapping and marking episode followed by no further manipulations. They then assumed that all individuals subsequently caught above a certain body weight were immigrants while those caught first at a lower weight were born on the plot. This is a clever way to circumvent some of the difficulties mentioned above, but creates other problems. Almost surely, some individuals born on the area will escape capture until they are above the arbitrary weight level, and some residents may have escaped the initial trapping effort. Such individuals will then be incorrectly classified as dispersers. This is particularly a problem since it is well known that trappability is often affected by age as well as size (e.g. Jensen *et al.*, 1991b). Similarly, some individuals may disperse before they reach the threshold weight, and then be incorrectly classified as resident recruits. The magnitude of these errors has been estimated for *Microtus pennsylvanicus* by Pugh and Tamarin (1991), by comparing estimates of dispersal by this method in an enclosure with a comparable unenclosed grid. The method has been critically reviewed by Tamarin (1984).

Jensen *et al.* (1991a) took another approach to minimize this array of problems. In a 2 ha area they had a removal area in the central portion of the grid. After live-trapping the entire area, pulse removals were performed on the central part of the grid. The border zone of live traps provided a means of determining where the immigrants into the vacated area came from. However, even in this experiment the authors were unable to distinguish unequivocally between proper immigrants and local recruits to the trappable population.

Experimental protocols based on removal grids (providing a dispersal sink) combined with radioisotope marking in the disperser source area could possibly solve some of the uncertainties of where the source population is, and hence what reference density to use when estimating dispersal rates by this method.

(c) Enclosures

Another approach is to use leaky enclosures (Gaines *et al.*, 1979b; Riggs, 1979; Verner, 1979). If exits are difficult to transgress, then it can be assumed that successful voles are highly motivated to disperse. If the escape routes are monitored, dispersing individuals can be identified. For example, Riggs (1979) used long exit tubes leading from the corners of large enclosures. The tubes were guarded by an earth barrier, a bare area, and sometimes a pool of water. Gaines *et al.* (1979b) employed exit traps placed at 15.2 m intervals around three 0.8 ha enclosures. Entrance to the traps was via a pipe opening into the centre of a 1 m buffer strip of suboptimal habitat maintained around the edges of the enclosures. Verner (1979) also used dispersal exits in his study.

This approach does not absolutely exclude residents from exploring the exit paths. However, it is possible to define an individual as a disperser only after it has entered an exit trap several times. Even leaky enclosures also risk the artificial conditions associated with enclosed populations (the fence effect; see p. 298.

(d) Drift fences

When Scandinavian hunters some thousand years ago were hunting reindeer, they used drift fences to lead dispersing (or moving) reindeer into pitfall traps (e.g. Blehr, 1973). Modern New Guineans do the same for leading forest-dwelling bandicoots into snares. This method has been

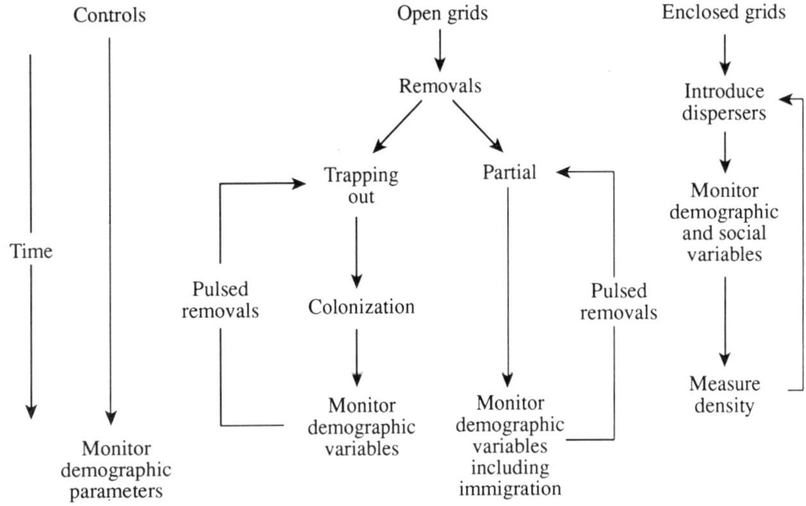

Figure A1.1 A generalized recommended protocol for the study of dispersal in small mammals that incorporates three modes (control, open, and enclosed grids), and that provides for investigating emigration, immigration, and colonization at the individual and population levels of analysis.

applied (we assume independently) by students of small mammal dispersal. Garsd and Howard (1981, 1982) reported on a long-term study in which such drift fences equipped with drop-bottles filled with alcohol were used. Drift fences in combination with live traps have been used in a study of dispersal in *Peromyscus californicus* (Ribble, 1991). This method requires a minimum of effort, but unless the drift fences are placed beyond a substantial strip of non-habitat area, it is very difficult to distinguish residents from dispersers. It would be most rewarding in situations where marked individuals from known home ranges were caught.

(e) Recommended protocol for the study of dispersal in small mammals

Now that we have surveyed the available methodologies, we can, on the basis of the survey of data and needs for more information provided by earlier parts of this book, synthesize all of this effort and suggest a general protocol for future research (Figure A1.1). Our scheme builds on the experimental design used by Johnson and Gaines (1985: figure 1), and incorporates three components: controls, open grids, and enclosed populations. It concerns questions about the dispersal process at the population level as well as in some respects at the individual level.

First, there are control populations in which survival, reproduction, emigration, and immigration are monitored with minimal disturbance.

These may be either open, enclosed, or both, depending on the rest of the experiment. Second, we recommend open grids in which removals are carried out periodically ('pulsed'). Two major variants of this approach are suggested. If removals are modest and at relatively short intervals, we simulate emigration, and can study the effects of this on a source population (the emigration effect). If, on the other hand, removals are total or close to it, we can monitor the colonization process as has been done by Johnson and Gaines (1985) and others.

The remaining component is to investigate the effects of immigration on a recipient population (the immigration effect). This is the least studied effect of the dispersal process. Open grids are generally unsuitable for this, because immigration cannot be satisfactorily manipulated under such conditions. Enclosures, especially large ones, are therefore recommended for this purpose. These would permit the introduction of dispersers as well as the variation of recipient population densities. Introduced individuals could be either a random selection of known dispersers or some combination of selected individuals chosen on the basis of sex, age, or genotype. Unmanipulated enclosed populations may be required for controls in most cases.

The advantage of this general protocol is that it calls attention to and provides for the need to investigate all aspects of dispersal including emigration, immigration, and colonization. Moreover, it suggests a large variety of permutations and modifications that will allow testing of more specific and detailed hypotheses. It does not, of course, concern investigation of dispersal biology at the level of physiology or at the more complex community and landscape levels.

A1.3 METHODS OF ANALYSIS

A1.3.1 Data on home range movements and home range shifts

Fundamental to any study of dispersal is the concept of home range. For dispersal to be meaningful, we must assume that organisms do not live randomly across the earth, but at any given time have a spatial domain with which they are familiar and that supplies the resource base for their survival. For sedentary organisms this home area extends only to the boundaries reached by their trophic structures and perceived by their sensory organs. Mobile creatures make frequent local movements within their home range, and these must be distinguished from movements outside. Therefore, in order to determine if an individual has become a disperser (i.e. left home), it may be necessary to know its home range first. Only then can home range shifts be confidently documented. Techniques that lead to more precise knowledge about home ranges are thus important tools for dispersal biology.

Dale M. Madison and his associates have obtained some of the best data on the details of home range structure and dynamics for small mammals (Appendix 2 and references therein). Data on the movements within the home ranges of individual animals, either derived on the basis of capture–recapture studies or on the basis of radiotelemetry studies, are often difficult to analyse. Mainly this is due to the difficulties of arriving at meaningful summary statistics. However, successful analysis also depends on whether home range boundaries are treated in a discrete or probabilistic manner.

These problems are briefly mentioned in Appendix 2, where references to computer program packages are given. Packages are also available for analysing capture–recapture data (Ford and Krumme, 1979) and for the large number of data points often derived from radiotelemetry (Dixon and Chapman, 1980). A non-parametric method for analysing such data sets was recently developed by Worton (1989). Ostfeld (1986) has developed a quantitative index of home range overlap among individuals.

Often these methods produce a single measure of some home range parameter and do not provide any estimates of uncertainty or confidence limits. This therefore limits their utility for testing results against some null hypotheses or for comparisons between two groups of organisms. It is feasible, however, to estimate the variance of the various home range parameters that we might be interested in by creating artificial data sets from the original data set (Stenseth *et al.*, manuscript). The basic method is to sample randomly the real data set *n* times, where *n* is the number of data points in the set, but with replacement of each data point after it is recorded. Thus we have a series of artificial data sets based on real information and all of the same size as the original set. We can continue making such artificial data sets until the average value of the parameter of interest stabilizes. Then this collection of artificial data sets gives us the estimated variance that we need for statistical comparisons and estimates of reliability.

A1.3.2 Distance moved – what is the H_0 distribution under the hypothesis of random settling?

It is often claimed that dispersing animals settle at non-random distances from their point of departure (Howard, 1960; Allred and Beck, 1963; French *et al.*, 1968; Îto and Miyashita, 1965). Frequently in such studies, insufficient attention is paid to what the appropriate null distribution should be.

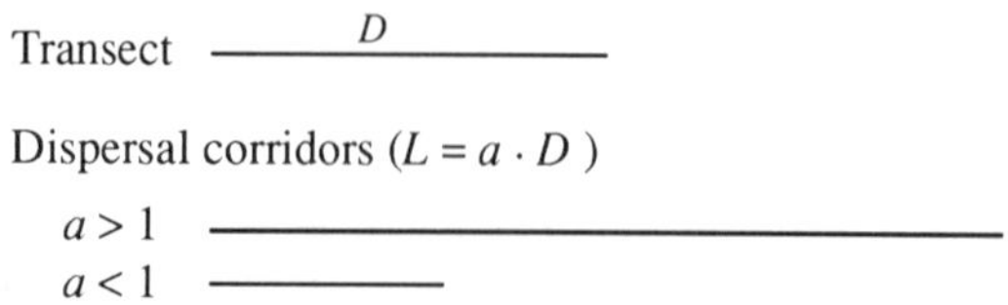

Figure A1.2 A one-dimensional model for calculating dispersal distances based on random settling. D is the transect length, L is the length of the dispersal corridor ($=aD$). a is the ratio of the length of the dispersal corridor (L) relative to the maximum detectable dispersal distance (D).

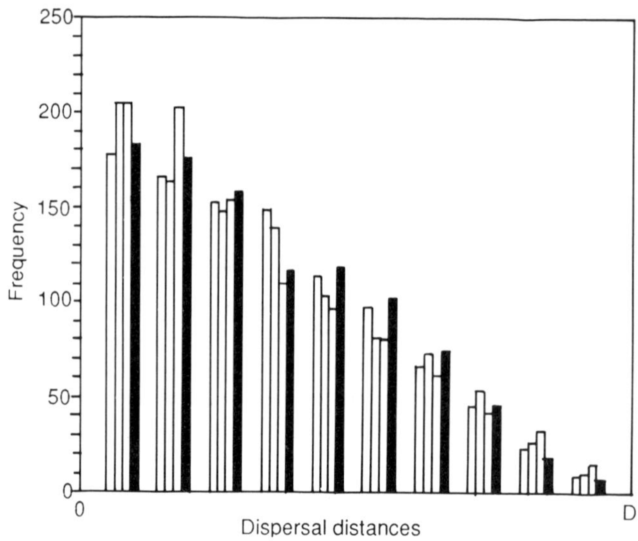

Figure A1.3 Distribution of detected recapture distances resulting from four simulations of the model in which $a = 1$. Actual dispersal distances are equally likely at all distances up to L; see text for further explanation.

The difficulty is this: in essentially all cases where dispersal distances are measured, there is a greater chance of observing short distances than long distances. Hence, there will in general be a bias toward observing the shorter dispersal distances. This is a particular case of size-biased sampling (Ord *et al.*, 1980).

To demonstrate this point, Figures A1.3–A1.5 give the results of hypothetical numerical examples for which the underlying assumptions are depicted in Figure A1.2. As is generally the case in the real world, a finite length of survey transect is assumed. Consider animals dispersing along

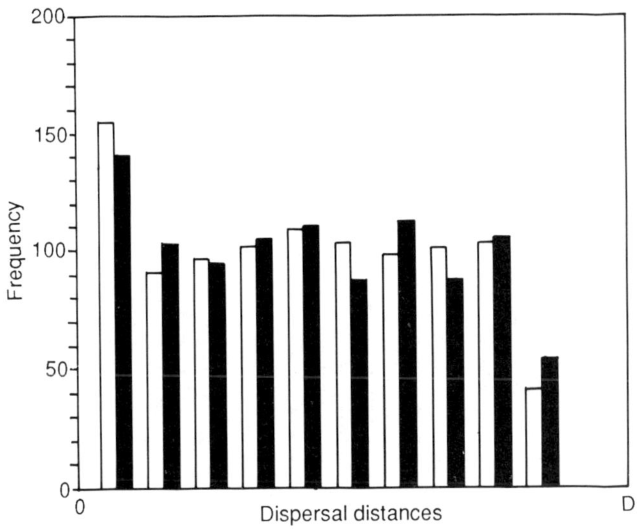

Figure A1.4 Distribution of detected recapture distances resulting from two simulations of the model in which $a = 0.1$. See Figure A1.3 and text for further explanation.

some sort of corridor (e.g. fishes in a stream); our example could, of course, have been extended to two dimensions but without providing any further insight. Animals are assumed to disperse from a random location along this transect; they settle down at randomly determined distances from this point of origin. Actual dispersal distances can be both shorter as well as longer than the transect.

Let the length of the transect be given by D. This is determined by the researcher. A dispersing individual can settle in a corridor of length L. This is determined by the animal's dispersal tendencies, and in our one-dimensional model is termed the 'dispersal corridor' (Figure A1.2). Now, let L equal aD. Thus a measures the length of the dispersal corridor relative to the maximum detectable dispersal distance (i.e. the transect length). As can be seen, the null distribution for the distances observed is a skewed one which in both cases differs from the generating distribution (which in our hypothetical example is assumed to be a rectangular distribution).

If the dispersal corridor (L) is longer than the transect, $a > 1$, individuals will settle both along the transect and beyond it (Figure A1.2). If, however, $a < 1$ individuals initially caught along the transect will mostly settle somewhere along it. The ability of the transect to detect dispersal distances will depend strongly on the value of a. If a is equal to 1

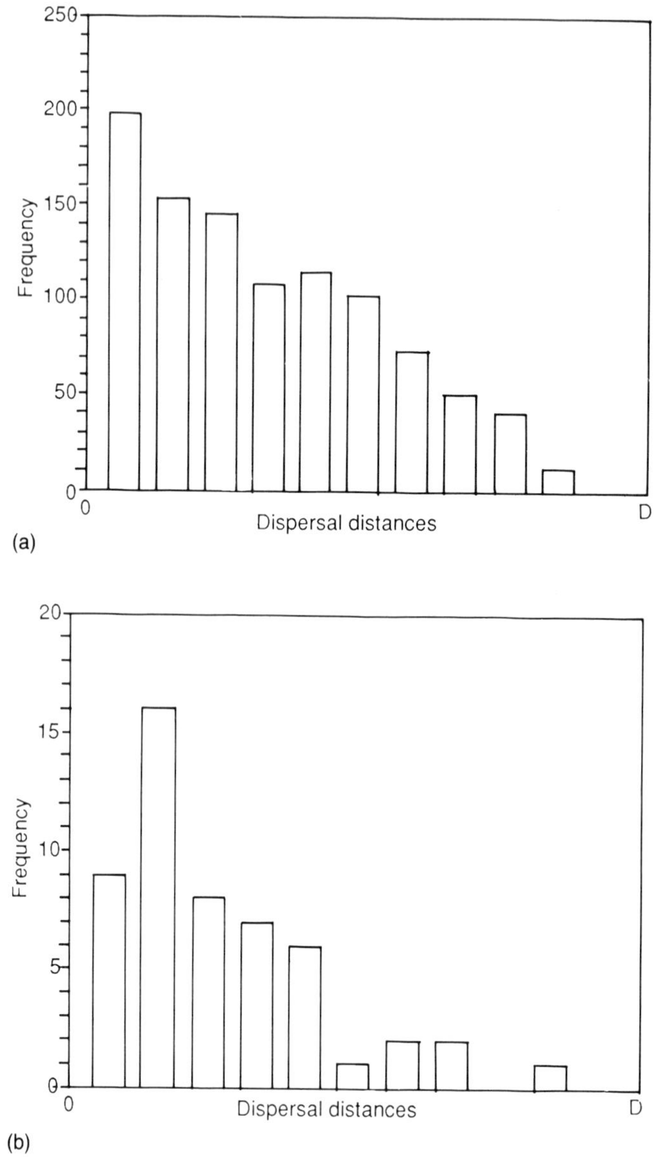

Figure A1.5 Distribution of detected recapture distances resulting when: (a) a = 0.5, and (b) a = 20. See Figure A1.3 and text for further explanation.

the two distances are equal. If a is 20, the maximal dispersal distance is 20 times longer than the transect. Out of 1000 individuals first caught on the study transect (and, for instance, marked there) only a small fraction is captured on that study transect. As a increases, a decreasing fraction is recaptured.

When $a = 1$, a strongly skewed distribution results even though all distances of dispersal are equally likely (Figure A1.3). Figure A1.4 depicts the situation where most dispersal events are detected by the transect (e.g. $a = 0.1$). In this case also the smaller dispersal distances are over-represented and the larger underrepresented, although the overall distribution resembles a rectangular one much more than in the other cases. An intermediate situation ($a = 0.5$) is shown in Figure A1.5a. Note that the distribution resembles that when $a = 1$. An extreme case ($a = 20$) is shown in Figure A1.5b which may be close to reality for many actual studies. The distribution of recapture distances is in this case strongly deviant from the real rectangular one and could even be mistaken for a Poisson distribution.

Figure A1.6a displays two hypothetical data sets representing dispersal distances in two dimensions (Murray, 1967). Although Murray is unclear with respect to the assumptions of his model, they all seem to correspond to ours. As can be seen, the same qualitative results emerge. This further supports the generality of our hypothetical examples.

Boonstra *et al.* (1987) provide two real data sets from *Microtus pennsylvanicus*. These results (Figure A1.6b) may easily, but possibly erroneously, be interpreted as a Poisson-type distribution. As we have shown in our simulations as well as in Murray's example, this distribution pattern of recapture distances likely represents a very different generating distribution. The basic problem arises from the way data are obtained that leads to size-biased sampling. Such frequency distributions of recapture distances can nevertheless give meaningful and interesting results (Figure A1.6c), showing sex differences in dispersal distances in a monogamous rodent (Ribble, 1991).

The above analysis does in fact provide guidelines for how to derive null distributions to be used when interpreting data on dispersal distances. In the examples used, we have assumed that dispersal distances in fact are randomly distributed according to a rectangular distribution. Other distributions could, of course, have been assumed: in the next section we do in fact discuss another possibility.

A1.3.3 Diffusion models

Dispersal out from some core area of good quality habitat may be viewed as a diffusion process (Levin, 1974; Okubo, 1980). We discuss one ex-

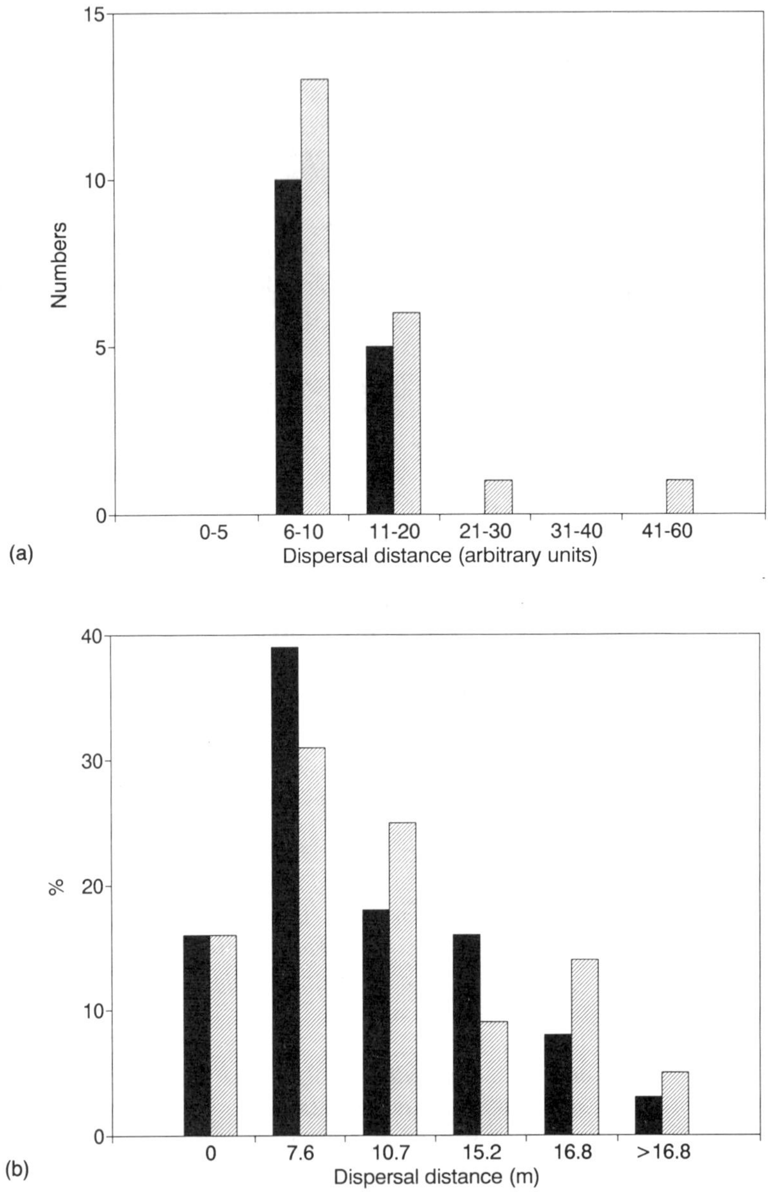

Figure A1.6 (a) Natal dispersal distances for males in two hypothetical data sets based on a 2-dimensional model, taken from Murray, 1967. Solid bars are for a population in which survival of young to breeding age was 23% and the shaded bars denote a population in which survival was 62%. In the first case, eight individuals left the 'study area' and 41 did so in the second case, thus markedly skewing the dispersal distance distributions. (b) Recapture distances from place of birth to first trap entered in *Microtus pennsylvanicus* from two grids in southern Ontario; n = 38 for the left hand bars and 43 for the right hand. Data are from Boonstra *et al.*, 1987. (c) Dispersal distances in male and female *Peromyscus califormicus*, redrawn from Ribble (1991).

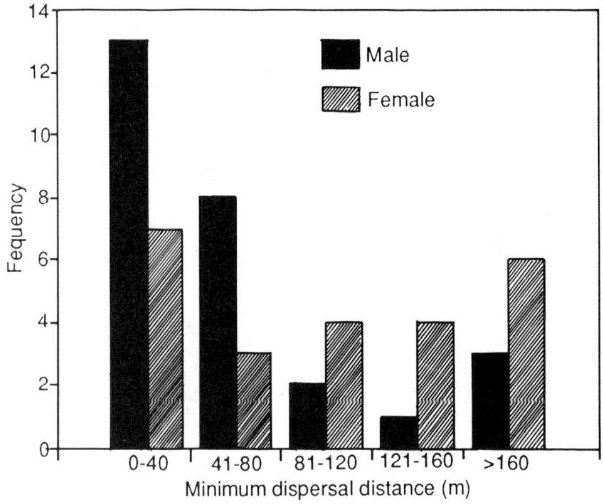

(c)

ample utilizing this approach in a species of small mammal in Appendix 3 as the technique of radioactive isotope marking of the animals was used. In this section we describe the idea behind diffusion models as well as explain, by use of an actual example (dispersal in the bark beetle, *Ips typographus*; Helland *et al.*, 1984, 1989), how particular hypotheses about dispersal, attraction to some particular site, and about settling can be incorporated into the analysis. These processes are of course fundamental to dispersal in all kinds of organisms.

(a) Background

The flight of insects is a very complex phenomenon, and any detailed mathematical model of such a flight must necessarily be an idealization. Helland *et al.* (1984, 1989) modelled the flight of bark beetles that are attracted to a pheromone source. The basic mathematical model used was a diffusion model (random walk with small steps) where the beetles have a drift (bias in the random walk) directed towards the pheromone source. The magnitude of this drift term was assumed inversely proportional to the distance from the source. The main advantage of this model is that practically every quantity of interest can be calculated explicitly from the model equations. In particular, one can predict the expected catch, under the assumed hypothesis, in a passive trap at any position relative to the release point and the attractive pheromone trap.

(b) The experiment

To test the models and estimate the parameters of the models, Helland *et al.* (1984) performed a release–recapture experiment. The experimental set-up consisted of one central plastic cylinder trap baited with a pheromone source. Passive trap stations were placed in eight concentric rings around the central pheromone trap, the distance between the rings being 3 m. The number of trap stations per ring was eight, placed at ±22.4°, ±67.5°, ±112.5° and ±157.4° with respect to the line between the release point and the pheromone trap. The set-up is shown in Figure A1.7, where also the catch in the passive traps in one experiment is indicated.

Approximately 6000 beetles were released from each of four platforms (NE, NW, SW, and SE), at 12 m distances apart. The beetles were marked with fluorescent powder using a different colour for each release platform, and released simultaneously. The catch in the central trap was registered every 2, 5, or 10 min, and the passive trap catch was recorded 3.5 h after release, when the swarming activity had ceased.

(c) Experimental results and model fitting

It is possible to include the effect of wind in the diffusion models that we discuss here. However, this required numerical analysis, and was judged an unnecessary complication by Helland *et al.* (1984). Capture points of the beetles were recorded as the distance from the central trap and the angle between the release point and the capture point, as seen from the central trap. Catches of marked beetles from all four release platforms with the same distance and angle were added together. When comparing these data with model predictions, we must take into account that the pheromone trap is not 100% efficient. Only 10.01 and 7.31% of the beetles released were caught in the two performed experiments. From experiments with male pea moths, *Cydia nigricana*, Wall and Perry (1982) report that even when they got within 1 m of the trap, only 40% of the insects actually got caught. Assuming that beetles reaching the central trap without being caught continue their flight with a new diffusion coefficient (phase two), one can arrive at predictions for the catch in the passive traps.

Helland *et al.* (1984) found that the fit of the model to the data was improved if they allowed for 'flight exercise' in the model, i.e. a state where the beetles moved according to a diffusion equation without drift before they started to respond to the pheromone.

The basic diffusion model used by Helland *et al.* (1984) is about the simplest one that can be imagined if we want to take drift towards a pheromone source into account (Figure A1.8a). It turned out that several modifications of this basic model were necessary in order to get a reasonable fit to the data. In Helland (1983) several other diffusion models were discussed, for instance models with a drift term inversely proportional to the *square* of the distance to the pheromone trap and one with a threshold response to the pheromone attractant (Figure A1.8b,c). For all these models they could calculate the expected total catch per ring when many

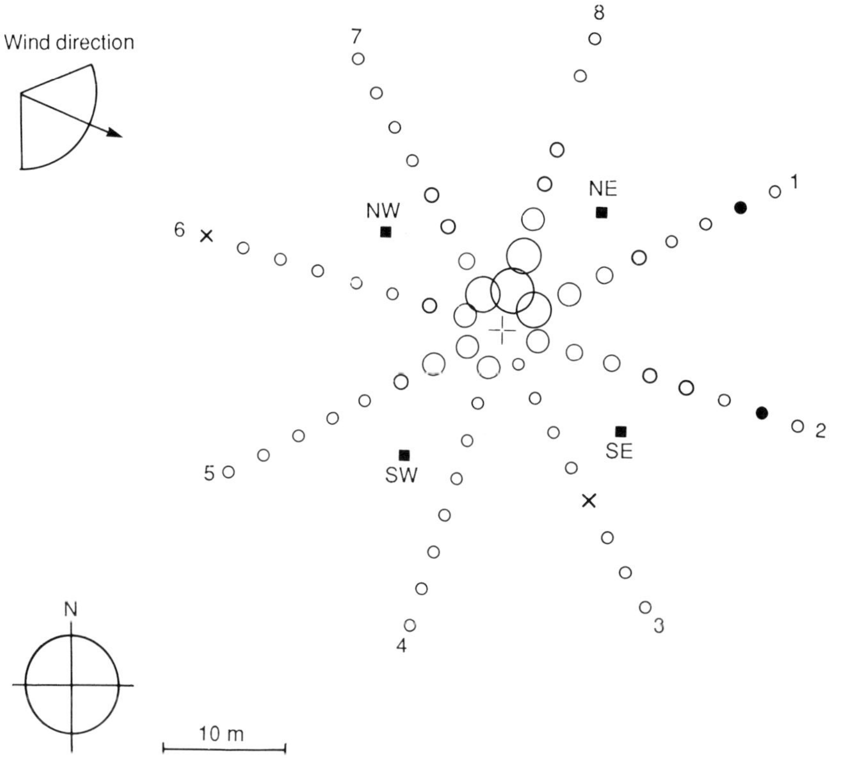

Figure A1.7 The experimental set-up used by Helland *et al.* (1984) to measure movements of the bark beetle *Ips typographicus* released at four sites (NW, NE, SW, SE), and recaptured at a pheromone bait in the centre (the dispersal sink) and in an array of passive traps located along eight transects radiating from the centre (distance between traps = 12 m). The catch in the passive traps for one experiment is also shown. × = 0, • = 1, ○ = 50, ◯ = 100 beetles.

passive traps are distributed along each concentric ring. In Figure A1.8 predictions for some of these models are shown together with the experimental results from Helland *et al.* (1984; figure A1.8d,e).

An attraction model of quite different type was proposed by Perry and Wall (1984). This was a simulation model where the shape of the so-called threshold contour for pheromones was taken into account. The model had several parameters that partly depended upon the micrometeorological conditions. The model parameters used by Perry and Wall were based on the observation by David *et al.* (1982) that moths fly upwind if they detect

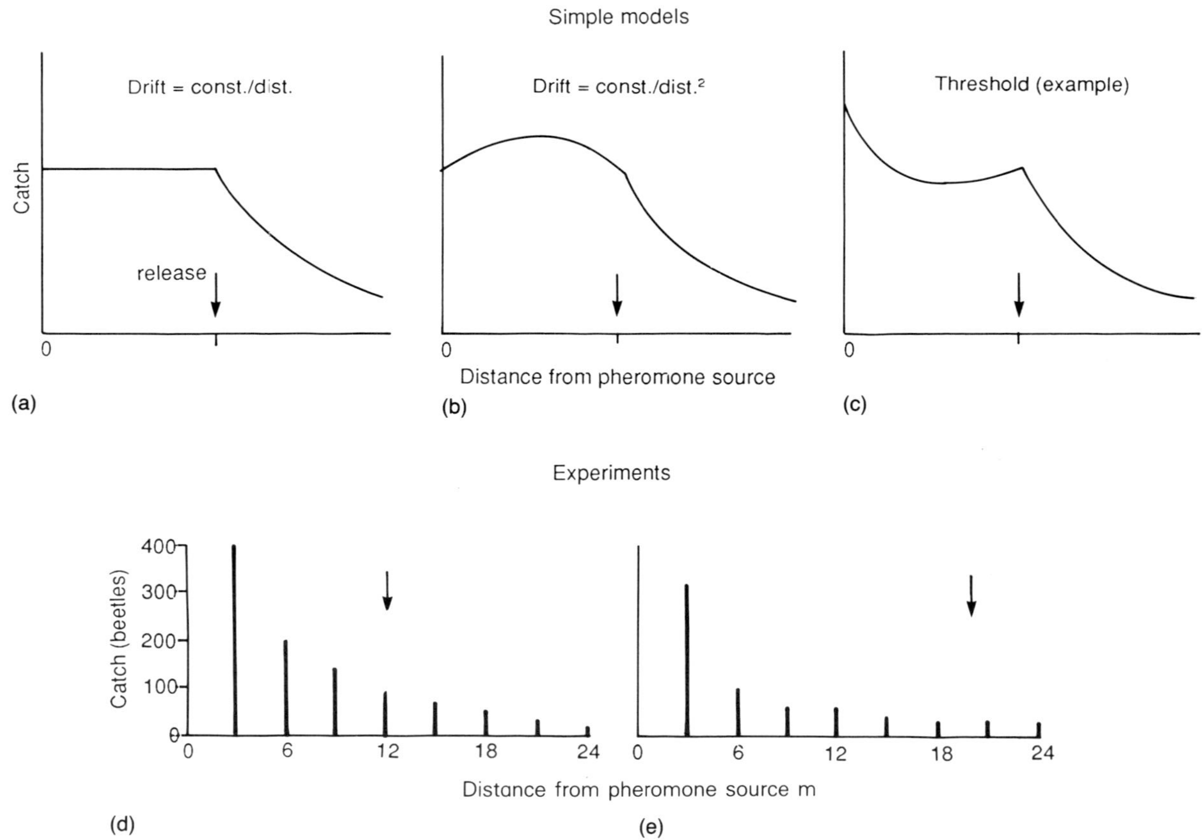

Figure A1.8 Three models of expected recapture distributions for the experimental array shown in Figure A1.7a,b,c. The pheromone source is at the origin and the distance to the release point is shown by the arrow. (a) The expected catch if drift is equal to a constant divided by the distance from the pheromone source. (b) Expected catch if drift is equal to a constant divided by the squared distance to the pheromone source. (c) Expected catch if there is an upper threshold of response to the pheromone above which no further increase in drift is observed. (d) and (e) Actual results from two experiments reported by Helland *et al.* (1984).

pheromone and crosswind otherwise. For beetles flying over open ground with constant wind direction, a model of this type will (if pheromones are detected) predict movement along straight lines in the upwind direction.

Application of diffusion models in order to study dispersal is discussed in Appendix 3. Mass migration in Norwegian lemmings and dispersal in voles could be studied in this way. Since voles and lemmings do not rely on pheromones in the same way that bark beetles do, the details of the model will, of course, be different. Obviously, there is much room here for profitable work by theoreticians and statisticians interested in biological problems. Such research should, we believe, be carried out in close collaboration with empiricists. The study by Helland *et al.* (1984, 1989) is a splendid example of such a profitable collaboration.

A1.4 FUTURE DIRECTIONS

Science can always benefit from new techniques and new methods of analysis. Such new approaches arise either from particularly innovative individuals who perceive a need in the context of some research question, or they are developed in other disciplines and are subsequently adapted to the problem at hand. In the latter case it is necessary that someone with interdisciplinary skills and perspective learn of the new technique and recognize its utility for other purposes. It is, therefore, difficult to prescribe new directions for technical developments.

Of course, we can always direct attention to the desirability of improvements in existing methodologies: lighter and more powerful radio transmitters for small creatures, better live traps, more discriminating exit traps for enclosures, and better methods for assessing genetic relatedness. In the latter case, the technique of DNA-fingerprinting (Jeffreys *et al.*, 1985; Burke *et al.*, 1989; Gyllensten *et al.*, 1990) seems to offer great potential, and many researchers are now experimenting with this method.

Even with the prospect of exciting new technologies ahead, we feel that much more can be accomplished with existing methods. We need to be skillful in matching our hypotheses to be tested with the most appropriate technologies. In the long run it is the theoretical framework of our science and its empirical base that will direct us to new methodologies, and not the converse.

REFERENCES

Allred, D.M. and Beck, D.E. (1963) Range of movement and dispersal of some rodents at the Nevada atomic test site. *Journal of Mammalogy*, **44**, 190–200.

Baird, D.D. and Birney, E.C. (1982). Pattern of colonization in *Microtus pennsylvanicus*. *Journal of Mammalogy*, **63**, 290–3.

Beacham, T.D. (1979) Size and growth characteristics of dispersing voles, *Microtus townsendii. Oecologia*, **42**, 1–10.

Beacham, T.D. (1980a) Dispersal during population fluctuations of the vole, *Microtus townsendii. Journal of Animal Ecology*, **49**, 867–77.

Beacham, T.D. (1980b) Demography of declining populations of the vole, *Microtus townsendii. Journal of Animal Ecology*, **49**, 453–64.

Beacham, T.D. (1981) Some demographic aspects of dispersers in fluctuating populations of the vole *Microtus townsendii. Oikos*, **36**, 273–80.

Blehr, O. (1973) Traditional reindeer hunting and social change in local communities surrounding Hardanger-vidda. *Norwegian Archeological Review*, **6**, 102–12.

Bondrup-Nielsen, S. (1983) Density estimation as a function of livetrapping grid and home range size. *Canadian Journal of Zoology*, **61**, 2361–5.

Bondrup-Nielsen, S. (1984) The role of habitat heterogeneity and female spacing behaviour in density regulation of Clethrionomys gapperi. Unpublished Ph.D. Dissertation, University of Alberta, Edmonton.

Boonstra, R. and Krebs, C.J. (1977) A fencing experiment on a high-density population of *Microtus townsendii. Canadian Journal of Zoology*, **55**, 1166–75.

Boonstra, R. and Krebs, C.J. (1978) Pitfall trapping of *Microtus townsendii. Journal of Mammalogy*, **59**, 136–48.

Boonstra, R., Krebs, C.J., Gaines, M.S. *et al.* (1987) Natal philopatry and breeding systems in voles (*Microtus* spp.). *Journal of Animal Ecology*, **56**, 655–73.

Burke, T., Davies, N.B., Bruford, M.W. and Hatchwell, B.J. (1989) Parental care and mating behaviour of polyandrous dunnocks *Prunella modularis* related to paternity by DNA fingerprinting. *Nature*, **338**, 249–51.

Calhoun, J.B. and Webb, W.L. (1953) Induced emigrations among small mammals. *Science*, **117**, 358–60.

Caughley, G. (1977) *Analysis of Vertebrate Populations*, John Wiley, New York, 234 pp.

Cormack, R.M. (1989) Loglinear models for capture–recapture. *Biometrics*, **45**, 395–413.

David, C.T., Kennedy, J.S., Ludlow, A.R. *et al.* (1982) A reappraisal of insect flight towards a distant point source of wind-borne odour. *Journal of Chemical Ecology*, **8**, 1207–15.

Dice, L.R. and Howard, W.E. (1951) Distance of dispersal by prairie deermice from birthplaces to breeding sites. Contributions of the Laboratory of Vertebrate Biology, University of Michigan, No. 50, 15 pp.

Dixon, K.R. and Chapman, J.A. (1980) Hormonic mean measure of animal activity area. *Ecology*, **61**, 1040–4.

Dueser, R.D., Rose, R.K. and Porter, J.H. (1984) A body-weight criterion to identify dispersing small mammals. *Journal of Mammalogy*, **65**, 727–9.

Dueser, R.D., Wilson, M.L. and Rose, R.K. (1981) Attributes of dispersing meadow voles in open-grid populations. *Acta Theriologica*, **26**, 139–62.

Fairbairn, D.J. (1977) The spring decline in deermice: death or dispersal? *Canadian Journal of Zoology*, **55**, 84–92.

Fairbairn, D.J. (1978) Dispersal of deermice, *Peromyscus maniculatus. Oecologia*, **32**, 171–93.

Ford, R.G. and Krumme, D.W. (1979) The analysis of space use patterns. *Journal of Theoretical Biology*, **76**, 125–55.

French, N.R., Tagami, T.Y. and Hayden, P. (1968) Dispersal in a population of desert rodents. *Journal of Mammalogy*, **49**, 272–80.

Gaines, M.S., Baker, C.L. and Vivas, A.M. (1979a) Demographic attributes of dispersing southern bog lemmings (*Synaptomys cooperi*) in eastern Kansas. *Oecologia*, **49**, 91–101.

Gaines, M.S., Vivas, A.M. and Baker, C.L. (1979b) An experimental analysis of dispersal in fluctuating vole populations: demographic parameters. *Ecology*, **60**, 814–28.

Garsd, A. and Howard, W.E. (1981) A 19-year study of microtine population fluctuations using time-series analysis. *Ecology*, **62**, 930–7.

Garsd, A. and Howard, W.E. (1982) Microtine fluctuation: an ecosystem approach based on time-series analysis. *Journal of Animal Ecology*, **51**, 225–34.

Gipps, J.H.W. and Jewell, P.A. (1979) Maintaining populations of bank voles, *Clethrionomys glareolus*, in large outdoor enclosures, and measuring the response of population variables to the castration of males. *Journal of Animal Ecology*, **48**, 535–55.

Gyllensten, U.B., Jakobsson, S. and Temrin, H. (1990) No evidence for illegitimate young in monogamous and polygamous warblers. *Nature*, **343**, 168–70.

Harris, S., Cresswell, W.F., Forde, P.G. *et al.* (1990) Home range analysis using radio-tracking data – a review of problems and techniques particularly applied to the study of mammals. *Mammalian Review*, **20**, 97–123.

Hansson, L. (1977) Spatial dynamics of field voles *Microtus agrestis* in heterogeneous landscapes. *Oikos*, **29**, 539–44.

Helland, I.S. (1983) Diffusion models for the dispersal of insects near an attractive center. *Journal of Mathematical Biology*, **18**, 103–22.

Helland, I.S., Anderbrant, O. and Hoff, J.M. (1989) Modelling bark beetle flight: a review. *Holarctic Ecology*, **12**, 427–31.

Helland, I.S., Hoff, J.M. and Anderbrant, O. (1984) Attraction of bark beetles (Coleoptera: Scolytidae) to a pheromone trap. Experiment and mathematical models. *Journal of Chemical Ecology*, **10**, 723–52.

Henttonen, H. and Järvinen, A. (1981) Lemmings in 1978 at Kilpisjärvi: population characteristics of a small peak. *Memoirs of the Society Fauna Flora Fennici*, 57, 25–30.

Henttonen, H., Oksanen, T., Jortikka, A. and Haukisalmi, V. (1987) How much do weasels shape microtine cycles in the northern Fennoscandian taiga? *Oikos*, 50, 352–65.

Hestbeck, J.B. (1986) Multiple regulation states in populations of the California vole, *Microtus californicus*. *Ecological Monographs*, **56**, 161–81.

Hilborn, R. and Krebs, C.J. (1976) Fates of disappearing individuals in fluctuating populations of *Microtus townsendii*. *Canadian Journal of Zoology*, **54**, 1507–18.

Howard, W.E. (1949) Dispersal, amount of inbreeding, and longevity in a local population of deermice on the George Reserve, Southern Michigan. *Contributions of the Laboratory of Vertebrate Biology*, University of Michigan, No. 43, 50 pp.

Howard, W.E. (1960) Innate and environmental dispersal in individual vertebrates. *American Midland Naturalist*, **63**, 152–61.

Hurlbert, S.H. (1984) Pseudoreplication and the design of field experiments. *Ecological Monographs*, **54**, 187–211.

Îto, Y. and Miyashita, K. (1965) Studies on the dispersal of leaf- and planthoppers. III. An examination of the distance–dispersal rate curves. *Japanese Journal of Ecology*, **15**, 85–9.

Jeffreys, A.J., Wilson, V. and Thein, S.L. (1985) Hypervariable 'minisatellite' regions in human DNA. *Nature*, **314**, 67–73.

Jensen, P.M., Stenseth, N.C. and Framstad, E. (1991a) An experimental study on dispersal of *Lemmus lemmus* (L.). *Biological Journal of the Linnean Society*, in press.

Jensen, P.M., Stenseth, N.C. and Framstad, E. (1991b) Trappability of the Norwegian lemming (*Lemmus lemmus* (L.)). *Biological Journal of the Linnean Society*, in press.

Johnson, M.L. and Gaines, M.S. (1985) Selective basis for emigration of the prairie vole, *Microtus ochrogaster*: open field experiment. *Journal of Animal Ecology*, **54**, 399–410.

Kalela, O. (1961) Seasonal change of habitat in the Norwegian lemming, *Lemmus lemmus* (L.). *Annales Academia Scientium Fennica 4A*, IV, 55, 1–72.

Kalela, O. (1971) Seasonal differences in habitats of the Norwegian lemming, *Lemmus lemmus* (L.), in 1959 and 1960 at Kilpisjärvi, Finnish Lapland. *Annales Academia Scientium Fennica, A IV*, **178**, 1–22.

Keith, T.P. and Tamarin, R.H. (1981) Genetic and demographic differences between dispersers and residents in cycling and non-cycling vole populations. *Journal of Mammalogy*, **62**, 713–25.

Kenward, R. (1987) *Wildlife Radio Tagging: Equipment, field techniques and data analysis*. Academic Press, London.

Krebs, C.J. (1989) *Ecological Methodology*, Harper and Row, New York, 654 pp.

Krebs, C.J., Keller, B.L. and Tamarin, R.H. (1969) *Microtus* population biology: demographic changes in fluctuating populations of *M. ochrogaster* and *M. pennsylvanicus* in southern Indiana. *Ecology*, **50**, 587–607.

Krebs, C.J., Redfield, J.A. and Taitt, M.J. (1978) A pulsed-removal experiment on the vole *Microtus townsendii*. *Canadian Journal of Zoology*, **56**, 2253–82.

Krebs, C.J., Wingate, I., LeDuc, J. *et al.* (1976) *Microtus* population biology: dispersal in fluctuating populations of *M. townsendii*. Canadian Journal of Zoology, **54**, 79–95.

Levin, S.A. (1974) Dispersion and population interactions. *American Naturalist*, **108**, 207–28.

Lidicker, W.Z. Jr (1975) The role of dispersal in the demography of small mammals, in *Small Mammals: their productivity and population dynamics* (eds F.B. Golley, K. Petrusewicz and L. Ryszkowski), Cambridge University Press, London, pp. 103–28.

Lidicker, W.Z. Jr (1985a) Population structuring as a factor in understanding microtine cycles. *Acta Zoologica Fennica*, **173**, 23–7.

Lidicker, W.Z. Jr (1985b) Dispersal, *Biology of New World* Microtus (ed. R.H.

Tamarin), Special Publication No. 8, American Society of Mammalogists, pp. 420–54.

Lidicker, W.Z. Jr (1988) Solving the enigma of microtine 'cycles'. *Journal of Mammalogy*, **69**, 225–35.

MacArthur, R.H. (1972) *Geographical Ecology: patterns in the distribution of species*, Harper and Row, New York, 269 pp.

Murray, B.G. Jr (1967) Dispersal in vertebrates. *Ecology*, **48**, 975–8.

Myers, J.H. and Krebs, C.J. (1971) Genetic, behavioral and reproductive aspects of dispersing field voles *Microtus pennsylvanicus* and *Microtus ochrogaster*. *Ecological Monographs*, **41**, 53–78.

Okubo, A. (1980) *Diffusion and Ecological Problems: mathematical models*, Springer-Verlag, New York.

Ord, J.K., Patil, G.P. and C. Taillie (eds) (1980) *Statistical Distributions in Ecological Work*. Inturn Co-operative Publishing House, Fairland, Maryland.

Ostfeld, R.S. (1986) Territoriality and mating system of California voles. *Journal of Animal Ecology*, **55**, 691–706.

Ostfeld, R.S., Lidicker, W.Z. Jr and Heske, E.J. (1985) The relationship between habitat heterogeneity, space one, and demography in a population of California voles. *Oikos*, **45**, 433–42.

Otis, D.L., Burnham, K.P., White, G.C. and Anderson, D.R. (1978) Statistical inference from capture data on closed animal populations. Wildlife Monograph No. 62, 135 pp.

Perry, J.N. and Wall, C. (1984) A mathematical model for the flight of pea moth to pheromone trap through a crop. *Philosophical Transactions of the Royal Society of London,* **B306**, 19–48.

Pollock, K.H., Nichols, J.D., Brownie, C. and Hines, J.E. (1990) Statistical inferences for capture–recapture experiments. Wildlife Monograph No. 107, 97 pp.

Pugh, S.R. and Tamarin, R.H. (1991) Demographic characteristics and reproductive success of resident and immigrant meadow voles, *Microtus pennsylvanicus*. *Canadian Journal of Zoology*, in press.

Pulliam, R. (1988) Sources, sinks, and population regulation. *American Naturalist*, **132**, 652–61.

Ribble, D.O. (1991) Dispersal in a monogamous rodent, *Peromyscus californicus*. *Ecology*, in press.

Riggs, L.A. (1979) Experimental studies of dispersal in the California vole, *Microtus californicus*. Ph.D. Dissertation, University of California, Berkeley, 237 pp.

Seber, G.A.F. (1982) *The Estimation of Animal Abundance and Related Parameters*, Chas. Griffin, London, 654 pp.

Seber, G.A.F. (1986) A review of estimating animal abundance. *Biometrics*, **42**, 267–92.

Smith, M.H., Gardner, R.H., Gentry, J.B. *et al.* (1975) Density estimations of small mammal populations, in *Small Mammals: their productivity and population dynamics* (eds F.B. Golley, K. Petrusewicz and L. Ryszkowski), Cambridge University Press, London, pp. 25–53.

Snedecor, G.W. and Cochran, W.G. (1967) *Statistical Methods*, Iowa State University Press, Ames, 593 pp.

Sokal, R.R. and Rohlf, F.J. (1981) *Biometry*, W.H. Freeman, San Francisco, 859 pp.

Southwood, T.R.E. (1987) *Ecological Methods with Particular Reference to the Study of Insect Populations*, 2nd edn, Chapman & Hall, London, 524 pp.

Staples, P.P. and Terman, C.R. (1977) An experimental study of movement in natural populations of *Mus musculus*, *Microtus pennsylvanicus*, and *Microtus pinetorum*. *Researches on Population Ecology* **18**, 267–83.

Stenseth, N.C. (1983) Causes and consequences of dispersal in small mammals, in *The Ecology of Animal Movement* (eds I. Swingland and P.J. Greenwood), Oxford University Press, Oxford, pp. 63–101.

Stenseth, N.C. and Hansson, L. (1979) Correcting for the edge effect in density estimation: explorations around a new method. *Oikos*, **32**, 337–48.

Stickel, L.F. (1946) The source of animals moving into a depopulated area. *Journal of Mammalogy*, **27**, 301–7.

Stoddart, D.M. (1970) Individual range, dispersion and dispersal in a population of water voles (*Arvicola terrestris* (L.)). *Journal of Animal Ecology*, **39**, 403–25.

Stoddart, D.M (1982) Demonstration of olfactory discrimination by the short-tailed vole, *Microtus agrestis* L. *Animal Behaviour* **30**, 293–4.

Sullivan, T.P. (1977) Demography and dispersal in island and mainland populations of the deer mouse, *Peromyscus maniculatus*. *Ecology*, **58**, 964–78.

Tamarin, R.H. (1977) Dispersal in island and mainland voles. *Ecology*, **58**, 1044–54.

Tamarin, R.H. (1980) Dispersal and population regulation in rodents, in *Biosocial Mechanisms of Population Regulation* (eds M.N. Cohen, R.S. Malpass and H.G Klein), Yale University Press, New Haven, pp. 117–33.

Tamarin, R.H. (1984) Body mass as a criterion of dispersal in voles: a critique. *Journal of Mammalogy*, **65**, 691–2.

Tamarin, R.H., Reich, L.M. and Moyer, C.A. (1984) Meadow vole cycles within fences. *Canadian Journal of Zoology*, **62**, 1796–804.

Verner, L. (1979) The significance of dispersal in fluctuating populations of *Microtus ochrogaster* and *Microtus pennsylvanicus*. Unpublished Ph.D. Dissertation, University of Illinois, Urbana.

Wall, C. and Perry, J.N. (1982) The behaviour of moths responding to pheromone sources in the field: a basis for discussion. *Le Colloques de l'INRA*, 7, 171–88.

White, G.C., Anderson, D.R., Burnham, K.P. and Otis, D.L. (1982) Capture–recapture and removal methods for sampling closed populations. Los Alamos Nat. Lab., LA 8787-NERP, Los Alamos, NM, 235 pp.

Wolff, J.O. and Holleman, D.F. (1978) Use of radioisotope labels to establish genetic relationships in free-ranging small mammals. *Journal of Mammalogy*, **59**, 859–60.

Wolff, J.O. and Lidicker, W.Z. Jr (1980) Population ecology of the taiga vole, *Microtus xanthognathus*, in interior Alaska. *Canadian Journal of Zoology*, **58**, 1800–12.

Worton, B.J. (1989) Kernel methods for estimating the utilization distribution in home-range studies. *Ecology*, 70, 164–8.

Appendix 2. Alternative approaches to the study of small mammal dispersal: insights from radiotelemetry

W.J. McSHEA AND D.M. MADISON

A2.1 INTRODUCTION

Dispersal is the permanent movement of an organism away from its home range, and includes movements both within and between populations. Radiotelemetry is the most accurate means available to measure movements in small mammals and therefore deserves serious consideration in a study of dispersal. This chapter will examine dispersal data derived from the use of radiotelemetry, relate the findings to proposed theory, and make suggestions for future avenues of investigation.

Dispersal studies have concentrated on the relationship between dispersal and population fluctuations (Krebs *et al.*, 1969; Gaines *et al.*, 1979; Tamarin, 1980), and the relationship between phenotypic attributes and the tendency to disperse (Savidge, 1974; Hilborn, 1975; Bekoff, 1977; Beacham, 1979; Baird and Birney, 1982). Radiotelemetry has the potential to assist in the investigation of both these areas.

Radiotelemetry has both technical and practical limitations. Transmitter size, and signal duration and distance can influence effectiveness (Madison *et al.*, 1985; Rado and Terkel, 1989), but money and time usually set the practical limits on a researcher's ability to monitor a large population of small mammals. Transmitters are expensive, they are placed on animals with high mortality rates, and one researcher can locate no more than 20–30 individuals during a short period: these limitations make it difficult to effectively monitor a significant portion of most rodent populations. We feel telemetry is a poor choice to study dispersal rates within a general population. However, telemetry can characterize movements and locate select individuals, and, therefore, can be used to determine the proximate causes and pattern of dispersal. The selective use of radiotelemetry to study specific subsets or experimental groups within a population holds the most promise for future research.

A2.2 VERIFICATION OF DISPERSAL

A major problem in the study of dispersal is the identification of dispersers (Gaines and McClenaghan, 1980; Dobson, 1981; Baird and Birney, 1982; Krohne *et al.*, 1984; Lidicker, 1985; Brochu *et al.*, 1988; Gregory and Cameron, 1988). Techniques for studying dispersal include extensive live-trap configurations, removal grids, and 'selective' enclosures. These methods involve the assumptions that dispersing animals have a trappability equal to resident animals, that there is no significant effect of the experimental manipulations on attracting or repelling dispersers, and that those animals captured within a removal grid represent an accurate subsample of those animals that actually disperse. The only means to positively identify dispersers is to record when a specific animal's movements no longer encompass its previously determined home range.

Radiotelemetry can provide specific and detailed information on movement that is derived without experimental manipulation and is independent of trapping schedules or spacing. Dispersal events have been recorded during studies on a wide variety of small rodent species (e.g. *Microtus pennsylvanicus*, Madison, 1980, McShea, 1990a; *M. pinetorum*, FitzGerald and Madison, 1983; *M. arvalis*, Boyce and Boyce, 1988; *Neotoma fuscipes*, Cranford, 1977; *Lemmus trimucronatus*, Banks *et al., 1975; Arvicola terrestris*, Leuze, 1976; *Peromyscus leucopus*, Mineau and Madison, 1977; *Dipodomys merriami*, Jones, 1989). However, only recently have studies been designed to specifically study dispersal (Jones, 1989; McShea, 1990a,b). Radiotelemetry studies of dispersal in microtine populations are feasible and would prove productive, particularly for the study of natal dispersal.

A2.3 DESCRIPTION OF MOVEMENTS

We define dispersal as the permanent shift of an animal away from its home range. However, the simple capture of an animal in a previously unoccupied area does not necessarily indicate dispersal. Burt (1943) considered a home range 'that area traversed by the individual in its normal activities of food gathering, mating, and caring for young'. Before an animal can be considered a disperser, it is necessary to determine the animal's home range and this area can be flexible and should be expected to change over time. The time interval over which an animal's position is measured will have a great effect on whether its use of an area is labelled as dispersal, or simply a return to a seldom used portion of its home range. The investigator's choice of a time interval should depend on the habits of the species, but should take into consideration the time span of events within the animal's life. For example, shifts within home ranges occur over the 21-day lactation period of female meadow voles (Madison, 1978).

Table A2.1 The frequency of home range shifts for reproductively active female meadow voles over two reproductive seasons (1983–84). A shift was considered to have occurred when a home range perimeter for one week was outside the area encompassed by the individual's movements during the previous week. Home range estimates were based on 10 positions taken at hourly intervals during the day (see Madison *et al.*, 1984, for details)

No. of dispersals	*No. of 3-week intervals individual was montiored*			
	1	2	3	4
0	8	7	3	3
1	0	7	2	2
2	2	3	3	1
3	0	2	1	1
4	0	0	1	2

The calculation of home range size and location is dependent on the sampling means, sampling interval, and the procedure chosen for dealing with 'outlier' positions (Dixon and Chapman, 1980; Swihart and Slade, 1985a,b; Ackerman *et al.*, 1990). Radiotelemetry gives a more accurate representation of an animal's movements than does trapping conducted over a similar period (Hackett and Trevor-Deutsch, 1982; Jones and Sherman, 1983). For *Microtus ochrogaster* at least eight trap captures of an individual are necessary to approximate the home range size derived from radiotracking (Desy *et al.*, 1989). However, even radiotelemetry can underestimate home range size due to autocorrelations between successive positions (Swihart and Slade, 1985a). The length of time an animal is monitored also increases the probability of recording a dispersal event (Table A2.1). Despite these problems, most labelling of animals as residents or dispersers does not take past and future dispersal into account. Positions obtained outside of areas of high utilization are often not included in estimates of home range size (Dixon and Chapman, 1980), yet these movements are important in deciding if home range shifts occur. Computer programs are available to convert radiotelemetry data into estimates of home range size (Stuwe and Blohowiak, 1985; Cary, 1990; Ackerman *et al.*, 1990). However, actual home range size is not as important as determining what area is familiar to an animal and thereby correctly identifying dispersal movements.

Movement in small mammals can assume several patterns (Figure A2.1). These patterns include movements within a home range (stationary, variable, and local reconnaissance) and movements away from a home

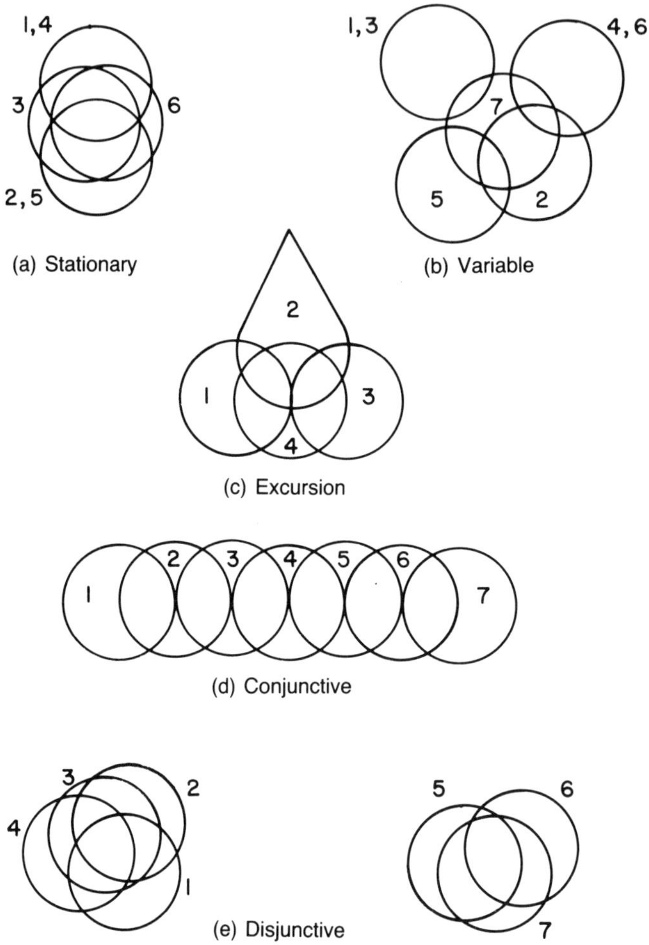

Figure A2.1 A representation of home range and dispersal patterns seen in microtines (see Madison, 1985, for more details). Stable home ranges include: stationary (a), where each home range perimeter overlaps previous estimates; and variable (b) where successive home range boundaries do not necessarily overlap, yet repeated monitoring reveals no net shift in location. Both of these home range types may include movements beyond the home range boundaries (excursions, c). Dispersal constitutes a shift in home range and may occur slowly (conjunctive, d) or rapidly (disjunctive, e). The numbers shown indicate successive estimates of home range location at intervals meaningful to the study species.

range (excursions, and dispersal, both conjunctive and disjunctive) (terminology following Madison, 1980, 1985). Excursions, 'distant reconnaissance', or 'sallies', have been described in several species (*Rattus norvegicus*, Hardy and Taylor, 1980; *Microtus pennsylvanicus*, Madison, 1978; *Dipodomys merriami*, Jones, 1989). In meadow voles, movements of over 100 m beyond the normal home range are usually followed by a subsequent return to the original home range. For female meadow voles 10% of all home range estimates include movements that can be considered excursions (based on 624 home range estimates of 124 females over a 2-year period). Male meadow vole excursion rates may be twice as frequent (Madison, 1980).

Cranford (1977) monitored woodrats (*Neotoma fuscipes*) that shifted home ranges in response to spring flooding, yet later returned to their original home range. *M. pennyslvanicus* also demonstrate home range shifts that last several days before the animal returns to its previous range (McShea, personal observation). Such excursions are not true dispersal movements, yet are indistinguishable from dispersal, as categorized by trap captures over a short period of time. 'True' dispersal is disjunctive dispersal (Figure A2.1e), which involves the relatively rapid shift in the entire home range. In order to differentiate this dispersal from the slower conjunctive dispersal, the time interval for sampling is important. The distinction between the two types of dispersal is important; conjunctive dispersal should have a much higher survival rate and may be a dispersal pattern for particular subsets of the population (i.e. pregnant females).

Transients are animals that do not appear to have a stable home range (Banks *et al.*, 1975; Jones, 1990). For some individuals dispersal may not be a one time event, but rather a regular, or chronic movement pattern. Jannett (1980) identifies several species of microtines where females disperse regularly in response to weaning their litters. Individuals within our populations have also differed in their propensity to move (Table A2.1). Constant, or periodic, dispersal is a movement pattern that is almost impossible to study through the use of static trap grids, yet this behaviour pattern may be more common than many suppose.

A2.4 THE LOCATION OF DISPERSERS

The fate of dispersers is important, especially in the light of Lidicker's (1975) contention that presaturation dispersers can increase their fitness by leaving the 'population'. For most species we have no idea how far animals move or what is the fitness of presaturation dispersers relative to non-dispersing animals. For the larger ground squirrels it is possible to visually monitor dispersal (Armitage and Downhower, 1974; Michener and Michener, 1977) or combine visual monitoring with radiotelemetry (Garrett and Franklin, 1988). With smaller species, large trapping grids can monitor some movements beyond the original home range (Krohne *et al.*, 1984; Extine and Scott, 1987), yet animals can still move beyond the grid boundaries. Radioactive tags can locate dispersing animals (Hilborn and Krebs, 1976; Appendix 3), but are only detectable over short distances. Radiotelemetry can provide reliable locations of highly mobile small mammals and has been successfully used to study dispersal. In water voles (*Arvicola terrestris*) dispersers have a higher rate of mortality than residents, and individuals differed in their dispersal movements; juvenile females moved long distances each day, while juvenile males moved at a very slow rate (Leuze, 1976). In *M. arvalis*, young from single mother nests dispersed farther than young from communal nests (Boyce and Boyce, 1988). Dispersal between winter communal groups of *M. pennsylvanicus* was predominately by males (Madison *et al.*, 1984; McShea, 1990b). These studies targeted specific segments of the population (i.e. juveniles, nestlings, winter groups) and were able to contrast movement patterns within the chosen group.

A2.5 DISPERSAL DISTANCE

The distance of home range shifts in voles, based on radiotelemetry, indicate that most dispersal does not involve movement over long distances (Madison, 1980, 1985; FitzGerald and Madison, 1983; Figure A2.2). These findings agree with those for Merriam's kangaroo rat (Jones, 1989). However, anecdotal notes from both radiotelemetry (1257 m in *Arvicola terrestris*, Leuze, 1980) and trapping studies (800 m in *Microtus xanthognathus*, Wolff and Lidicker, 1980) show that animals are capable of dispersing long distances. Laboratory studies demonstrate the same potential for long distance dispersal (mean maximum distance 5.8 km per running bout for adult *M. pennsylvanicus*, Lomolino, 1983).

Upon the disappearance of a monitored animal, we usually initiate a systematic search of the surrounding area to retrieve the valuable transmitter. Despite our frequent success at locating transmitters carried up to 300 m by predators, we have never located a dispersed vole farther than 100 m from its original home range (Figure A2.2). Therefore, for meadow voles, either most dispersal involves minor shifts in home range location, or our data represent just part of a bimodal distribution for dispersal distances (e.g. <100 m and >400 m). Over 85% of breeding dispersals recorded in a study of female *M. arvalis* were less then 160 m, with the farthest dispersal distance being 537 m (Boyce and Boyce, 1988). A 400+ mode cannot be practically evaluated because of the rather limited range of the transmitters. These range limitations can be reduced through the use of larger transmitter packages (possible in larger species such as sciurids),

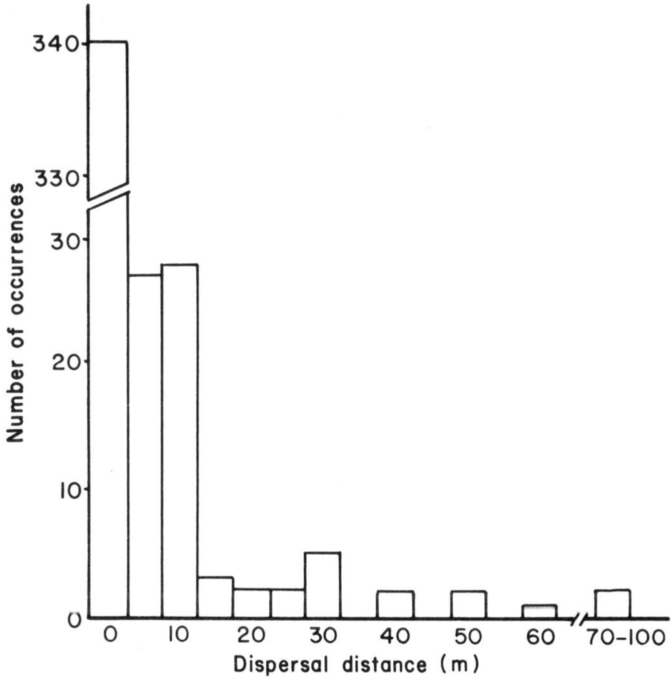

Figure A2.2 The distance moved between weekly estimates of activity centres for female meadow voles.

systematic search patterns and a shorter time span between subsequent position fixes.

A2.6 DISPERSAL FREQUENCY

Radiotelemetry studies to date indicate non-natal dispersal to be a rare event. Pine voles have only been recorded moving into adjacent home ranges (FitzGerald and Madison, 1983). Dispersal occurred in only three out of 56 adult meadow voles monitored for three or more weeks (Madison, 1985). Less than 5% of home range estimates for female meadow voles show shifts from the previous week (Figure A2.2). For woodrats (*Neotoma fuscipes*) 34 out of 37 animals monitored for a year showed no shift in their home ranges (Cranford, 1977). As the data now stand, stability and constancy of home range use is the dominant pattern for those age and gender groups studied.

This apparent stability may be the result of the animals sampled and the periods of study. Snow cover has been shown to have a releasing effect on meadow vole movement (Madison *et al.*, 1984), yet most studies detail only summer movements. Studies of small mammals have also concentrated on adults due to limitations of transmitter weight (Mineau and Madison, 1977; Cranford, 1977; McShea, 1990a), whereas studies that included juveniles indicate their dispersal movements are more frequent (Madison *et al.*, 1984; Boyce and Boyce, 1988). Boyce and Boyce (1988) also found social organization to influence breeding dispersal in *M. arvalis*, with single females dispersing and communal females not dispersing with each parturition.

A2.7 FUTURE STUDIES

Lidicker (1975, 1985) recognized that not all dispersal has the same proximate cause. He outlined four mechanisms for presaturation dispersal: ontogenetic, seasonal, colonizing and interference. Dispersal rates should be the sum total of these factors. Mutual tolerance, as measured through movement synchrony and co-habitation, is not readily observable through live trapping, yet is important for recording several of Lidicker's dispersal mechanisms. Radiotelemetry has an advantage over trapping by allowing the measurement of dynamic interactions between individuals. If a population or a local subgroup of animals has social cohesion (Bekoff, 1977), then dispersers should be identifiable prior to dispersal by exhibiting reduced interactions and movement synchrony with other members of the group. Any act that decreases cohesion (e.g. sexual maturation, removal of familiar neighbours) should change the movement patterns of animals and ultimately result in dispersal. The synchrony of movements between two animals has been termed 'dynamic interactions' and can be analysed via the correlation between subsequent positions for each animal (Dunn, 1979; MacDonald *et al.*, 1980).

The dispersal mechanisms of Lidicker (1985) can be further broken down in regard to their applicability to study through radiotelemetry; dispersals that involve a transition stage and dispersals that are catastrophic. During transitional dispersals, individuals should pass through a progression of stages that culminate in dispersal. Catastrophic dispersal events would not be prefaced by exploratory movements and should result in a dramatic, but brief, change in movement patterns. The impact of a predator on a population is usually catastrophic and should result in a sudden increase in the movements of individuals. Apparent dispersals in response to predators have been recorded during overwintering studies of meadow voles (Madison *et al.*, 1984; McShea, 1990a,b).

Transitional dispersals, such as ontogenetic dispersal from family units, occur over a longer time period. Young are originally found within their mother's home range, but show increased dispersion with increased age

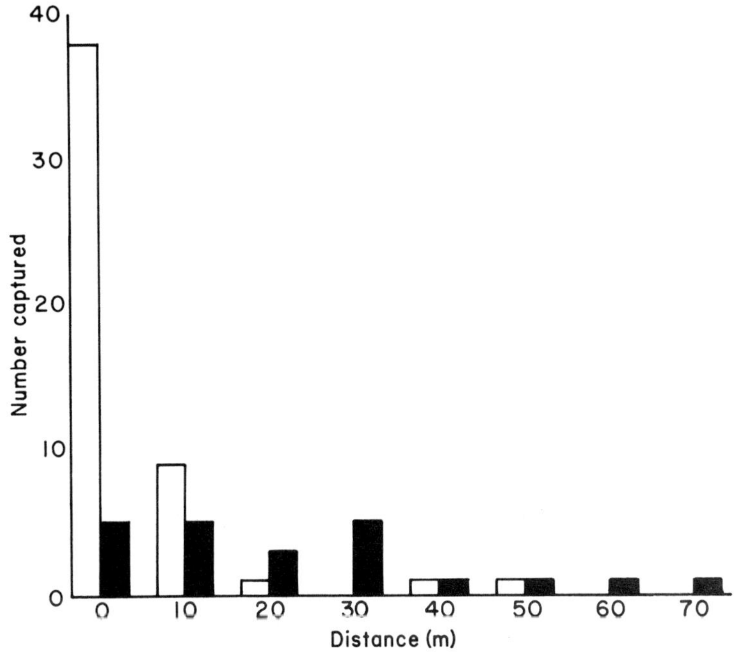

Figure A2.3 The position of young relative to their mother when first captured at less than 16 g (open bars) and when subsequently captured at over 20 g (dark bars). The female's mean home range radius (± 2 SD) was centred upon the trap closest to her centre of activity, and offspring capture locations were referenced to this position. The captured juveniles were originally marked as pups within the nests.

(Figure A2.3). These data are derived from the trap captures of juveniles, who were marked while pups within the nests of monitored females. It gives only a hint of the data that could be collected if both mother and offspring were monitored, but it does show the variation that exists between individuals. Overwintering associations between mother and offspring also revealed individual differences in dispersal, with male offspring delaying dispersal (Madison *et al.*, 1984).

Seasonal dispersal can be both transitional and catastrophic. It can result from the breakup of communal wintering groups with the onset of breeding or with the sudden flooding of overwintering home ranges. The addition of snow cover can serve as a trigger for dispersal into 'suboptimal' areas (Madison *et al.*, 1984; Figure A2.4). Spring floods have resulted in dispersal during several studies (Cranford, 1977; Webster, 1979).

The distinction between transitional and catastrophic is that transitional

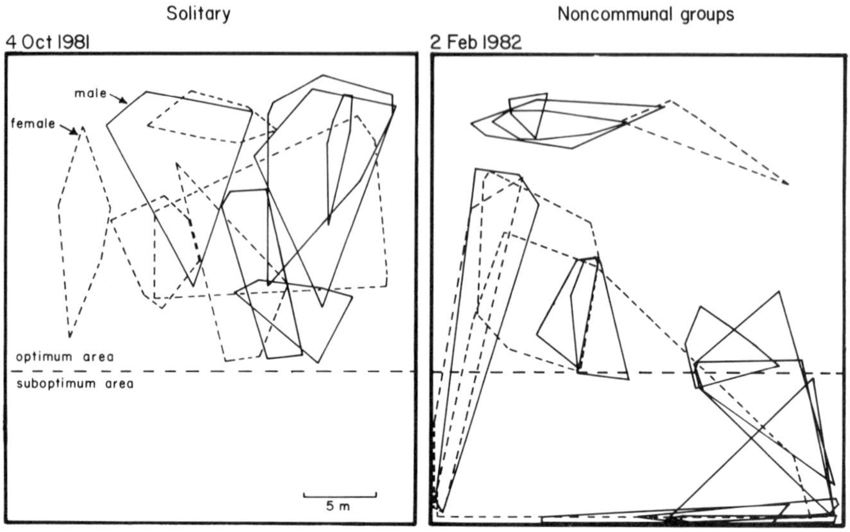

Figure A2.4 The movement of meadow voles into suboptimal area within an enclosure in Apalachin, New York (1981–82). Movement into the suboptimal area coincided with snow fall (see Madison *et al.*, 1984, for details).

events can be more easily predicted and monitored. Transitional dispersals can be described, and tests of proximate causes for dispersal (such as the cohesion hypothesis of Bekoff, 1977) can be examined through the rate that animals move through these stages. Catastrophic dispersers may not exhibit genetic and phenotypic differences from the entire population because the proximate cause of dispersal (i.e. a flooded field or a hungry predator) is independent of the animal's attributes.

A2.8 CONCLUSIONS

Radiotelemetry offers an excellent technique for the study of dispersal in unmanipulated populations. Present data indicate that dispersal is rare and usually occurs over short distances. However, these results may be due to limited studies, or a bimodal distribution of dispersal distances. The actual dispersal event itself is often overlooked in the discussion of small mammal dispersal, but we believe 'how' an animal disperses will provide information on the cause of the dispersal and influence the ability of the animal to survive its shift in home range. We argue against the random monitoring of animals and suggest a focus on transitional dispersals as a means to obtain information on the causes of dispersal in small mammal populations.

REFERENCES

Ackerman, B., Leban, F., Samuel, M.O. and Garton, E.O. (1990) Users manual for program home range. University of Idaho Forestry, Wildlife and Range Experimental Station Technical Report 15, 80 pp.

Armitage, K.B. and Downhower, J.F. (1974) Demography of yellow-bellied marmot populations. *Ecology*, **55**, 1233–45.

Baird, D.D. and Birney, E.C. (1982) Characteristics of dispersing meadow voles *Microtus pennsylvanicus*. *American Midland Naturalist*, **107**, 262–83.

Banks, E., Brooks, R. and Schnell, J. (1975) A radiotracking study of home range and activity of the brown lemming (*Lemmus trimucronatus*). *Journal of Mammalogy*, **56**, 888–901.

Beacham, T.D. (1979) Size and growth characteristics of dispersing voles, *Microtus townsendii*. *Oecologia*, **42**, 1–10.

Bekoff, M. (1977) Mammalian dispersal and the ontogeny of individual behavioral phenotypes. *American Naturalist*, **111**, 715–32.

Boyce, C.C.K. and Boyce, J.L. III (1988) Population ecology of *Microtus arvalis* II. Natal and bredding dispersal of females. *Journal of Animal Ecology*, **57**, 723–36.

Brochu, L., Caron, L. and Bergeron, J.-M. (1988) Diet quality and body condition of dispersing and resident voles (*Microtus pennsylvanicus*). *Journal of Mammalogy*, **69**, 704–11.

Burt, W.H. (1943) Territoriality and home range concepts as applied to mammals. *Journal of Mammalogy*, **24**, 346–52.

Cary, J. (1990) SEAS – spacial ecology analysis programs. Department of Wildlife and Ecology, University of Wisconsin, Madison, WI.

Cranford, J.A. (1977) Home range and habitat utilization by *Neotoma fuscipes* as determined by radiotelemetry. *Journal of Mammalogy*, **58**, 165–72.

Desy, E.A., Batzli, G.O. and Liu, J. (1989) Comparison of vole movements assessed by live trapping and radiotracking. *Journal of Mammalogy*, **70**, 652–6.

Dixon, K.R. and Chapman, J.A. (1980) Harmonic mean measure of animal activity area. *Ecology*, **61**, 1040–4.

Dobson, S.F. (1981) An experimental examination of an artificial dispersal sink. *Journal of Mammalogy*, **62**, 74–81.

Dunn, J.E. (1979) A complete test for dynamic territorial interaction, in *Proceedings of the Second International Conference on Wildlife Biotelemetry* (ed. F.M. Long), Laramie, Wyoming.

Extine, D.D. and Scott, I.J. (1987) Dispersion and habitat occupancy of the beach mouse, *Peromyscus polionotus niveiventris*. *Journal of Mammalogy*, **68**, 297–304.

FitzGerald, R.W. and Madison, D.M. (1983) Social organization of a free-ranging population of pine voles, *Microtus pinetorum*. *Behavioural Ecology and Sociobiology*, **13**, 183–7.

Gaines, M.S. and McClenaghan, L.R. Jr (1980) Dispersal in small mammals. *Annual Review of Ecology and Systematics*, **11**, 163–96.

Gaines, M.S., Visas, A.M. and Baker, C.L. (1979) An experimental analysis of dispersal in fluctuating vole populations: demographic parameters. *Ecology*, **60** 814–28.

Garrett, M.G. and Franklin, W.L. (1988) Behavioral ecology of dispersal in the black-tailed prairie dog. *Journal of Mammalogy*, **69**, 236–50.

Gregory, M.J. and Cameron, G.N. (1988) Examination of socially induced dispersal in *Sigmodon hispidus*. *Journal of Mammalogy*, **69**, 251–60.

Hackett, D.F. and Trevor-Deutsch, B. (1982) Radiotelemetric assessment of grid trapping technique in the study of the eastern chipmunk (*Tamias striatus*). *Biotelemetry Patient* Morit., **9**, 213–26.

Hardy, A.R. and Taylor, K.D. (1980) Radio tracking of *Rattus norvegicus* on farms, in *A Handbook on Biotelemetry and Radio Tracking* (eds C.J. Amlaner and D.W. Macdonald), Pergamon Press, Oxford, pp. 657–66.

Hilborn, R. (1975) Similarities in dispersal tendencies among siblings in four species of voles (*Microtus*). *Ecology*, **56**, 1221–25.

Hilborn, R. and Krebs, C.J. (1976) Fates of dispersing individuals in fluctuating populations of *Microtus townsendii*. *Canadian Journal of Zoology*, **54**, 1507–18.

Jannett, F.J. Jr (1980) Social dynamics of the montane vole, *Microtus montanus*, as a paradigm. *The Biologist*, **62**, 3–19.

Jones, E.N. (1990) Effects of forage availability on home range and population density of *Microtus pennsylvanicus*. *Journal of Mammalogy*, **71**, 382–9.

Jones, E.N. and Sherman, L.J. (1983) A comparison of meadow vole home ranges derived from grid trapping and radiotelemetry. *Journal of Wildlife Management*, **47**, 558–61.

Jones, W.T. (1989) Dipersal distance and the range of nightly movements in Merriam's kangaroo rats. *Journal of Mammalogy*, **70**, 27–34.

Krebs, C.J., Keller, B.L. and Tamarin, R.H. (1969) *Microtus* population biology: demographic changes in fluctuating populations of *M. ochrogaster* and *M. pennsylvanicus* in southern Indiana. *Ecology*, **50**, 587–607.

Krohne, D.T., Dubbs, B.A. and Baccus, R. (1984) An analysis of dispersal in an unmanipulated population of *Peromyscus leucopus*. *American Midland Naturalist*, **112**, 146–56.

Leuze, C.C.K. (1976) Social behavior and dispersion in the water voles, *Arvicola terrestris* (Lacepede). Unpublished Ph.D. Thesis, University of Aberdeen.

Leuze, C.C.K. (1980) The application of radio tracking and its effect on the behavioral ecology of the water voles, *Arvicola terrestris* (Lacepede), in *A Handbook on Biotelemetry and Radio Tracking* (eds C.J. Amlander Jr and D.W. Macdonald), Pergamon Press, Oxford, pp. 361–66.

Lidicker, W.Z. Jr (1975) The role of dispersal in the demography of small mammals, in *Small Mammals: their productivity and population dynamics* (eds F.B. Golley, K. Petrusewicz and L. Ryszkowski), Cambridge University Press, London, pp. 103–34.

Lidicker, W.Z. Jr (1985) An overview of dispersal in non-volant small mammals, in *Migration: mechanisms and adaptive significance* (ed. M.A. Rankin), Contributions in Marine Science, Suppl. 27, pp. 359–75.

Lomolino, M.V. (1983) Island biogeography, immigrant selection and body size of mammals on islands. Unpublished Ph.D. Dissertation, State University of New York, Binghamton, 250 pp.

Macdonald, D.W., Ball, F.G. and Hough, N.G. (1980) The evaluation of home range size and configuration using radio tracking data, in *A Handbook of Biotelemetry and Radio Tracking* (eds C.J. Amlander and D.W. Macdonald), Pergamon Press, Oxford, 804 pp.

Madison, D.M. (1978) Movement indicators of reproductive events among female meadow voles, as revealed by radiotelemetry. *Journal of Mammalogy*, **59**, 835–43.

Madison, D.M. (1980) Movement types and weather correlates in free-ranging meadow voles, in *Proceedings of the Fourth Eastem Pine and Meadow Vole Symposium* (ed. R.E. Byers), pp. 34–42.

Madison, D.M. (1985) Activity rhythm and spacing, in *Biology of New World Microtus* (ed. R.H. Tamarin), American Society of Mammalogists, Special Publication No. 8, pp. 373–419.

Madison, D.M., FitzGerald, R.W. and McShea, W.J. (1984) Dynamics of social nesting in overwintering meadow voles (*Microtus pennsylvanicus*): possible consequences for population cycling. *Behavioural Ecology and Sociobiology*, **15**, 9–17.

Madison, D.M., FitzGerald, R.W. and McShea, W.J. (1985) A user's guide to the successful radiotracking of small mammals in the field, in *Proceedings of the Fifth International Conference on Wildlife Radiotelemetry* (eds R.W. Weeks and F.M. Long), Laramie, Wyoming, pp. 28–39.

McShea, W.J. (1990a) Social tolerance and proximate mechanisms of dispersal between overwintering groups of meadow voles (*Microtus pennsylvanicus*). *Animal Behaviour*, **39**, 346–51.

McShea, W.J. (1990b) Predation and its potential impact on the behavior of microtine rodents, in *Social Systems and Population Cycles in Voles* (eds R.H. Tamarin, R. Ostfeld, S. Pugh and G. Bujalska), Birkhauser Verlag, Basel, pp. 101–10.

Michener, G.R. and Michener, D.R. (1977) Population structure and dispersal in Richardson's ground squirrels. *Ecology*, **58**, 359–68.

Mineau, P. and Madison, D.M. (1977) Radiotracking of *Peromyscus leucopus*. *Canadian Journal of Zoology*, **55**, 465–8.

Rado, R. and Terkel, J. (1989) A radio-tracking system for subterranean rodents. *Journal of Wildlife Management*, **53**, 946--9.

Savidge, I.R. (1974) Social factors in dispersal of deer mice (*Peromyscus maniculatus*) from their natal site. *American Midland Naturalist*, **91**, 395–405.

Stuwe, M. and Blohowiak, C. (1985) McPAAL – Micro-computer programs for the analysis of animal locations. Conservation and Research Center, Smithsonian Institution, Front Royal, VA.

Swihart, R.K. and Slade, N.A. (1985a) Influence of sampling interval on estimates of home range size. *Journal of Wildlife Management*, **49**, 1019–25.

Swihart, R.K. and Slade, N.A. (1985b) Testing for independence of observations in animal movements. *Ecology*, **66**, 1176–84.

Tamarin, R.H. (1980) Dispersal and population regulation in rodents, in *Biosocial*

Mechanisms of Population Regulation (eds M.N. Cohen, R.S. Malpass and H.G. Klein), Yale University Press, New Haven, Connecticut, pp. 117–33.

Webster, A.B. (1979) A radiotelemetry study of social behavior and activity of free-ranging meadow voles, *Microtus pennsylvanicus*. Unpublished M.Sc. Thesis, University of Guelph, Guelph, Canada.

Wolff, J.O. and Lidicker, W.Z. Jr (1980) Population ecology of the taiga voles. *Behavioural Ecology and Sociobiology*, 9, 237–40.

Appendix 3. The use of radioisotopes in the study of dispersal: with a case study

N.C. STENSETH AND W.Z. LIDICKER JR

A3.1 INTRODUCTION

To mark a sufficient number of individuals whose movements can be followed without this effort becoming prohibitively large, represents a major problem in the study of animal dispersal. Ear tags, leg rings or toe clips can be used for obtaining a sufficient sample size of marked individuals (indeed these represent the most commonly used techniques in population studies on small mammals). However, after dispersal, marked individuals will be difficult or impossible to locate due to the dilution effect of distance. Furthermore, those using capture–recapture techniques are generally forced to focus their attention on study areas of very limited size, a situation not favourable for obtaining quantitatively reliable information on dispersal. (See Appendix 1 for a thorough discussion of a situation in which this sort of bias can be severe.) Finally, with these techniques individuals have to repeatedly enter traps and be subject to stressful handling.

Radiotelemetry (Appendix 2) has the advantage of making it easy to locate 'marked' individuals, and of not requiring re-trapping and additional handling. However, the method is not designed for large-scale dispersal studies, and is often characterized by such small sample sizes that the information obtained cannot be used readily for estimating dispersal rates or for deriving frequency distributions of dispersal distances. Those using the technique of radiotelemetry are therefore often limited to an intensive examination of individual movements. Yet it is generally the case that we want to be able to say something about classes of individuals. Hence, we need techniques that permit the evaluation of many individuals.

Marking entire populations or substantial segments thereof by radioisotopes circumvents some of the difficulties inherent in the use of live traps or radiotelemetry. Indeed, field studies of animals using radioactive isotope markers have been popular since Griffin (1952) outlined the technique, called attention to the potential of radioisotope marking and provided guidelines for designing relevant studies. Pendleton (1956) also reviewed the use of radioisotopes in ecological studies. A comprehensive and extremely useful review of isotope marking in small mammals was published by Bailey *et al.* (1973).

An important attribute of studies using radioactive markers is that a large sample can fairly readily be obtained making it easier to infer the desired characteristics of classes of individuals. A second important advantage is that marked individuals are relatively easy to locate even after dispersal.

Those planning to use radioisotopes can take advantage of a wide variety of isotopes offering different levels of toxicity and detectability and having widely differing physical half-lives (International Committee on Radiation Protection, 1959; Table A3.1). The physical half-life becomes a secondary consideration, however, in those cases where organisms are labelled internally. In such circumstances, the radionuclides enter into various physiological and biochemical processes in the body, and hence, their retention times in the living organism (measured as biological half-life) may be very different from the physical half-life of the isotope. If the biological half-lives of specific isotopes are unknown for the subject species, they should be measured *a priori* or estimated from data on closely related species. Some isotopes will localize in specific organs so that only these organs need be collected in the field from dead individuals if it is necessary to test for the presence of the experimental isotope in the laboratory. Finally, the extent to which an isotope is assimilated by an organism depends on whether it is ingested in food or injected directly into the body cavity or circulating system. In other words, it is necessary to determine the behaviour of specific isotopes in specific species before any field use of the isotopes is attempted.

Organisms can also be labelled with combinations of isotopes. Using combinations of two, three, or more isotopes in the same individual greatly increases the number of individuals that can be uniquely marked in a given population. For gamma-emitting isotopes, identification of combinations of radionuclides in the same individual is made possible by the fact that each isotope has a characteristic spectral signature. That is, each isotope shows peak energy emissions at different MeV (or KeV) levels (Table A3.1). The method has become feasible for field biologists because whole body γ-emission counters are available for use under field conditions (Tamarin *et al.*, 1983). Injecting rather than feeding the isotopes will assure that appropriate levels of each for detection will be assimilated.

If the number of isotopes available is equal to n and k is the number used in each individual, the maximum number of possible unique combinations will be

Table A3.1 Radioactive isotopes used in studies of movements in small mammals. Physical half-lives and diagnostic main energy emission peaks are also given along with references to users. d, days; y, years

Isotope	*Half-life*	*Energy peaks (KeV)*	*References*
Antimony-124	60.2 d	602,1690,723	Punt and van Nieuwenhoven (1957)
Antimony-125	2.73 y	428,463,601,636	Sheridan and Tamarin (1986), Scott and Tan (1985), Crabtree *et al.* (1989)
Cadmium-109	463 d	88	Pelton and Marcum (1977), Crabtree *et al.* (1989)
Calcium-45	163 d	*	Twigg and Miller (1963), Rongstad (1965), Meslow and Keith (1968)
Carbon-14	5730 y	*	Soldatkin *et al.* (1961)
Cerium-144	285 d	133	Pelton and Marcum (1977)
Cesium-134	2.06 y	563,569,605,795	Pelton and Marcum (1977), Crabtree *et al.* (1989), Eie (1991)
Cesium-137	30 y	661	Tamarin *et al.* (1983), Scott and Tan (1985)
Chromium-51	27.7 d	320	Myllymäki *et al.* (1971), Tamarin *et al.* (1983), Sheridan and Tamarin (1986)
Cobalt-57	271.8 d	14,122,136	Crabtree *et al.* (1989), Eie (1991)
Cobalt-58	70.9 d	810	Sheridan and Tamarin (1986), Scott and Tan (1985)
Cobalt-60	5.27 y	1173,1332	Godfrey (1953), Linn and Shillito (1960), Barbour (1963), Hamar *et al.* (1964), Pelton and Marcum (1977), Tamarin *et al.* (1983), Scott and Tan (1985), Sheridan and Tamarin (1986)
Gold-198	2.3 d	97,180,204,215	Kaye (1960, 1961), Cope *et al.* (1961)
Iodine-131	8 d	364	Gifford and Griffin (1960), Johanningsmeier and Goodnight (1962), Gentry *et al.* (1971), Myllymäki *et al.* (1971) Bailey *et al.* (1973), Hoff (1981), Kingsley *et al.* (1987)
Iron-59	44.3 d	1079,1292	Griffin (1952), Gentry *et al.* (1971), Tamarin *et al.* (1983), Scott and Tan (1985), Sheridan and Tamarin (1986), Eie (1991)

Table A3.1 (cont'd)

Isotope	*Half-life*	*Energy peaks (KeV)*	*References*
Manganese-54	312.2 d	835	Pelton and Marcum (1977), Tamarin *et al.* (1983), Scott and Tan (1985), Sheridan and Tamarin (1986), Crabtree *et al.* (1989)
Niobium-95	35 d	766	Sheridan and Tamarin (1986)
Phosphorus-32	14.3 d	*	Jenkins (1954), Miller (1957), Hamar *et al.* (1963), Shura-Bura *et al.* (1960), Rudenchik (1963), Stoddart (1970), Hoff (1981)
Scandium-46	83.8 d	889,1121	Tamarin *et al.* (1983), Scott and Tan (1985), Sheridan and Tamarin (1986)
Selenium-75	120 d	121,136,265, 280,401	Frigerio and Eisler (1968), Hoff (1981), Tamarin *et al.* (1983), Scott and Tan (1985), Sheridan and Tamarin (1986), Eie (1991)
Silver-110m	249 d	658,885	Linn and Shillito (1960), Tamarin *et al.* (1983), Scott and Tan (1985)
Strontium-85	64.8 d	514	Scott and Tan (1985), Sheridan and Tamarin (1986)
Strontium-89	51 d	910 (weak)	Shura-Bura and Kharlamov (1961)
Sulphur-35	87.3 d	*	Rudenchik *et al.* (1967), Korneyev (1967a,b), Dickman *et al.* (1983)
Tantalum-182	114.4 d	68,1121,1221	Graham and Ambrose (1967), Hilborn and Krebs (1976), Jannett (1978), Hoff (1981), Genoud (1984), Salvioni (1988)
Yterbium-88	106.6 d	898,1836	Scott and Tan (1985), Sheridan and Tamarin (1986)
Zinc-65	244 d	1115.5	Griffin (1952), Nellis *et al.* (1967), Gentry *et al.* (1971), Pelton and Marcum (1977), Wolff and Holleman (1978), Kruuk *et al.* (1980), Tamarin *et al.* (1983), Scott and Tan (1985), Sheridan and Tamarin (1986), Crabtree *et al.* (1989), Eie (1991)

* Beta-emitting isotope.

$$\binom{n}{k} = \frac{n!}{k!\,(n-k)!} \tag{A3.1}$$

All of these may not, of course, be distinguishable spectrographically.

This chapter briefly surveys the use of radioisotope marking in population studies of small mammals. Then we describe in some detail a Fennoscandinavian study on dispersal using radioisotope marking. In this case study we attempt to provide sufficient details as well as guides to the relevant literature to enable others to perform similar types of analysis.

A3.2 A BRIEF SURVEY OF RADIOISOTOPE USE IN POPULATION STUDIES

A3.2.1 General

Most of the early studies using radioactive tags on small mammals focused on the determination of home ranges, on measuring activity rhythms, and on assessing population densities. Mihok *et al.* (1985a) provide a brief guide to the Russian literature in which similar radioactive marking techniques have been used.

(a) Home range measurements

One of the earliest attempts to determine home range size using radioactive tags was that of Godfrey (1953, 1954a). She used leg rings tagged with ^{60}Co to find nests and to follow movements of 23 *Microtus agrestis*. The technique was then extended (Godfrey, 1954b) to the study of subterrenean moles (*Talpa europea*). Also at that time, Jenkins (1954) discovered that by feeding ^{32}P to a female lemming, its faeces were labelled permitting the tracking of its movements, and what is more, its offspring became radioactive. These two findings were subsequently widely exploited by students of animal movements. A recent example of a home range study is provided by Jannett (1978), who investigated aspects of territorial behaviour in *Microtus montanus* with the use of ^{182}Ta wires placed subcutaneously.

Eie (1991), together with his collaborators, used an innovative approach to study home range and movements in the European marten (*Martes martes*), a species notoriously difficult to live-trap. Because it was not feasible to trap and radioactively label a large enough sample of individuals to make reasonable estimates of movements, he allowed the martens to label themselves. This represents an alternative to the usual situation where the investigator first catches and then chooses labels for each individual. They set out combinations of five carefully selected radionuclides (Table A3.1) placed in baits and located in five different sections of a large grid. Later, when the animals were killed by fur trappers, it was possible to analyse the carcasses for the presence of specific radionuclides which would in turn identify the section of the grid on which that particular animal had fed.

(b) Daily activity rhythms

Radionuclides have also been used effectively to study daily activity rhythms. By monitoring the passage of labelled individuals near radiation-detecting instruments, good records of when individuals are active through 24 h cycles can be registered. Examples of relatively recent investigations of this type include Hamar *et al.* (1974), Karulin *et al.* (1976), Airoldi (1979), Genoud (1984), and Salvioni (1988).

(c) Density estimation

A third area of investigation utilizing radioactive tags is that of density estimation. The basic approach is to mark a sample of individuals, allow them to return to the population, and then capture a second sample to determine the proportion of labelled to unlabelled individuals in the population, that is, a standard Lincoln Index approach (e.g. Seber, 1982). Myllymäki *et al.* (1971) used triple isotope marking to calibrate standard grid-trapping techniques used for estimating population size in *Microtus agrestis*. Similarly, Kruuk *et al.* (1980) used ^{65}Zn to label the faeces of European badgers (*Meles meles*). By collecting faeces and finding the proportion of labelled to unlabelled, they could estimate badger numbers.

(d) Mother–offspring relations

In recent years, researchers have taken advantage of the knowledge that radioactive labels are passed from mothers to their offspring either during pregnancy or lactation. By knowing the family relationships of juveniles, much can be learned about social behaviour and reproductive success of different females. For a variety of gamma-emitting radionuclides, Morimoto *et al.* (1985) determined appropriate dosages, whole body elimination rates, and transfer rates between mothers and litters in captive *Microtus pennsylvanicus*.

Transfer of radioactive labels between adult males and females in seminal fluid has also been successfully used by Scott and Tan (1985) to assess paternity and mating success in the marsupial *Antechinus stuartii*. They labelled 12 adult males with 12 different radionuclides at the beginning of the breeding season, which is brief in this species, and then determined copulation frequency for each male by detecting the isotopes in adult females.

Tamarin *et al.* (1983) used eight different radionuclides alone and in combination to determine matrilineal kinship of recruits into a population to *M. pennsylvanicus*, and hence to measure differential success among adult females. They labelled six pregnant females and were able to detect seven litters under field conditions. Sheridan and Tamarin (1986) further developed the multiple isotope approach to include 13 isotopes with combinations of two being injected into pregnant and lactating females. A total of 78 unique combinations would thereby be potentially available. They were able to label 35 females, 18 of which successfully recruited 40 litters varying in size from 1 to 8. These juveniles represented 62% of all young recruited in one breeding season. From our current perspective, it is unfortunate that they apparently did not take advantage of this pool of marked young to study dispersal.

A3.2.2 Dispersal

Hilborn and Krebs (1976) were the first to address dispersal explicitly by using radioactive isotopes. They glued ^{182}Ta wires to the ear tags of 219 *Microtus townsendii*, and assessed dispersal rates by whether or not the radioactive tags remained on their study plot. They found that tags could be detected as much as two feet underground, permitting the location of some, but possibly not all, dead voles. However, their method could not distinguish between losses from predation and dispersal. Crabtree *et al.* (1989), describe a method for radioactively marking the scats of carnivores in order to trace movement patterns. They used six isotopes in subcutaneous implants and claimed that scats were labelled for a 6-month period.

Wolff and Holleman (1978) marked three pregnant females of *M. xanthognathus* with ^{65}Zn and then were able to follow movements of the radioactively tagged juveniles. Even though this particular study was limited by very small sample sizes (three pregnant females), it represented an important start in the use of radionuclides for investigating mother–progeny relations, and indicated its potential for the study of dispersal. Dickman *et al.* (1983) applied ^{35}S in a similar way in order to study dispersal of juveniles from known mothers in two rodent species (*Mus musculus*, *Rattus fuscipes*) and one marsupial (*Antechinus stuartii*).

Myllymäki, Hansson, and Hoff completed a major study on dispersal in *Microtus agrestis* in which radioactive labelling was employed. This study is discussed in detail in the next section, and also serves to illustrate a diffusion model for dispersal behaviour.

Using a slightly different approach, Iversen and Turner (1976a,b) and Mihok *et al.* (1985a,b) studied the small rodent community (*Clethrionomys gapperi* and *Microtus pennsylvanicus*) on a field exposed to ^{137}Cs source. Even though they did not study dispersal, such a set-up could potentially be used for that purpose. Live-trapping around such a radioactive source establishes the level of radioactivity found in individuals living at various distances from the radioactive source. Given this, dispersers may be identified by the fact that they will be radioactive at a level appropriate to somewhere else, and an approximate movement distance can then be determined.

A3.3 THE MYLLYMÄKI–HANSSON–HOFF STUDY

A3.3.1 The study

Arvo Myllymäki and Lennart Hansson performed what we believe to be the first large-scale experiment applying radioisotopes in order to study dispersal in small mammals. Their data were analysed by Hoff (1981). Unfortunately that report is inaccessible to non-Scandinavian speaking people; hence, we describe their findings in some detail below. We take this opportunity to thank Arvo Myllymäki, Lennart Hansson, and Jann Morten Hoff for permission to summarize their findings.

Their project involved an investigation of dispersal in *Microtus agrestis* and was designed so as to utilize a source–sink perspective (Pulliam, 1988). Although several other species of small mammals were present on their study area, none was present in sufficient numbers to suspect that interspecific interactions affected the results.

A3.3.2 Study area

The experiment was carried out at Hiirola, 20 km north of Mikkeli in southeastern Finland during a peak rodent year, 1976 (Figure A3.1). The experimental field consisted of three different habitat types: (1) an abandoned field of 2.4 ha, on mineral soil with continuous cover and abundant nutritious forbs. It was considered to be optimal habitat (patch A in Figure A3.1). (2) Close to this optimal habitat was an abandoned previously grazed meadow on peat soil covering an area of 2.2 ha and dominated by *Deschampsia caespitosa*; this was considered suboptimal habitat for field voles (e.g. Myllymäki, 1977; Hansson, 1977) (patch C in Figure A3.1). (3) The optimal and suboptimal habitats were surrounded by early successional clearcuts with grasses and sedges, dominated by *Calamagrostis* spp. and young pines (habitat B in Figure A3.1). This habitat type resulted from clearcutting of spruce forest and was considered marginal habitat or transition habitat (*sensu* Stenseth, 1983), since reproduction was probably insufficient to maintain a consistent population density above zero in the absence of immigration.

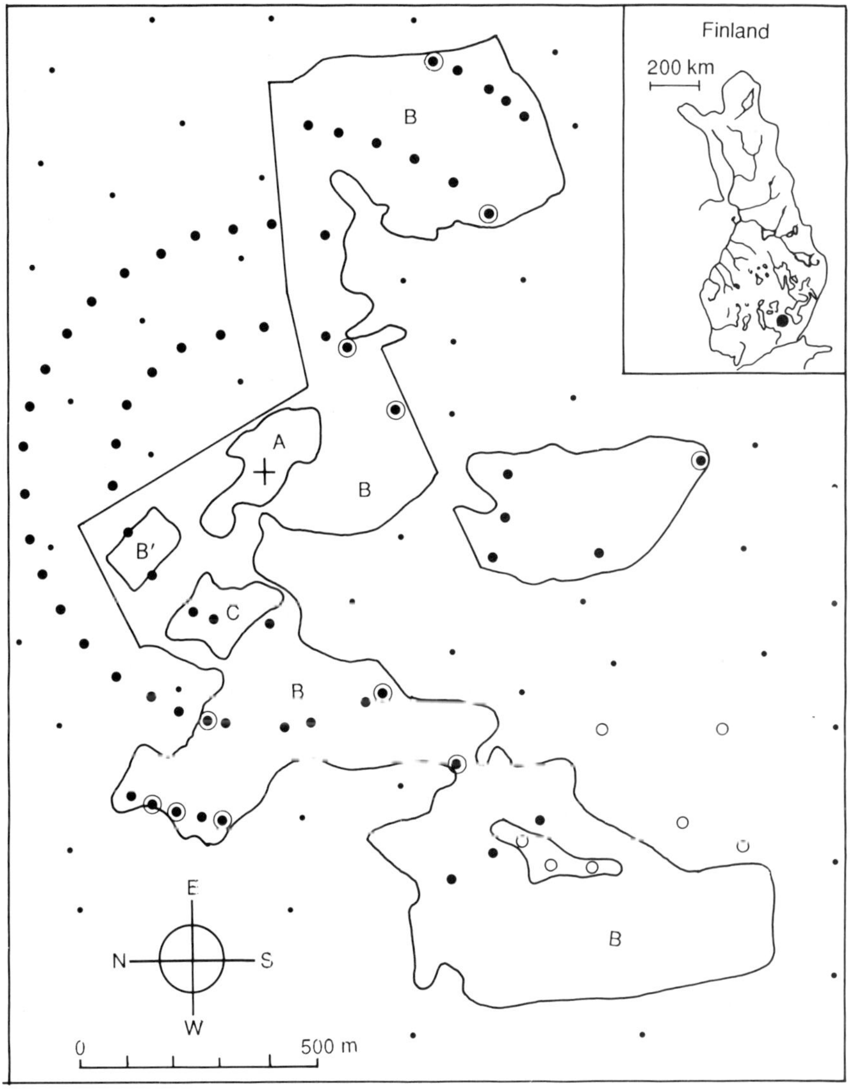

Figure A3.1 Study area at Hiirola, 20 km north of Mikkeli, Finland, used in the Myllymäki–Hansson–Hoff investigations. Locations of trapping grids, distribution of major community types, and placement of assessment traps are shown. A, optimal habitat patches and grid; B, marginal habitat (clearcuts dominated by grasses, especially *Calamagrostis* spp.); B′, trap grid in marginal habitat; C, sub-optimal habitat dominated by *Deschampsia caespitosa*; •, snap-trap station not on a forest edge; ◉, snap-trap station at a forest edge; ○, deciduous forest; *, spruce forest.

A3.3.3 Method of sampling and marking the population

Live-trapping using Ugglan traps (e.g. Hansson, 1967) was performed (6–28 July) in the entire optimal and suboptimal habitats (patches A and C in Figure A3.1) in addition to a 2 ha large clearcut area (patch B′ in Figure A3.1). Traps were placed in a grid-like pattern with 15 m between each station. Traps were checked every 2 h during the daytime, and between trapping periods, they were locked open. Each individual caught was weighed and marked by toe clipping. Location on the grid was recorded upon each capture.

Animals caught in the optimal habitat (patch A in Figure A3.1) were additionally marked by using radioactive isotopes (7–15 July) of iodine (^{131}I), selenium (^{75}Se), phosphorus (^{32}P) and tantalum (^{182}Ta). Thus, only the source population (optimal habitat) was radioactively labelled. A major part of the population in this habitat was marked by radioactive I and P by spreading pieces of apples soaked in a radioactive liquid over the area. Non-lactating pregnant females were marked with Se isotopes by feeding them radioactive food in cages for one day. Lactating females were marked with a wire of radioactive tantalum placed underneath the skin of their backs. The ^{131}I, ^{75}Se, and ^{182}Ta isotopes could be detected throughout the experimental period, whereas individuals marked with the ^{32}P isotope could only be rediscovered for a short length of time; after a month there was insufficient activity left in the animals (see half-life values in Table A3.1).

In order to study the spreading of the individuals out from the optimal habitat (patch A in Figure A3.1), four (incomplete) concentric circles with snap-traps were located around the optimal habitat at distances of 300, 500, 700, and 900 m, respectively (Figure A3.1). Each snap-trap station was a small quadrant of 15 × 15 m with three snap-traps in each corner (Myllymäki *et al.*, 1971). Pieces of apple were used as bait. Traps were left set between each trapping period (Table A3.2) in order to maximize the chances of intercepting dispersers. Trapping on these lines was performed from 26 July to 9 September. The entire trapping schedule is summarized in Table A3.2.

At the end of the experiment, the optimal habitat patch was trapped intensively using snap-traps (1–6 September). A grid with 7 m between stations and four traps at each station was used. The results of this experiment relevant in our context are given in Table A3.3.

A3.3.4 Pattern of dispersal

(a) Methods

The rate of dispersal may be estimated on the basis of how individuals marked in the optimal habitat are distributed among the catches on the

Table A3.2 Trapping schedule for the Myllymäki–Hansson–Hoff experiment at Hiirola

Trapping session	*Dates*
A. Kill-trapping in optimal habitat	
1	1–2 Sept.
2	2–3 Sept.
3	3–4 Sept.
4	4 Sept.
5	5 Sept.
6	6 Sept.
B. Kill-trapping on concentric trap-lines	
Period 1	
1	26 July
2	27 July
3	28 July
4	29 July
5	31 July
6	1 Aug.
7	2 Aug.
8	3 Aug.
9	4 Aug.
10	7 Aug.
Period 2	
11	9 Aug.
12	16 Aug.
13	1–3 Sept.
14	4 Sept.
15	5 Sept.
16	7 Sept.
17	9 Sept.

concentric trap lines of kill-traps. This presumes, of course, that catches reflect densities; hence, we make this assumption.

Assuming (as we do) that dispersal out from the optimal habitat can be viewed as a diffusion process, the rate of dispersal will then be a parameter estimated by this model and based on the trapping data. Previously, dispersal in small mammals has been viewed as such a diffusion process by Skellam (1951); for reference to general technical treatments, see Ricciardi (1977), Okubo (1980), and Appendix 1.

Individuals starting from a central point and moving outwards independently of each other over a small distance, ε, in a random direction during a small interval of time, $\triangle t$, constitute a diffusion process when $\triangle t$ and ε

Table A3.3 Distribution of sexes and reproductive classes among the trapped voles (*Microtus agrestis*), Hiirola study site

	Males			*Females*			*Ages combined*			*Sex ratios*
Habitat	*Juv.*	*Subad.*	*Ad.*	*Juv.*	*Subad.*	*Ad.*	*Males*	*Fem.*	*Total*	*(% males)*
Optimal, period 1	86	14	40	75	1	217	140	293	433	32.3
Optimal, period 2	45	16	37	34	4	194	98	232	330	29.7
Suboptimal	16	6	37	20	3	53	59	76	135	43.7
Marginal	11	1	15	12	2	27	27	41	68	39.7
Optimal, kill-trap	110	138	24	83	150	102*	272	335	607	44.7
Marginal kill-trap circles	56	122	180	65	37	177	358	279	637†	56.2

* 15 of these were post-reproductive.
† 170 additional individuals were caught but their sex, etc., could not be determined due to decomposition.

approach zero. If individuals have the same probability of moving in any direction, Pielou (1977), among others, has shown that the probability distribution of the individual distances, F_s, moving from the starting point is, after time t, given by

$$\varphi\,(F_s;\ t) = \frac{2F_s}{a^2 t}\, e^{-F_s^2/(a^2 t)} \tag{A3.2}$$

where a^2 is the variance of distance moved from the centre per unit of time.

The parameter a can be estimated as follows: let N_0 be the number of individuals marked on the optimal habitat, t_j the time between marking and recapture number j at the circles with snap-traps, n_j the number of marked individuals recaptured at t_j; F_{kj} the number of marked individuals recaptured per trap-station at circle k and at time t_j; r_k the distance between circle k and the centre; and $f_{kj} = F_{kj}k_iF_{ij}$. Then,

$$E(F_{kj}) = \varphi(r_k;\ t_j) \tag{A3.3}$$

$$E(F_{kj}) = N_0\ \varphi(r_k;\ t_j) \tag{A3.4}$$

and

$$(F_{kj}/F_{ij}) = \varphi\,(r_k; t_j)/\varphi\,(r_i;\ t_j) = (r_k/r_i)\, e^{-(r_k^2 - r_i^2)/(a^2 t)} \tag{A3.5}$$

for $k \neq i$. The letter E in equations (A3.3) and (A3.4) denotes the expectations of capture at each trap station. From these formulae we have

$$\hat{a} = [(r_k^2 - r_i^2)/t_j \ (\ln r_k - \ln r_i - \ln F_{kj} + \ln F_{ij})]^{1/2}. \qquad \text{(A3.6)}$$

It is sufficient to have two trap rings in order to estimate a. Since there are four rings, we will have six different estimates of a for each period. The final estimate of a, therefore, is given as a weighted mean of the six original estimates; weighting factors are $F_{kj} + F_{if}$. This will give most weight to the a values estimated on the basis of the largest data sample. If a is estimated on the basis of several visitations, the final estimate will be the weighted mean of those from each visitation; we therefore use n_j as weighting factors.

From Pielou (1977) we know that the fraction p of N_0 being at a distance greater than R from the centre after a time t has elapsed, is given by

$$p = e^{-R^2/(a^2 t)} \qquad \text{(A3.7)}$$

from which we have

$$R = a - (-\mathrm{t} - \ln p)^{1/2}. \qquad \text{(A3.8)}$$

We can test how the diffusion model fits data by comparing observed and expected distributions of isotope-marked individuals. For this purpose it is necessary to correct for the different number of traps in the various trap circles. The corrected catches are given by

$$Q_{kj} = n_j f_{kj}, \ h = 1, \ldots, 4. \qquad \text{(A3.9)}$$

The corresponding expected distribution is given by

$$EQ_{kj} = n_j E f_{kj} = [n_j^2 r_k / (a^2 t_j)] \ e^{-r_k^2/(\hat{a}^2 t_j)}$$
$$k = 1, \ldots, 4. \qquad \text{(A3.10)}$$

The hypothesis that dispersal out from a source population follows a diffusion process may then be tested against the data by applying a G-test (e.g. Sokal and Rohlf, 1981). This test statistic is given by

$$G = 2 - \sum_{k=1}^{l} \sum_{j=1}^{m} Q_{kj} \ln \frac{Q_{kj}}{EQ_{kj}} \qquad \text{(A3.11)}$$

where l is the number of trapping rings and m is the number of visitations. The G–statistic is approximately χ^2–distributed with the degrees of freedom given by $m\ (l - 1)$ minus the number of parameters estimated in order to arrive at EQ_{kj} (which in our example is given by l_j; equations A3.3 and A3.4). Hansson, Myllymäki, and Hoff assumed that the diffusion process starts on 15 July (after which no more isotope marking was done) (Hoff, 1981).

Table A3.4 Captures per trap station of *Microtus agrestis* in various habitats at the Hiirola study site

	Forest	*Forest edge*	*Clearcuts*
Average	0.026	1.112	1.158
SD	0.055	0.572	0.503

(b) Results

In spite of more than 600 animals being caught on the concentric circles, samples had to be combined. The material was collected into two classes: one included catches from the 10 first kill trappings and the other included the latter seven trappings (Table A3.2). No marked individuals were caught at the forest stations. Furthermore, since few unmarked field voles were caught in the forest (Table A3.4), these stations were excluded from further analysis. This was done in order to avoid biases when calculating catch per trap station in the various kill-trap circles; as can be seen from Figure A3.1, there was an uneven number of 'forest stations' in the various trap lines. Data from trapping after 1 September were also disregarded since the traps then had been activated for 17 or more days. Captures per trap station in various habitat types are given in Table A3.4.

Observed and expected distributions of marked animals are given in Table A3.5. As can be seen, the fit is indeed very good. The rate of spread, a, does not seem to depend on the season; this conclusion should, however, be treated with some caution since the entire experiment only covered a relatively short period and since we only have two (combined) sampling periods.

By assuming $a = 113$ (Table A3.5) and $p = 0.5$ in equation (A3.7), we obtain a median dispersal distance per day equal to 94 m. Furthermore, after one day, 95% of the marked individuals will be inside a circle with a radius equal to 196 m from the centre of origin.

The estimated distances may be atypically high because the vole habitats were surrounded by extensive forested habitat. The high value of a suggests that field voles move quickly through the forest which is only a transitional habitat for them (Hansson, 1977); see also the low capture rate in forests shown in Table A3.4. Thus, the good fit between the diffusion model and the observations most plausibly is the result of movements being largely through transition habitat. This kind of dispersal has rarely been measured.

On the basis of this analysis, it can therefore be concluded that dispersal of field voles out from an optimal habitat surrounded by transitional habitat may be described as a diffusion process with a median value of dispersal distance per day equal to approximately 100 m.

Table A3.5 Observed and expected distribution of marked individuals recaptured in the concentric trap circles. All individuals were marked on 15 July. Time between visitations: $t_1 = 17$, $t_2 = 42$. Number of animals caught: $n_1 = 25$, $n_2 = 23$. Estimated standard deviation of dispersal distance from origin per day is given by a. A: a is estimated by using data from each visitation. B: a is estimated on the basis of the entire material combined

	Visitation no.	*Ring* 1	2	3	4	a (m day^{-1})
Observed	1	11.4	8.2	4.1	1.4	
	2	6.2	7.9	4.6	4.4	
Expected						
A	1	10.5	8.8	4.4	1.4	117 ($G = 0.19$, $p = 0.89$)
	2	5.8	7.0	6.1	4.2	110 ($G = 0.51$, $p = 0.78$)
B	1	11.0	8.8	4.1	1.2	113 ($G = 0.75$, $p = 0.94$)
	2	5.6	6.9	6.2	4.4	113 ($G = 0.75$, $p = 0.94$)

A3.4 CONCLUSIONS

In this appendix we have in addition to a brief review of the literature provided an example of dispersal analysed as a diffusion process for animals marked with radioactive isotopes. We have provided some technical details of how to analyse such data so that ecologists can apply the techniques even though they may have little knowledge about diffusion processes.

The use of radioactive materials clearly has great potential in population studies, including those involving dispersal. Like all methods, however, the technique does carry limitations on the number of animals that can be marked individually, and on the distances over which movements can be detected. The possibility of labelling individuals with combinations of isotopes has dramatically alleviated the limitations on the number of unique marks possible in a given time and place. Since this is insufficiently appreciated, we offer the following calculation. If only 10 isotopes are carefully selected so that five can be distinguished in a single individual, the number of unique labels, using combinations of five, four, three, two, and one isotopes, is 637 (equation A3.1).

Of course, there are important safety issues to be considered in the use of radioactive materials, not only for the investigator, but for the animals labelled, and for potential predators. These matters are discussed fully in the various review papers cited above, and we urge anyone contemplating use of these techniques to become familiar with all of this literature before initiating any study involving radioactive materials. In spite of these considerations, we remain enthusiastic about the prospective utility of radioactive tagging.

We are confident that radioactive isotopes have a legitimate place among the tools available to those doing research on dispersal in small mammals. We have pointed to the circumstances under which they may be the technique of choice while at the same time insisting that we keep their limitations in mind. Finally, we hope that the example given of the use of a diffusion model for analysing dispersal data will encourage others to try this approach.

ACKNOWLEDGEMENTS

We thank Jon Atle Eie for his contributions to our literature search. And we repeat our appreciation to Lennart Hansson, Arvo Myllymäki and Jann Morten Hoff for allowing us to use their unpublished data to provide an excellent illustration of the radioisotope approach to the study of dispersal.

REFERENCES

Airoldi, J.P. (1979) Etude du rythme d'activité du compangnol terrestre, *Arvicola terrestris scherman* Shaw. *Mammalia*, **43**, 25–52.

Bailey, G.N.A., Linn, I.J. and Walker, P.J. (1973) Radioactive marking of small mammals. *Mammal Review*, **3**, 11–23.

Barbour, R.W. (1963) *Microtus*: a simple method of recording time spent in the nest. *Science*, **141**, 41.

Cope, J.B., Churchwell, E. and Koontz, K. (1961) A method of tagging bats with radioactive gold-198 in homing experiments. Proceedings of the Indiana Academy of Science, 70, 267–9.

Crabtree, R.L., Burton, F.G., Garland, T.R. *et al.* (1989) Slow-release radioisotope implants as individual markers for carnivores. *Journal of Wildlife Management*, 53, 949–54.

Dickman, C.R., King, D.H., Happold, D.C.D. and Howell, M.J. (1983) Identification of the filial relationships of free-living small mammals by ^{35}Sulphur. *Australian Journal of Zoology*, **31**, 467–74.

Eie, J.A. (1991) Social organization in martens. Unpublished M.Sc. Dissertation, University of Oslo.

Frigerio, N.A. and Eisler, W.J. (1968) Low cost, automatic, nest and burrow monitor using radioactive tagging. *Ecology*, **49**, 788–91.

Genoud, M. (1984) Activity of *Sorex coronatus* (Insectivora, Soricidae) in the field. *Zeitschrift für Saugetierkunde*, **49**, 74–8.

Gentry, J.B., Smith, M.H. and Byers, R.J. (1971) Use of radioactively tagged bait to study movement patterns in small mammal populations. *Annales Zoologica Fennici*, **8**, 17–21.

Gifford, C.E. and Griffin, D.R. (1960) Notes on homing and migratory behaviour of bats. *Ecology*, **41**, 378–81.

Godfrey, G.K. (1953) A technique for finding *Microtus* nests. *Journal of Mammalogy*, **34**, 503–5.

Godfrey, G.K. (1954a) Tracing field voles (*Microtus agrestis*) with a Geiger–Muller counter. *Ecology*, **35**, 5–10.

Godfrey, G.K. (1954b) Use of radioactive isotopes in small-mammal ecology. *Nature*, **174**, 951–2.

Graham, W.J. and Ambrose, H.W. (1967) A technique for continuously locating small mammals in field enclosures. *Journal of Mammalogy*, **48**, 639–42.

Griffin, D.R. (1952) Radioactive tagging of animals under natural conditions. *Ecology*, **33**, 329–35.

Hamar, M., Şuten, G.H. and Şutova, M. (1963) 'Home range' studies in rodents by marking with P^{32}. *Reviue de Biologie Académia République Populatire Roumaine*, **8**, 431–46.

Hamar, M., Şuten, G.H. and Şutova, M. (1964) Home range and activity studies of the mole rat (*Spalax leucodon* Nordm.) by Co^{60} marking. *Rev. Roum. Biol. (Sér. Zool.)*, **9**, 421–33.

Hamar, M., Şuten, G.H., Şutova, M. and Tuta, A. (1974) Determination of the structure and use of underground galleries of *Arvicola terrestris scherman* Shaw and *Pitymys subterraneus* de S.L. by ^{60}Co labeling. *EPPO Publications Series A*, **58**, 129–35.

Hansson, L. (1967) Index line catches as a basis of population studies on small mammals. *Oikos*, **18**, 261–76.

Hansson, L. (1977) Spatial dynamics of field voles *Microtus agrestis* in heterogeneous landscapes. *Oikos*, **29**, 539–44.

Hilborn, R. and Krebs, C.J. (1976) Fates of disappearing individuals in fluctuating populations of *Microtus townsendii*. *Canadian Journal of Zoology*, **54**, 1507–18.

Hoff, J.M. (1981) Teoretiske studier over migrasjonens betydning for populasjonsdynamikken hos smågnagere: heterogene områder, med speciell vekt på markmusa (*Microtus agrestis*). M.Sc. Dissertation, University of Oslo, 177 pp.

International Commission on Radiation Protection (1959) *Report of Committee II on Permissible Dose for Internal Radiation*, Pergamon Press, Oxford, 233 pp.

Iverson, S.L. and Turner, B.N. (1976a) Effect of acute irradiation on survival of captive and free-ranging meadow voles, in *Radioecology and Energy Resources* (ed. C.E. Cushing Jr), Ecological Society of America, Special Publication No. 1, pp. 359–62.

Iverson, S.L. and Turner, B.N. (1976b) Small mammal radioecology: natural reproductive patterns of seven species. AECL-5393. Whiteshell Nuclear Research Establishment, Atomic Energy of Canada Limited.

Jannett, F.L. Jr (1978) The density-dependent formation of extended maternal families of the montane vole, *Microtus montanus nanus*. *Behavioural Ecology and Sociobiology*, **3**, 245–63.

Jenkins, D.W. (1954) Advances in medical entomology using radioisotopes. *Experimental Parasitology*, **3**, 474–90.

Johanningsmeier, A.G. and Goodnight, C.J. (1962) Use of iodine-131 to measure movements of small mammals. *Science*, **138**, 147–8.

Karulin, B.E., Litvin, V.Y., Nikitina, N.A. *et al.* (1976) A study of activity, mobility and diurnal range in *Microtus oeconomus* on the Yamal Peninsula by means of marking with radioactive cobalt. *Zoologicheskii Zhurnal*, **55**, 1052–60 (in Russian with English summary).

Kaye, S.V. (1960) Gold-198 wires used to study movements of small mammals. *Science*, **131**, 824.

Kaye, S.V. (1961) Movements of harvest mice tagged with gold-198. *Journal of Mammalogy*, **42**, 323–37.

Kingsley, C., Morimoto, D.C. and Tamarin, R.H. (1987) Elimination of ^{131}I in meadow voles (*Microtus* pennsylvanicus) measured by whole-body γ spectrometry. *Health Physics*, **52**, 481–3.

Korneyev, G.A. (1967a) On the nature of the utilization of great gerbil (*Rhombomys opimus*) colonies by midday gerbils (*Meriones meridianus*) as revealed by radioactive labelling of the rodents. *Nauk Doklady Vyssh. Shk.* (*Biol.*), **1**, 26–30 (in Russian).

Korneyev, G.A. (1967b) On interspecies ectoparasitic contacts of some species of mammals in colonies of great gerbils. *Parazitologiya*, **1**, 233–7 (in Russian).

Kruuk, H., Gorman, M. and Parrish, T. (1980) The use of ^{65}Zn for estimating populations of carnivores. *Oikos*, **34**, 206–8.

Linn, I. and Shillito, J. (1960) Rings for marking very small mammals. *Proceedings of the Zoological Society of London*, **134**, 489–95.

Meslow, E.C. and Keith, L.B. (1968) Demographic parameters of a snowshoe hare population. *Journal of Wildlife Management*, **32**, 812–34.

Mihok, S., Schwartz, B. and Iverson, S.L. (1985a) Ecology of red-backed voles (*Clethrionomys gapperi*) in a gradient of gamma radiation. *Annales Zoologica Fennici*, **22**, 257–71.

Mihok, S., Turner, B.N. and Iverson, S.L. (1985b) The characterization of vole population dynamics. *Ecological Monographs*, **55**, 399–420.

Miller, L.S. (1957) Tracing vole movements by radioactive excretory products. *Ecology*, **38**, 132–6.

Morimoto, D., Tamarin, R.H. and Levy, C. (1985) Whole-body biological elimination rates of gamma-emitting radionuclides in captive meadow voles, *Microtus pennsylvanicus*. *Health Physics*, **49**, 1217–28.

Myllymäki, A. (1977) Demographic mechanisms in the fluctuating populations of the field vole *Microtus agrestis*. *Oikos*, **29**, 468–93.

Myllymäki A., Paasikallio, A. and Hakkinen, U. (1971) Analysis of a 'standard trapping' of *Microtus agrestis* (L.) with triple isotope marking outside the quadrat. *Annales Zoologica Fennici*, **8**, 22–34.

Nellis, D.W., Jenkins, J.H. and Marshall, A.D. (1967) Radioactive zinc as a feces tag in rabbits, foxes and bobcats. *Proceedings of the 21st Conference of the Southeastern Association of Game and Fish Commissions* (Columbia SC), pp. 205–7.

Okubo, A. (1980) *Diffusion and Ecological Problems: mathematical models*, Springer, Heidelberg, 254 pp.

Pelton, M.R. and Marcum, L.C. (1977) The potential use of radioisotopes for determining densities of black bears and other carnivores, in *Proceedings of the 1975 Predator Symposium* (eds R.L. Phillips and C. Jonkel), Montana Forestry Conservation Experimental Station, University of Montana, Missoula, pp. 221–36.

Pendleton, R.C. (1956) Uses of marking animals in ecological studies: labeling animals with radioisotopes. *Ecology*, **37**, 686–9.

Pielou, E.C. (1977) *Mathematical Ecology*, Wiley, New York, 385 pp.

Pulliam, H.R. (1988) Sources, sinks and population regulation. *American Naturalist*, **132**, 652–61.

Punt, A. and van Nieuwenhoven, P.J. (1957) The use of radioactive bands in tracing hibernating bats. *Experientia*, **13**, 51–4.

Ricciardi, L.M. (1977) *Diffusion Processes and Related Topics in Biology*, Springer, Berlin, 200 pp.

Rongstad, O. (1965) Calcium-45 labeling of mammals for use in population studies. *Health Physics*, **11**, 1543–56.

Rudenchik, Yu. V. (1963) The study of mobility of great gerbils with the aid of radioactive indicators, in *Proceedings of the Science Conference on Natural Foci and Prophylaxis of Plague* (ed. M.A. Aikimbaev), Alma-Ata, pp. 198–200 (in Russian).

Rudenchik, Yu., V., Soldatkin, I.S., Severova, E.A. *et al.* (1967) Quantitative evaluation of the possibilities of a territorial advance of plague epizootics in a population of great gerbils (North Kysylkumy) *Zoologicheskii Zhurnal*, **46**, 117–23 (in Russian).

Salvioni, M. (1988) Home range and social behavior of three species of European *Pitymys* (Mammalia, Rodentia). *Behavioural Ecology and Sociobiology*, **22**, 203–10.

Scott, M.P. and Tan, T.N. (1985) A radiotracer technique for the determination of male mating success in natural populations. *Behavioural Ecology and Sociobiology*, **17**, 29–33.

Seber, G.A.F. (1982) *The Estimation of Animal Abundance and Related Parameters*, 2nd edn, Griffin, London, 654 pp.

Sheridan, M. and Tamarin, R.H. (1986) Kinships in a natural meadow vole population. *Behavioural Ecology and Sociobiology*, **19**, 207–11.

Shura-Bura, B.L. and Kharlamov, V.P. (1961) Audioradiography as a method to reveal labelled rodents and their ectoparasites when studying migration problems. *Zoologicheskii Zhurnal*, **40**, 258–63 (in Russian).

Shura-Bura, B.L., Tararin, R.A. and Mel'nikov, B.K. (1960) On the methods of radioactive labelling rats with the aim of studying migration problems. *Zoologicheskii Zhurnal*, **39**, 1700–6 (in Russian).

Skellam, J.G. (1951) Random dispersal in theoretical populations. *Biometrika*, **38**, 196–218.

Sokal, R.R. and Rohlf, F.J. (1981) *Biometry*, 2nd edn, W.H. Freeman, San Francisco, 859 pp.

Soldatkin, I.S., Novokrashchenova, N.S., Rudenchik, Yu. V. *et al.* (1961) An experiment in studying activity of feeding of the fleas parasitizing great gerbils under natural conditions with the use of radioactive tracers. *Zoologicheskii Zhurnal*, **40**, 1647–50 (in Russian).

Stenseth, N.C. (1983) Causes and consequences of dispersal in small mammals, in *The Ecology of Animal Movement* (eds I. Swingland and P.J. Greenwood), Oxford University Press, Oxford, pp. 63–101.

Stoddart, D.M. (1970) Individual range, dispersion and dispersal in a population of water voles (*Arvicola terrestris* (L.)). *Journal of Animal Ecology*, **39**, 403–25.

Tamarin, R.H., Sheridan, M. and Levy, C.K. (1983) Determining matrilineal kinship in natural populations of rodents using radionuclides. *Canadian Journal of Zoology*, **61**, 271–4.

Twigg, G.I. and Miller, H. (1963) The use of calcium^{45} as an agent for labeling rat populations. *Journal of Mammalogy*, **44**, 335–7.

Wolff, J.O. and Holleman, D.F. (1978) Use of radioisotope labels to establish genetic relationships in free-ranging small mammals. *Journal of Mammalogy*, **59**, 859–60.

Epilogue

Looking back over the chapters in this book, it is apparent that we have accomplished at least five things regarding our understanding of dispersal.

1. We have a fairly good theoretical basis for making predictions regarding which circumstances will favour dispersal (Parts I and II, Chapter 9).
2. There is a fairly good, although somewhat more fragmentary, theoretical basis for predicting the demographic consequences of dispersal (Part V).
3. We have compiled an extensive amount of data making it possible to see patterns concerning which individuals are dispersers under which circumstances (Part II), and the population dynamic consequences of dispersal (Part III).
4. We have used this knowledge in applied fields such as in conservation biology and pest management, and have discovered that these applied fields have also contributed in turn to basic research (Part IV).
5. We have reviewed, for the novice and the experienced practitioner, a wide variety of techniques now in use for studying movement patterns of small mammals (Appendices 1–3).

We have also attempted in this book to build bridges between disparate subdisciplines, groups that typically operate in nearly complete isolation from each other. As in other fields of science, there are gaps, even gulfs, between theorists and empiricists, as well as between investigators who work with different taxonomic groups of organisms.

The two of us represent distant points on the theorist–empiricist continuum, and having now edited this book we can attest to finding profit in our collaboration. In fact, it is not merely the adding together of ideas that results from such interaction, but the intellectual exchanges leading to mutually improved understanding that yield synergistic effects. The result has indeed been more than the sum of the parts. As the book is designed with this in mind, we hope that these synergistic effects will be experienced by the readers as well.

We have also tried to bridge the gap between investigators studying different groups of organisms (Part VI). We think our invited commentaries add a very important, possibly unique, dimension to this book, and have allowed us to present a much more unified approach to the study of dispersal.

Our focus, however, has been on small mammals as model organisms. Indeed, we think it is very profitable to pick model organisms with great care. One should always ask which species and which system are most appropriate for studying a particular process. As we have argued elsewhere in this book, small mammals serve as very good model organisms in which a wide variety of general processes can be studied easily. Ecologists should, we believe, more often be critically selective of their organisms of study (in a similar manner to that done in several other fields of biology).

What next? We see three areas which should be given more emphasis in the near future.

1. So far there has been a strong and productive focus on population level phenomena. Partly this has been encouraged by a series of technical improvements. But, if we want to put our investigations into an evolutionary framework, which we should do, then we need individual-level studies to a much greater extent. This is because the individual organism represents the major unit of selection in biology. It is therefore essential to be able, for example, to compare the fitnesses of individuals which disperse and those that do not in order to test models for the evolution of dispersal. With the development of DNA-fingerprinting techniques (p. 313 in Appendix 1), this goal may now be somewhat easier to achieve than previously.

2. Another emphasis in the studies on dispersal is one on the emigration process. We also need now to focus more on immigration, both the process itself and on the effects of invaders moving into an established population. This bias in attention has not, we believe, been a result of technical constraints. Rather, it is a result of a bias introduced because of our tendency to think most about those individuals leaving our study plot, the individuals we know best. We forget (or choose to neglect) those coming into our study plot. Besides, we tend to establish our study sites in places with especially favourable habitats so that emigration is inherently a more important and conspicuous process than immigration.

3. The third suggestion for the future is to incorporate habitat texture or heterogeneity into our research thinking and protocols. That is, we need to pay more attention to the role of habitat patchiness and habitat fragmentation in the dispersal process. Variations in patch quality, size, shape, and connectedness all need to be brought into our collective consciousness. In doing this we will become aware of the importance of spatial and temporal scale in forming our questions and integrating our results. This in itself will make it easier for those working with organisms of quite different sizes to integrate their thinking and findings (Part VI).

There are many things we still do not know about dispersal, and it remains an exciting and critically important area of research. Yet, there are many things we do know about it, and this too needs to be acknowledged. This book is a documentation of both these facts. Indeed, we now are at a stage where we know approximately where we ought to go in order to find out more about our subject. We cannot say exactly where and how, as this is the frontier of scientific investigation, but we do have a solid platform for further scientific work on dispersal, especially that using small mammals as model organisms. We trust that this book represents such a foundation.

Nils Christian Stenseth
William Z. Lidicker Jr

Index